MAGILL'S ENCYCLOPEDIA OF SCIENCE

ANIMAL LIFE

MAGILL'S ENCYCLOPEDIA OF SCIENCE

ANIMAL LIFE

Volume 4

Respiratory System–Zoos
Index

Editor
Carl W. Hoagstrom, Ph.D.
Ohio Northern University

Project Editor
Tracy Irons-Georges

SALEM PRESS, INC.
Pasadena, California
Hackensack, New Jersey

Editor in Chief: Dawn P. Dawson

Project Editor: Tracy Irons-Georges
Copy Editor: Leslie Ellen Jones
Research Supervisor: Jeffry Jensen
Research Assistant: Jeff Stephens
Aquisitions Editor: Mark Rehn
Photograph Editor: Philip Bader
Production Editor: Joyce I. Buchea
Page Design: James Hutson
Layout: Eddie Murillo
Additional Layout: William Zimmerman

Graphics by: Electronic Illustrators Group, Morgan Hill, Calif.
Cover Design: Moritz Design, Los Angeles, Calif.

Some of the updated and revised essays in this work originally appeared in *Magill's Survey of Science: Life Science*, edited by Frank N. Magill (Pasadena, Calif.: Salem Press, Inc., 1991), and in *Magill's Survey of Science: Life Science Supplement*, edited by Laura L. Mays Hoopes (Pasadena, Calif.: Salem Press, Inc., 1998).

∞ The paper used in these volumes conforms to the American National Standard for Permanence of Paper for Printed Library Materials, Z39.48-1992 (R1997).

Library of Congress Cataloging-in-Publication Data

Magill's encyclopedia of science : animal life / editor, Carl W. Hoagstrom.
p. cm.
Includes bibliographical references (p.).
ISBN 1-58765-019-3 (set : alk. paper) — ISBN 1-58765-020-7 (v. 1 : alk. paper) — ISBN 1-58765-021-5 (v. 2 : alk. paper) — ISBN 1-58765-022-3 (v. 3 : alk. paper) — ISBN 1-58765-023-1 (v. 4 : alk. paper)
1. Zoology—Encyclopedias. I. Hoagstrom, Carl W.

QL7 .M34 2002
590'.3—dc21

2001049799

Second Printing

PRINTED IN THE UNITED STATES OF AMERICA

TABLE OF CONTENTS

MAGILL'S ENCYCLOPEDIA OF SCIENCE

ANIMAL LIFE

RESPIRATORY SYSTEM

Types of animal science: Anatomy, geography, physiology
Fields of study: Anatomy, biochemistry, cell biology, environmental science, histology, immunology, pathology, physiology

The respiratory system maintains a constant flow of oxygen into the blood while removing carbon dioxide. Other functions of the respiratory system include maintaining blood acid-base balance, reducing body temperature, communicating by means of sounds, and removing inhaled microbes.

Principal Terms

ALVEOLUS: the thin-walled, saclike lung structure where gas exchange takes place

CHEMORECEPTOR: specialized nervous tissue that senses changes in pH (hydrogen ions) and oxygen

COUNTERCURRENT EXCHANGER: the process where a medium (air or water) flowing in one direction over a tissue surface encounters blood flowing through the tissue in the opposite direction; this improves the gas diffusion by maintaining a concentration gradient

DIFFUSION: the process by which gas molecules move from a higher to a lower concentration through a medium or across a permeable barrier; the rate at which gases cross a barrier is increased by the surface area, and gas concentration gradient is decreased by the thickness of the barrier; gas solubility determines the amount that crosses the barrier

GILL: an evaginated organ structure where the membrane wall turns out and forms an elevated, protruding structure; typically used for water respiration

LUNG: an invaginated organ structure where the membrane wall turns in and forms a pouch or saclike structure

Animals generally meet their energy needs by oxidation of food, and the respiratory system supplies the oxygen necessary for cell metabolism while removing its waste product, carbon dioxide. Oxygen is available either dissolved in water or as a component of the air, and animals have evolved special organ structures to effectively obtain oxygen from their environment.

Organs of Gas Exchange

Single-cell and simple organisms, such as flatworms and protozoa, can obtain sufficient oxygen to meet their energy demands by simple diffusion through their body surface. Some amphibians utilize gas exchange through their skin to supplement their lung respiration, but generally, larger, more complex animals require specialized organ systems with a large surface area for gas exchange and a circulatory system for distribution of oxygen to each cell. The basic mechanism, however, for gas exchange between the environment and the blood and between the blood and cells is by diffusion. The three major types of gas exchange organs are the gill for water respiration, the lung for air and in some special cases water respiration, and the tracheas system of tubules for air respiration in insects.

Gills consist of several gill arches located in the operculum or gill cover on each side of the fish's head. A gill arch contains two rows of gill filaments, and each filament has a row of parallel platelike structures on its surface called lamellae. The lamellae are everted structures that rise up

Respiratory Pigments

The amount of oxygen dissolved into circulatory fluids is insufficient to meet the oxygen requirements for almost all vertebrate animals. Respiratory pigments, colored proteins that contain a metal (usually iron or copper) have evolved to increase the amount of oxygen carried by the blood one hundredfold (20 milliliters of oxygen per 100 milliliters of blood, compared to 0.2 milliliters of oxygen per 100 milliliters of blood). These pigments have the unique property of binding oxygen at the gas exchange surface of the gill or lung, transporting the oxygen in circulation, and then releasing the oxygen to the cells, where it can be utilized for energy metabolism.

The most common respiratory pigment is hemoglobin, which has a positively charged iron atom (divalent) attached to a circular protein ring (porphyrin ring). Oxygen is negatively charged and binds to the positive charge of the iron. This bond is easily broken at the tissue level, where oxygen is released. Copper-containing respiratory pigments are called hemocyanins and are found in mollusks and arthropods. Next to hemoglobin, they are the most widely used respiratory pigment.

Hemoglobin is bright red when it carries oxygen and turns a purple-blue color when it has released some or all of its oxygen molecules. Hemoglobin can be found either circulating free in the plasma, typically in insects called hemolymph, or contained in red bloods cells called erthyrocytes, circulating in the plasma.

from the filament surface and are only a fraction of a millimeter apart. Water flows between the lamellae, and oxygen diffuses from the water into the lamellar capillary blood. The lamellar blood flows in the opposite direction of the water flow and creates a countercurrent exchanger. The countercurrent maximizes the diffusion of oxygen into the lamellar capillary blood by maintaining a diffusion gradient over its entire length.

Lungs, in contrast to gills, are invaginations, where the surface turns in and forms a hollow or saclike structure. Lungs typically are divided into two functional areas: the conducting zone and the respiratory zone. The conducting zone branches from the trachea to the bronchioles and distributes air to the respiratory zone but is not involved in gas exchange. The respiratory zone comprises the majority of the lung and contains small respiratory bronchioles and ducts that lead to the primary gas exchange area, the alveolus. The alveoli vary from simple saclike structures in a pulmonate land snail to the complex alveolar wall structure of mammals. The alveolar wall is fifty micrometers thick, or about one fiftieth the thickness of a sheet of paper, and is composed of epithelial cells covering the alveolar surface, an interstitial space, and the endothelial cells that make up the capillaries. This thin-wall structure allows for the diffusion of oxygen and carbon dioxide between the air and blood.

The insect tracheas respiratory system is unique because it is both the gas exchange and distribution system. Pairs of openings on the insect's thorax and abdomen called spiracles regulate the movement of air in and out of a tubule system. The spiracles open and close in a pattern that allows unidirectional flow of air through the tubule system. The tubules branch and extend throughout the insect's body and deliver oxygen to the cells independent of the circulatory system.

Air and Water Environments

Important aspects of the atmosphere for respiration are the barometric pressure and concentration of gases, temperature, and humidity. The atmospheric gases important to animals are oxygen, carbon dioxide, and nitrogen, and the atmosphere is a constant 20.95 percent oxygen, 0.03 percent carbon dioxide, and 79 percent nitrogen (plus other inert gases). The rate of diffusion of oxygen from the inspired air into the circulation depends on the partial pressure of the oxygen. The barometric pressure, however, decreases with increasing altitude, and this decreases the partial pressure of oxygen, which decreases the diffusion of oxygen into the blood. Thus an animal's difficulty in obtaining adequate oxygen at higher altitudes is related to the reduction in atmospheric pressure and not to a change in the percentage of oxygen in the atmosphere.

The temperature and amount of water vapor or humidity in the atmosphere are variable, and during inspiration, the inspired air is warmed to body temperature and saturated with water vapor (100 percent humidity). The heat and moisture come from the airways and can potentially cool and dehydrate an animal. Therefore, a minimal amount of air is inspired to prevent excess heat and water loss. However, heat-stressed animals will use this respiratory heat loss or panting to cool their bodies.

Water poses several challenges for respiration compared to air: a lower oxygen content, slower gas diffusion rate, higher viscosity, and greater weight. The amount of oxygen available in the water is thirty times less than that found in air. Thus, more water has to flow over the gill surface for adequate oxygen delivery. The speed at which oxygen moves through the water is ten thousand times slower than oxygen moving through the air. Thus, the distance between the water and the gill surfaces can only be a fraction of a millimeter apart. In contrast, the lung gas exchange surfaces are a few millimeters apart.

Water's greater viscosity and weight compared to air require more energy to move water over the gill surface. Water-breathing animals compensate for this by having a unidirectional flow through the gill. This avoids water being moved, stopped, and then moved again in the opposite direction, which works well for air, but would be very energy costly for the heavier, more viscous water.

The gill structure depends on water to support and separate the rows of lamellar structures. Thus, when a fish is exposed to air, the gill structure collapses on itself and greatly reduces the surface area available for oxygen diffusion. Thus, the fish will suffocate if not returned to the water.

Breathing Water and Breathing Air

Water can be moved through the gill lamellae by either opercular pumping or ram ventilation. Opercular pumping involves the movement of the mouth and opercular covering to create pressure gradients for unidirectional flow of water through the mouth, across the gill surface, and out the opercular covering (unidirectional flow). Ram ventilation takes advantage of the fish's forward speed to flow water through the mouth and gill. Opercular pumping is used from rest to slow swimming speeds, and a fish switches over to ram ventilation when swimming at faster speeds.

For air breathers, inspiration (inflating the lungs) can be accomplished by either positive-pressure or negative-pressure breathing. Positive-pressure breathing requires air pressure to inflate the lungs, which is similar to inflating a balloon or tire with a compressed air. The pressure is considered positive because it is greater than atmospheric pressure. For example, frogs use positive-pressure breathing by closing their mouths and then elevating the floor of the mouth. This compresses and pressurizes the air and forces it into the lung. The elastic lung tissue is stretched like an inflated balloon by the increased volume. The process of the air moving out of the lung is called expiration. When the frog relaxes and opens its mouth the lung elastic recoil forces the air out similar to a balloon deflating.

With negative-pressure breathing, the lung is pulled open by contraction of the diaphragm. The pressure becomes negative (below atmospheric pressure), and air flows into the lung until it equalizes with the atmospheric pressure. If additional inflation is required, such as during exercise, accessory inspiratory muscles lift the ribs to inflate the lungs further. Expiration is accomplished by the relaxation of the inspiratory muscles, and the lung elastic recoil increases airway pressure and air flows out of the lung.

Inspiration is always an active process, whereas expiration results from the passive elastic recoil of the lung tissue. However, active expiration is possible by contracting muscles that pull the ribs down and by using abdominal muscles to push the diaphragm farther into the thoracic (chest) cavity.

August Krogh

Born: November 15, 1874; Grenå, Jutland, Denmark
Died: September 13, 1949; Copenhagen, Denmark
Fields of study: Anatomy, environmental science, invertebrate biology, physiology, zoology
Contribution: Krogh originally described how animals exchange oxygen and carbon dioxide by diffusion and invented many instruments needed to conduct experiments that led to his conclusions. He received the Nobel Prize in Physiology or Medicine in 1920 for his studies on capillary function and muscle metabolism. In this work, Krogh published the first account of regulation of blood perfusion in muscle and other organs.

August Krogh's work focused on gas exchange in mammalian respiratory systems. (Nobel Foundation)

In 1897, Schack August Steenberg Krogh began working in the lab of the famous physiology professor, Christian Bohr. Dr. Bohr had studied the solubility of oxygen in different tissues and fluids, as well as the mechanisms of muscle contraction. These experiments greatly influenced Krogh's early studies of gas exchange in snails, frogs, and fishes. In 1899, Krogh published the equivalent of a master's thesis, demonstrating that, in birds, oxygen moved by diffusion through the thin lung membranes into the blood. His dissertation, in 1903, studied gas exchange in the frog and showed that skin respiration remains fairly constant, whereas large variability occurs in lung respiration. Krogh reasoned this was an example of the oxygen secretion hypothesis proposed by Bohr. However, later he would doubt his conclusion and demonstrate that oxygen moves solely by diffusion through tissues.

Krogh participated in an expedition in 1902 to Disko, North Greenland, where he investigated the carbon dioxide and oxygen content in springwater, streams, and the sea. From these studies, Krogh described the important role of the oceans in regulation of atmospheric carbon dioxide. He applied these techniques of measuring dissolved gases in animal physiological studies in 1904.

Krogh won the prestigious Seegen Prize, awarded by the Austrian Academy of Sciences, in 1906 for investigating whether free nitrogen or nitrogenous gases were released as a normal by-product of metabolism. He showed that gaseous nitrogen remained constant by using his unique respiratory gas quantification methods. Krogh determined nitrogen dynamics with gas measurements instead of using the traditional German method of Liebig and Rubner, who measured nitrogen content in ingested food and liquids and excreted nitrogen in feces and urine.

Marie Jörgensen, a medical student and scientist, married August Krogh in 1905. Together they published seven papers on the quantification and diffusion of gases in the blood. This overturned the view held by Dr. Bohr and the scientific establishment that stated that oxygen and carbon dioxide were "secreted" by a glandlike structure in the lung.

In 1908, a special position as Associate Professor in Zoo-Physiology was created for Krogh at the University of Copenhagen, and in 1910 Krogh founded a zoo-physiology (animal) laboratory at the University of Copenhagen. The laboratory was moved and enlarged in 1928 with financing from the Rockefeller Foundation. Eight years later, Krogh was promoted to a chair, which he held until his retirement in 1945. The Krogh Institute is still active today.

—*Robert C. Tyler*

Setting Breathing Rate

In water-breathing animals, such as fish and lobsters, the level of oxygen sets the ventilation rate (volume of water moved through the gill per minute) such that as oxygen content in the water decreases, the frequency of breathing movements increases. During fast swimming, fish using ram ventilation regulate the mouth opening so that the amount of water flowing over the gills just meets tissue oxygen demand. A wider mouth opening than is necessary increases the fish's frictional drag through the water and thus decreases the energy efficiency. Carbon dioxide is highly soluble in water and easily diffuses from water-breathing animals. Thus, blood carbon dioxide levels in water-breathing animals are very low and not used to regulate respiration rate.

In air-breathing animals, the blood levels of carbon dioxide and oxygen regulate the ventilation rate (air volume moved in and out of the lungs per minute). Carbon dioxide quickly diffuses from the small capillaries in the brain circulation into the fluid surrounding the brain cells (cerebral spinal fluid). Here the carbon dioxide reacts with water and forms carbonic acid. The hydrogen ions released from the carbonic acid stimulate chemoreceptor cells that in turn stimulate the respiratory center in the medulla, located in the brain stem. Higher concentrations of carbon dioxide increase the hydrogen ion concentration and thus increase ventilation rate. Air-breathing animals primarily regulate ventilation rate by carbon dioxide produced from metabolism and not low blood oxygen levels.

However, oxygen can regulate ventilation in animals at high altitudes. Oxygen partial pressure is sensed by chemoreceptors in the aorta and the carotid artery. These peripheral chemoreceptors sense the partial pressure of oxygen in the blood plasma, and as the partial pressure of oxygen in the air decreases, such as with altitude, the partial pressure of oxygen in blood also decreases. This increases ventilation, which then compensates for the lower oxygen partial pressure. In addition to low oxygen partial pressure, the peripheral chemoreceptors are stimulated by blood acidosis. For example, lactic acid released from skeletal muscles during strenuous exercise stimulates the ventilation rate in animals and humans.

—*Robert C. Tyler*

See also: Anatomy; Beaks and bills; Bone and cartilage; Brain; Circulatory systems of invertebrates; Circulatory systems of vertebrates; Claws, nails, and hooves; Digestive tract; Ears; Endoskeletons; Exoskeletons; Eyes; Fins and flippers; Immune system; Kidneys and other excretory structures; Lungs, gills, and tracheas; Muscles in invertebrates; Muscles in vertebrates; Nervous systems of vertebrates; Noses; Physiology; Reproductive system of female mammals; Reproductive system of male mammals; Respiration and low oxygen; Respiration in birds; Sense organs; Skin; Tails; Teeth, fangs, and tusks; Tentacles; Wings.

Bibliography

Fish, F. E. "Biomechanics and Energetics in Aquatic and Semiaquatic Mammals: Platypus to Whale." *Physiological and Biochemical Zoology* 73, no. 6 (2000): 683-698. An overview of the mechanics of aquatic respiration.

Pough, F., H. Heiser, J. B. Heiser, and W. N. McFarland. *Vertebrate Life*. 5th ed. Upper Saddle River, N.J.: Prentice Hall, 1999. Emphasis on the differences in animals based on the sequences of evolution.

Schmidt-Nielson, Knut. *Animal Physiology: Adaptation and Environment*. 5th ed. New York: Cambridge University Press, 1997. This classic textbook in animal physiology is a standard for many high school and college courses and focuses on physiologic function and adaptation to different environmental conditions.

Weibel, Ewald R. *The Pathway for Oxygen*. Cambridge, Mass.: Harvard University Press, 1984. Ewald Weibel's research presented in this book established many of the concepts and facts currently known about respiratory system structure and function.

Willmer, Pat, Graham Stone, and Ian Johnston. *Environmental Physiology of Animals*. Malden, Mass.: Blackwell Science, 2000. This textbook has simple-to-understand diagrams and explanations of animal organ systems and their adaptation to different environments.

Withers, P. C. *Comparative Animal Physiology*. Fort Worth, Tex.: Saunders, 1992. Gives a broad overview of animal physiology comparing the characteristics of each species.

RHINOCEROSES

Types of animal science: Anatomy, classification
Fields of study: Anatomy, zoology

The five rhinoceros species, in the ungulate family Rhinocerotidae, are among the world's largest land animals.

Principal Terms

DIURNAL: active during the day
KERATIN: a tough, fibrous substance plentiful in hair and horns
NOCTURNAL: active at night
PERISSODACTYL: ungulates having an odd number of toes
PREHENSILE: able to grip things
RUMINANT: a herbivore that chews and swallows plants, which enter its stomach for partial digestion, are regurgitated, chewed again, and reenter the stomach for more digestion
UNGULATE: a hoofed mammal

Rhinoceroses (rhinos), which are among the world's largest land animals, belong to the ungulate family Rhinocerotidae. There are three Asian and two African species existing today; the fossil record shows several dozen extinct species as well. The name of the animal comes from Greek *rhino* + *ceros*, meaning "nose-horned."

Physical Characteristics of Rhinoceroses

Rhinos weigh up to four tons and have short, thick, supportive legs. Rhino skin is thick, gray to brown in color, hangs loosely on the body, and is almost hairless. In the Asian species, skin folds at the junctures of the neck and limbs make them look armored. The Asian species also have have incisors and canine teeth, which are missing in the African species. Rhinos have long, prehensile upper lips, for grasping branches and removing leaves, which they eat.

Depending on the species, rhinos have one or two nose horns. In two-horned species, the horn closest to the end of the snout is longer. The horns are made of keratin, a fibrous substance that also composes hair. The horns are used for digging food, for defense, and in mating combats.

Rhinos are ungulates with three toes per foot, each of which ends in hooflike nails. Each front foot has a vestigial fourth toe. Rhinos, which are ruminants related to horses, eat grass, bulbs, leafy twigs, and shrubs. Although they look clumsy, rhinos can run as fast as horses. They have sharp vision, very good smell, and excellent hearing. Their keen hearing is due in part to their funnel-shaped ears, that swivel in different directions.

Rhino Life Cycles

Most rhinos are both diurnal and nocturnal, active in daylight hours as well as after dark. They eat during the cool mornings and evenings, staying in mud wallows during hot afternoons. Rhinos have few enemies because of their size and their dangerous horns—an angry rhino charges its attackers. Humans are rhinos' great enemies, killing them for their horns, which are used in jewelry or medicinally.

Most rhinos, especially males, live alone except during mating. There are some exceptions to solitary living: mothers live with their offspring, and young males or females may form same-sex groups. Males have territories, which are marked and defended. They fight each other for mates. While rhinos may not live with others of their own species, they almost always have symbiotic birds, called oxpeckers, living on and around them. The birds eat insects from the rhinos' skins. This gives the bird food, and frees the rhino from the insects.

Mating takes place year round, and gestation lasts up to fifteen months. The female gives birth to a baby that weighs between 100 and 150 pounds. The young rhino stays with mother for 2.5 years, though it can feed itself in 2.5 months. Rhinos mate at seven to ten years of age. Females wait for approximately three years between gestations, only becoming pregnant after the previous offspring has left them. Rhinos live for up to forty-five years.

Rhinoceros Species

There are five rhino species: three in Asia and Malaya, and two in Africa. African rhinos are two-horned and classified as "black" or "white," though all are bluish-gray. Black rhinos live in habitats from mountain forests to scrub lands. Their maximum body length is 10 feet, their height is 5 feet at the shoulder, and they weigh 1.75 tons. Each has a front horn up to 3.5 feet long. The rear horn is shorter.

Rhinoceros Facts

Classification:
Kingdom: Animalia
Subkingdom: Bilateria
Phylum: Chordata
Subphylum: Vertebrata
Class: Mammalia
Order: Perissodactyla
Family: Rhinocerotidae (rhinoceroses)
Genus and species: Ceratotherium simum (white rhino); *Diceros bicornis* (black rhino); *Rhinoceros unicornis* (Indian rhino), *R. sondaicus* (Javan rhino); *Dicerorhinus sumatrensis* (Sumatran rhino)
Geographical location: Africa and Asia
Habitat: Forests, grasslands, and scrub lands
Gestational period: Eight to seventeen months
Life span: Twenty to forty-five years
Special anatomy: Incisors and canine teeth in Asian species; very thick skin; one or two horns; three-toed feet

Despite their size and awkward-looking bodies, rhinoceroses can run as fast as horses. (Corbis)

Black rhinos are mostly nocturnal, eating in the cool morning and evening hours and wallowing in river mud during the hot daylight hours. They eat grass, leaves, herbs, fruit, branches, and twigs. In the wild, a male has a marked territory, which he defends. When the territories of several males overlap, they form groups that share resources and defend the combined territory from strangers.

White rhinos, similar to black rhinos, are the largest land mammals except elephants. Their maximum length is 13 feet, shoulder height is 6.75 feet, and they weigh 4 tons. Females use their horns for digging, defense, and guiding their offspring. Nearly extinct, white rhinos exist only in preserves.

Indian rhinos—the largest Asian rhino species—average 10 feet in length and 5.5 feet in shoulder height, and weigh 2.75 tons. They have one thick, foot-long horn; their skin is sprinkled with knobs, and folds at the limb joints make them look as though they have armor. Females, although 75 percent the weight of males, have similar body heights and lengths. These rhinos live in marshy jungles and eat reeds, grass, twigs, and plant shoots. There are 1,500 Indian rhinos, all living in preserves and protected by legislation. Javan rhinos are similar to the Indian species, but smaller. They occur only in Western Java, though they once lived in forests of Bengal, Burma, Borneo, Java, and Sumatra.

Sumatran rhinos, the smallest rhinos, have two horns. They are approximately 4 feet tall, and weigh about 1 ton. Unlike the smooth-skinned African rhinos, they are hairy, especially on the tail and ears. The few living Sumatran rhinos are in Sumatra's forested hills.

Rhinoceroses are reputedly dangerous. However, they are usually peaceful and timid, except when threatened. Legally protected rhinos suffer from the market for rhino horn, reputed to be a medicine and aphrodisiac in traditional Asian medicinal practice. This market has been a major factor in driving four of the five rhino species into endangerment.

—Sanford S. Singer

See also: Fauna: Africa; Horns and antlers; Horses; Ruminants; Ungulates

Bibliography

Cunningham, Carol, and Joel Berger. *Horn of Darkness: Rhinos on the Edge*. New York: Oxford University Press, 1997. Written by a husband-and-wife naturalist team, provides information based on their own field research on black rhinos in Namibia and discusses the species' dwindling numbers.

Penny, Malcolm. *Black Rhino: Habitats, Life Cycle, Food Chains, Threats*. Austin, Tex.: Raintree Steck-Vaughn, 2001. Full coverage of black rhinos, written for a juvenile audience.

Walker, Sally M. *Rhinos*. Minneapolis: Carolrhoda Books, 1996. This book describes rhino physical characteristics, life cycle, behavior, and conservation.

Watt, Melanie. *Black Rhinos*. Austin, Tex.: Raintree Steck-Vaughn, 1998. A brief book describing rhino life, environment, habits, and endangerment.

RHYTHMS AND BEHAVIOR

Type of animal science: Ethology
Field of study: Neurobiology

Biological rhythms are cyclical variations in biological functions that schedule and time countless physiological processes, including those necessary for behavior. Rhythms and behavior are closely associated; rhythms in hormonal and neural activity often bring about the expression of the specific behavior.

Principal Terms

BIOLOGICAL RHYTHM: a cyclical variation in a biological process or behavior, often with a duration that is approximately daily, tidal, monthly, or yearly

CIRCADIAN RHYTHM: a cyclical variation in a biological process or behavior that has a duration of about a day—from twenty to twenty-eight hours

ENDOGENOUS: refers to rhythms that are expressions of only internal processes within the cell or organism

ENTRAINMENT: the synchronization of one biological rhythm to another rhythm, such as the twenty-four-hour rhythm of a light-dark cycle

EXOGENOUS: refers to rhythms that originate outside the organism in the environment

FREE-RUNNING: denotes a rhythm that is not entrained to an environmental signal such as a light-dark cycle

FREQUENCY: the number of repetitions of a rhythm per unit of time, such as a heart rate of seventy beats per minute

PERIOD: the length of one complete cycle of a rhythm

PHOTOPERIODISM: the responses of an organism to seasonally changing day length, that cause altered physiological states

ZEITGEBER: "time giver" in German, it is also referred to as a synchronizer or entraining agent

Circadian and other biological rhythms have been observed and described in so many processes and behaviors in so many diverse organisms that their presence in higher plants and animals is considered a basic characteristic of life. The term "circadian" (from the Latin *circa*, meaning "about," and *diem*, "day") was coined by Franz Halberg to describe these approximately twenty-four-hour rhythms, which in time were found to exist not only in plants but also in animals and in human beings. The "circa" prefix is used also with words denoting other time periods (such as "circannual").

Circadian and Circannual Rhythms

Circadian rhythms enable animals to time precisely their daily activities. Animals are broadly classified as diurnal if they are active by day, nocturnal if they are active at night, and crepuscular if they are active at both dawn and dusk. Many species schedule their activity to start within minutes of the same time each day. Thus, the Swiss psychiatrist Auguste Henri Forel, in 1906, noticed that bees adapted to his schedule of eating breakfast on the terrace: The bees came each morning at breakfast time to feed on the jam. One of Karl von Frisch's coworkers, Ingeborg Beling, found that she could train bees to visit a feeding station every twenty-four hours but that the bees could not be trained to come every nineteen hours. Individual species of flowers produce nectar only at certain times of the day, and bees have been observed to plan their visits according to the time of nectar flow.

The activity rhythms of caged flying squirrels have been studied in detail by Patricia DeCoursey. She found that the time of the onset of activity in this nocturnal rodent was very uniform from day to day but that the time gradually drifted during the year from about 4:30 P.M. in January to 7:30 P.M. in July and then back to 4:30 P.M. by the following January. Such a pattern is called a circannual rhythm. Circannual rhythms are particularly evident in migratory birds, which show seasonal changes in both their physiology and their behavior. Many mammals have distinct reproductive seasons. Mammals that hibernate, such as ground squirrels and woodchucks, gain fat in the fall, enter hibernation, and then wake up in the spring according to a circannual rhythm.

Although the annual changes in temperature might be expected to be the environmental factor that would signal a change in season to a plant or animal, it is now known that many plants and animals respond to changes in the length of the photoperiod. This response is called photoperiodism. Photoperiodism was found first in plants in 1920, in insects in 1923, in birds in 1926, and in mammals in 1932. In a typical experiment, light was artificially added to the short days of late fall to create a longer photoperiod—similar to that characteristic of spring. As a result, the organisms came into reproductive development months early.

Circadian rhythms have been found to play an essential role in photoperiodism. What later became known as the Bünning hypothesis postulated that a circadian rhythm was involved in the organism's mechanism which measures the length of the photoperiod. It was hypothesized that the first twelve hours of the circadian rhythm was a light-requiring phase and the last twelve hours was a dark-requiring phase. Short-day effects occurred when the light was limited to the light-requiring phase, but long-day effects occurred when light was present during the dark-requiring phase.

In this scheme light plays two roles: It is a zeitgeber to synchronize rhythms and an inducer to stimulate reproductive responses. Later experiments demonstrated that short photoperiods followed by a brief flash of light in the middle of the dark were interpreted by the organism as long photoperiods, and the organisms became reproductively developed months early. Thus, the important thing is really not how long the photoperiod is, but rather when light is present with respect to a circadian rhythm of sensitivity to light.

Marine Rhythms

Some of the most dramatic examples of biological rhythms are found in marine organisms. The periods or lengths of the rhythms are rather diverse and include circadian, circatidal, circalunar, and circannual rhythms, and various combinations of them. Perhaps most famous is the rhythm of reproductive activity of the South Pacific marine worm referred to as the palolo worm. This species spawns at the last quarter of the moon in October and November (spring in the Southern Hemisphere). The worm lives buried in coral reefs and, at spawning, the last twenty-five to forty centimeters of the worm, which bears the gametes, breaks off and rises to the surface of the sea. The gametes are released into the seawater, where fertilization takes place. The spawning always occurs at daybreak. The exact timing of the spawning is an adaption that increases the chances for successful reproduction in this species.

Similarly, the California grunion, a small smelt about fifteen centimeters long, spawns in the spring at about fifteen minutes after the time of the high tide each month. During the spawnings, or "grunion runs," the fish ride the waves onto the sandy beaches, where the females burrow the posterior end of their bodies into the sand. The male curls around the female's body and releases sperm as the female lays her eggs. The fish return to the sea and the eggs continue to develop until approximately fifteen days later, when the high tide returns and uncovers the hatching young. During the grunion runs, the adult fish are caught by fishermen (legally only by hand) and are eaten. Neither the palolo worm nor the grunion has been sufficiently studied to determine what environmental factor—moonlight, gravity, magnetism, or another factor—synchronizes their rhythms so precisely.

Chronobiology

The broad field of the study of biological rhythms is called chronobiology. A rhythm is the cyclical repetition of a property or behavior, whether it concerns the level of body temperature, enzyme activity, or hormone level in the blood, or describes an activity of the whole animal, such as feeding patterns, daily or seasonal migrations, or seasons of reproduction. The period of the rhythm is the time it takes to complete one full cycle. This could be measured from crest to crest or trough to trough. The frequency of the rhythm refers to how many cycles occur per unit of time (such as a heart rate of seventy beats per minute). The amplitude refers to the strength of the rhythm (for example, one-half of the height of the rhythm when shown on a graph).

The properties of biological rhythms are fascinating. They are ubiquitous, innate, probably endogenous, free-running, self-sustaining, entrainable, relatively temperature-independent, and relatively unsusceptible to chemical perturbations. Biological rhythms are said to be ubiquitous because they are found everywhere—at all levels of life from cell organelles to cells, tissues, organs, whole organisms, and populations. They are found in all kinds of living things, with the possible exception of the prokaryocytes. They are said to be innate because the rhythms are not learned and are largely programmed by the genetic makeup of the organism. Biological rhythms are probably endogenous, with an oscillator inside the cells of the organism, but it should be noted that Frank A. Brown has published extensive evidence that the timing information may be exogenously derived from geophysical fluctuations. Biological rhythms are entrainable, which means that they usually are kept in synchrony with day/night or other environmental schedules. Entrainment is maintained by an organism's responses to environmental factors called synchronizers, zeitgebers, entraining agents, or time cues. Light, temperature, noise, and feeding are some of the zeitgebers that have been identified. The rhythms are called self-sustaining because they continue in the absence of any obvious zeitgebers. Biological rhythms have been found to be relatively temperature-independent, which is important because they often function as clocks. Biological rhythms can free-run when isolated from zeitgebers. When a rhythm free-runs, its period is found to be slightly different from the entrained period.

Despite years of investigation, biological rhythms are poorly understood. The search continues to find biological bases for the rhythmic processes so commonly seen. The innermost rhythmic process is sometimes referred to as the "biological clock," since it represents the seat of the cell's or organism's timekeeping mechanism.

Studying Biological Rhythms

One of the earliest scientific observations of a biological rhythm was reported in 1729 by Jean Jacques d'Ortous de Mairan, a French astronomer. He made detailed observations of the reactions to constant darkness of a so-called sensitive plant that normally has its leaves unfolded during the daylight hours and folded during the night. De Mairan wondered whether the leaves respond directly to the presence of the sunlight and therefore open at dawn and close at dusk. Placing the plant in constant darkness to see how it responded, de Mairan found that the plant continued to show the rhythmic folding and unfolding of its leaves. The curious results were published in a brief report in the Proceedings of the Royal Academy of Paris.

Several years later, in 1758, Henri-Louis Duhamel repeated de Mairan's experiment and further observed that warm temperatures failed to alter the pattern of the rhythmic opening and closing of the leaves of the sensitive plant. Later studies, in the nineteenth century, revealed that in the sensitive plant *Mimosa pudica* the rhythmic opening and closing of leaves in constant dark completed a full cycle in 22 to 22.5 hours. It was found that plants supplied with lamps during the night and kept in darkness during the day adapted to the new schedule within a few days and unfolded their leaves only during the artificial day. Charles Darwin did experiments that convinced him that plants survived frosts more successfully when they could fold their leaves at night.

The extent to which circadian or other biological rhythms are endogenous (originate inside the organism) has been a subject of debate. Frank A. Brown spent most of his research career trying to resolve this question. Brown found that even when organisms were placed in heavy metal chambers that were airtight, it was virtually impossible to isolate an organism from its rhythmic, geophysical environment. Normally, circadian rhythms keep in synchrony with the day/night cycle (supposedly they are reset slightly each day, since they are not exactly twenty-four hours long). Brown studied in detail free-running rhythms—that is, rhythms that are found in organisms in a seemingly constant environment. When he averaged oxygen uptake data over many months, he found exact geophysical rhythms of twenty-four hours as well as exact lunar and annual rhythms in the metabolism of many different organisms, such as potatoes, carrots, hamsters, and rats. Furthermore, he showed that many animals, such as snails and flatworms, are influenced by subtle changes in the earth's magnetic field. Therefore, he concluded that the actual timing information that underlies circadian and other biological rhythms may well be exogenous and derived from the rhythmic, geophysical environment that pervades the organisms' everyday surroundings. Despite such evidence for exogenous influences, most biologists today regard biological rhythms as the product of essentially endogenous processes.

Endocrine and Nervous System Rhythms

When looking for some basis for endogenous rhythms, researchers often investigate the endocrine and nervous systems because of their large roles in integrating and controlling biological functions. An especially interesting study has been made by Albert H. Meier. He has found that endocrine rhythms play a role in regulating the seasonal changes of physiology and behavior in the migratory white-throated sparrow and other vertebrates. By injecting birds with the hormones corticosterone and prolactin in different time relationships, he was able to induce seasonal changes. In early studies, he found that injections of prolactin either caused fat gain or fat loss, depending simply on whether the injections were given in the morning or in the afternoon. Migratory birds gain fat before they migrate and use this fat as an energy source for their flights. If the birds are given daily injections of corticosterone and prolactin four hours apart, the birds gain fat, try to fly from the south side of their cages, and do not have well-developed gonads—all characteristics of the normal fall bird. If the birds are given daily injections of corticosterone and prolactin eight hours apart, they remain lean, do not show any directed flight, and do not have well-developed gonads—traits characteristic of the normal summer bird. On the other hand, if the birds are given daily injections of cortiscosterone and prolactin twelve hours apart, the birds gain weight, try to fly out the north sides of their cages, and have gonads that will grow in response to a lengthening photoperiod—traits characteristic of the normal spring bird. Some assays, using radioactive isotopes of corticosterone and prolactin, have been made in wild populations of white-throated sparrows, and the results show that the timing of the peaks of the hormones is roughly similar to the time relationships just discussed. Further research by this group centered on the modification of brain chemistry to bring about the seasonal changes in vertebrates.

The methods used to study biological rhythms range from the simple methods used by pioneers in the field to the latest innovations in molecular biology. The field is attracting many new researchers; new discoveries are being published almost daily. Yet much remains to be done before the essential nature of biological rhythms can be understood.

Implications of Biological Rhythms

There are many implications to the fact that plants and animals possess circadian and other biological rhythms. Some scientists have speculated about whether man can survive living in space vehicles that leave the geophysical environment of the earth-moon complex. Will it be necessary, they ask, to try to duplicate parts of the terrestrial geophysical environment—by, for example, installing a rhythmic magnetic field in the space vehicles?

More mundane applications of a better knowledge of rhythms are to be found in animal husbandry. The annual rhythm of the reproduction of farm animals can be manipulated to result in higher productivity. It is a standard practice to lengthen the photoperiod in the henhouse to increase egg production and minimize the winter decrease in production. Sheep are treated with the hormone melatonin, naturally produced in more abundance in the winter months, in the early fall to hasten the reproductive season. Many more benefits of a better understanding of biological rhythms await discovery.

—John T. Burns

See also: Ethology; Hormones and behavior; Instincts; Mating; Migration; Reproduction.

Bibliography

Arendt, J., D. S. Minors, and J. M. Waterhouse, eds. *Biological Rhythms in Clinical Practice*. London: Butterworth, 1989. Part 1 has an overview of the nature of biological rhythms understandable to the layperson; however, the rest of the book is a highly technical review of findings in medical chronobiology by seventeen of the best-known researchers in the field.

Brady, John, ed. *Biological Timekeeping*. Cambridge, England: Cambridge University Press, 1982. A somewhat technical review of rhythmic phenomena, photoperiodism, and circadian physiology and clock mechanisms. Each of ten topics is covered by an expert in the field. Rhythms in insects are better represented here than usually is the case. A glossary is included.

Cloudsley-Thompson, J. L. *Rhythmic Activity in Animal Physiology and Behaviour*. New York: Academic Press, 1961. A classic text at the college level. The book is written from the viewpoint of a physiological ecologist. The book has an excellent bibliography of early research papers.

Follett, Brian K., Susumu Ishii, and Asha Chandola, eds. *The Endocrine System and the Environment*. New York: Springer-Verlag, 1985. This collection of research papers contains an article by A. H. Meier and J. M. Wilson, "Resetting Annual Cycles with Neurotransmitter-Affecting Drugs," which is of interest for anyone studying rhythms and the function of the brain. Most of the articles are on birds and fish; no invertebrates are included. For the advanced college student or the graduate student seeking experimental details.

Moore-Ede, Martin C., Frank M. Sulzman, and Charles A. Fuller. *The Clocks That Time Us*. Cambridge, Mass.: Harvard University Press, 1982. One of the best overviews of the subject for a college-level reader. The chapters "The Neural Basis of Circadian Rhythmicity" and "Medical Implications of Circadian Rhythmicity" are especially noteworthy. Some sections are heavy with experimental details. The squirrel monkey is frequently used as an example, since the authors have done research on this species.

NATO Scientific Affairs Division. *Rhythms in Fishes*. NATO ASI Series. Series A, Life Sciences 236. New York: Plenum, 1992. Discusses fish biorhythms and their impact on aquaculture, as well as the effect of pollution on fish biorhythms. Aimed primarily at fish researchers.

Palmer, John D. *An Introduction to Biological Rhythms*. New York: Academic Press, 1976. An easily read introduction to the broad field of biological rhythms; well illustrated and filled with the excitement of scientific discovery. Includes essays by important researchers in the field.

Ward, Richie R. *The Living Clocks*. New York: Alfred A. Knopf, 1971. A saga of the personalities and discoveries that make up the story of the development of the field. Contains a fascinating chapter on Frank A. Brown's theories. One of the most easily read books on biological rhythms. Highly recommended for all readers.

RODENTS

Type of animal science: Classification
Fields of study: Anatomy, vertebrate biology, physiology

About two thousand species of rodents, the gnawing mammals, are classified in the order Rodentia. They are found worldwide, in trees, on the ground, underground, and in partly aquatic environments. Rodents are fur sources, pets, laboratory animals, and pests.

Principal Terms

CUSPID: a tearing tooth found in the mouth of a carnivorous animal

HERBIVORE: any animal that subsists entirely on plant foods

INCISOR: a cutting tooth which acts like scissors or a chisel

MOLAR: a flat tooth found at the back of the jaw and used to grind food

OMNIVORE: an animal which eats both plants and other animals

Rodents, comprising about two thousand species, form the largest, most abundant mammal order. They are found almost everywhere on the earth. Most are ground dwellers and many rodent species dwell underground in burrows or tunnel networks of varying complexity and size. However, rodents also dwell in tree nests (squirrels) or lodges in ponds and streams (beavers), or simply run in herds (capybaras). Judging from fossil remains, rodents were widespread and plentiful fifty million years ago. It is believed that they evolved from small, insect-eating mammals, and did not develop into large species until a million years ago. The largest ancient rodents were giant, bear-sized beavers. Contemporary rodents are usually small. However, the largest modern rodents are herbivorous capybaras, which grow to approximately 100 pounds as adults.

Rodents also show remarkable diversity in their diets. These range from the vegetarian capybaras to the all-encompassing diet of omnivorous rats, which will eat meat. Rodents have many roles relative to humans. Hamsters and other small rodents are pets, capybaras are eaten as food, chinchillas are fur sources, and a few, such as rats and mice, are pests that compete with humans for their food crop supplies. The tremendous adaptability of rodents, especially rats, explains their wide geographical distribution in areas differing hugely in climate.

Physical Characteristics of Rodents

Among the two thousand known rodent species, size varies widely. Some small adult mouse species weigh about a fifteenth of a pound. At the other extreme, capybaras, largest of contemporary rodents, are the size of pigs.

Regardless of size, all rodents possess pairs of large, chisel-like front teeth in both the upper and lower jaws. The roots of these incisor teeth are located far back in rodent jawbones and grow continuously. Rodents lack the tearing teeth (cuspids) of carnivores as well as several premolars. Therefore, a large space exists between their incisors and molars. This allows the incisors to operate well in gnawing. The design of rodent dentition also allows the gnawed food to be transferred easily to the molars for efficient grinding. In addition, the muscles of the rodent lower jaw are arranged so as to enable its easy movement backward, forward, and laterally. This optimizes grinding of gnawed food.

Rodent incisors are different from those in other animals. Their continued growth from the

Groundhogs, like many rodents, live in underground dens where they hibernate through the winter. (Corbis)

root is valuable, especially because only the front surfaces of these teeth are protected by enamel, the hardest material in teeth. Thus, gnawing food causes the rear surfaces of the teeth to wear down faster than their front surfaces. This wear pattern is the basis for development of the chisel-like incisor edges. It continues as long as a rodent eats regularly, keeping the incisors sharp. Another interesting aspect of rodent mouths is that cheek fur grows inside the mouth and fills up the space between incisors and molars. This hair acts as padding and filters out food chunks too large to be swallowed comfortably.

Other than the special development of "gnawing machinery" of the mouth and teeth, rodents are anatomically unspecialized, with no other ubiquitous anatomic features. Where any special characteristic has developed in some rodents, it appears to be due to environmental need. For example, claws and front paws of burrowing rodents, such as woodchucks and moles, make them efficient diggers. In addition, gliding adaptations in some squirrels allow them to "fly" (or actually glide) from tree to tree. Furthermore, leaping rodents such as the kangaroo rat use both hind feet together to enhance leaping capability. Yet another such adaptation is the webbed feet seen in beavers.

The Lives of Rodents

Rodents, like all other mammals, are warm-blooded. They carry offspring to term in a uterus where each fetus is connected to the mother via the placenta, give birth to them, and nurse them. Depending on the rodent, the sequence of events between fertilization and the end of the nursing period takes between 5.5 weeks for a small mouse, to well over a year for large rodents. The process is easiest to describe for rats, although it is quite similar for mice and hamsters.

After fertilization, rat eggs make their way into a complex uterus which can hold eight to sixteen fetuses. There, each attaches to the uterine wall and develops, over three weeks, into a rat pup. The pups are born pink, hairless, blind, and incompletely developed. They are then nurtured by their mothers, who have the instinct of all mammals to care for their offspring. Rats breast-feed their pups for three weeks. At the end of this time, they are fully covered in hair, have full vision, and have begun to eat foods other than milk.

In another month the pups are sexually mature and can breed. This makes it clear why omnivorous wild rats pose a threat to humans. Any pair of rats can produce up to eighty offspring per year. Furthermore, within six weeks after birth, any two offspring can, and do, reproduce.

Inbred laboratory rats live for two to three years, depending on the strain. Males are much larger than females (often twice their size) and may attain body weights up to two pounds. In the wild the life expectancy of rats varies greatly. However, reports of animals living for over five years occur. Some males have been reported to be as large as small cats or dogs. Wild rats live in complex tunnels as colonies of a hundred or more animals.

Other rodents live different versions of the life of rats. Litter size, gestation time, group organization all vary. For example, the larger rodents have only a few offspring per litter, and some rodents live in tree nests (squirrels) or lodges in ponds (beavers). Life expectancies may be ten years or more, assuming death by natural causes.

Destructive and Beneficial Rodents

Rats and mice interact extensively with humans in a destructive fashion. The problems involved are

Rodent Facts

Classification:
Kingdom: Animalia
Subkingdom: Bilateria
Phylum: Chordata
Subphylum: Vertebrata
Class: Mammalia
Order: Rodentia (rodents)
Suborders: Sciurognathi (squirrel-like rodents), Myomorpha (mouselike rodents), Caviomorpha (cavylike rodents)
Families: Castoridae (beavers, 1 genus, 2 species); Aplodontidae (mountain beaver); Sciuridae (squirrels, 49 genera, 267 species); Geomyidae (pocket gophers, 5 genera, 34 species); Anomaluridae (scaly-tailed squirrels, 3 genera, 7 species); Heteromyidae (pocket mice, 5 genera, 65 species); Pedetidae (springhares); Muridae (rats and mice, fifteen subfamilies, 241 genera, 1,082 species); Gliridae and Selevinidae (dormice, 8 genera, 11 species); Zapodidae (jumping mice and birchmice, 4 genera, 14 species); Dipodidae (jereboas, 11 genera, 31 species); Erethizontidae (northwestern porcupines, 4 genera, 10 species); Caviidae (cavies, 5 genera, 14 species); Hydrochoeridae (capybaras); Myocastoridae (coypus); Capromyidae (hutias, 4 genera, 13 species); Dinomyidae (pacaranas); Agoutidae (pacas, 1 genus, 2 species); Dasyproctidae (agoutis and acouchis, 2 genera, 13 species); Abrocomidae (chinchilla rats, 1 genus, 2 species); Echimyidae (spiny rats, 15 genera, 55 species); Chinchillidae (chinchillas and vicuñas, 3 genera, 6 species); Octodontidae (degus, 5 genera, 8 species); Ctenomyidae (tuco-tucos, 1 genus, 33 species); Thryonomyidae (cane rats, 1 genus, 2 species); Petromyidae (African rock rat species); Hystricidae (Old World porcupines, 4 genera, 11 species); Ctenodactylidae (gundis, 4 genera, 5 species); Bathyergidae (African mole rats, 5 genera, 9 species)

Geographical location: Every continent except Antarctica

Habitat: Mostly on land or underground, in forests, plains, and deserts; some live in a partly freshwater environment, using ponds or streams

Gestational period: Varies greatly, though generally two weeks in a mouse and about a month in a rat

Life span: Most smaller species live for one to three years, while large rodents survive for over ten years

Special anatomy: All rodents possess incisor teeth designed for gnawing, which grow continuously from the roots and wear away at their tips, giving them chisel-like edges that can gnaw through very hard materials; however, they lack cuspids (tearing teeth) seen in carnivores

Several Kinds of Rodents

BEAVERS, the largest Northern Hemisphere rodents, may weigh over fifty pounds. They live in dammed streams and in ponds, in lodges made of logs, rocks, and mud. Prized for their fur, beavers are vegetarians that eat tree bark and the roots, stems, and twigs of aquatic plants.

CAPYBARAS, the largest rodents, weigh about one hundred pounds. Found in Central and South America, they are land animals, travel in herds, and eat grass.

CHINCHILLAS, originally South American burrowers, are the source of valuable, silky fur. In the wild, they are colonial and venture above ground only at night to forage for vegetable foods.

CHIPMUNKS are small ground squirrels about ten inches in length. They are most often brown, with characteristic black and white stripes along their backs, and short tails.

GERBILS are desert rodents which dig extensive burrows. These gray, brown, or black rodents are related to rats but are more streamlined, and have prominent ears and furred tails.

MUSKRATS are semiaquatic rodents, named for musk glands in their hindquarters. Their luxuriant fur is used in fur garments.

PORCUPINES are arboreal or ground-living rodents with fur modified into defensive quills. When endangered, they backpedal at high speeds to impale predators with sharp quills that inflict painful wounds.

TREE SQUIRRELS are arboreal climbers and leapers. They are from four inches to several feet long and feed mostly on nuts.

RATS and MICE belong to the largest rodent family, which also includes hamsters. "Rat" denotes the large species, while smaller species are "mice." Often, they are crop-eating pests. Rats and mice are often used as lab animals.

competition for food, and disease transmission from rodents to humans. Rats and mice, viewed as pests, are known to eat 10 to 25 percent of grain crops grown, harvested, and stored worldwide. This percentage varies depending upon the extent of use of rodenticides, such as warfarin, in various nations and the extent of agricultural technology. Very careful use of rodenticides is important because they are quite toxic to humans.

Rodents are disease vectors, historically causing outbreaks of serious epidemics of the bubonic plague and tularemia. This was especially serious during the Middle Ages, when rats were responsible for the transmission of the Black Death. Currently, most sporadic outbreaks of rodent-derived infectious disease are handled by use of rodenticides to kill carriers and antibiotics to destroy rodent-borne microorganisms that infect humans. Most often it is not the rodents themselves that cause disease outbreaks. Rather, infection occurs as contaminated fleas and ticks move from rodents to humans. Rats are seen as the main disease vectors because they abound near and in human habitations. However, mice and any other infected rodents can be disease vectors.

Concerning beneficial use of rodents, one can point to the myriad rats, mice, hamsters, and guinea pigs utilized as laboratory animals in testing and developing pharmaceuticals, the identification of toxic cosmetic, paint, and food components, isolation of disease cures, and so on. This aspect of research is likely to become less common because a large segment of the population deems it morally inappropriate to submit animals to these testing procedures.

Another benefit of rodents that is becoming morally unacceptable is harvesting rodent fur. Beaver fur was once hugely important to the world fur trade. Presently, as beaver are nearly extinct, the use of rodents to provide fur for human use has shifted to muskrats, nutria, and chinchillas, which are valued for their attractive, luxuriant coats.

—Sanford S. Singer

See also: Beavers; Diseases; Gophers; Home building; Mice and rats; Porcupines; Squirrels.

Bibliography

Alderton, David. *Rodents of the World*. New York: Facts on File, 1996. Lots of information and numerous fine illustrations of rodents of the world, as well as useful bibliographical references.

Lacey, Eileen A., James P. Patton, and Guy N. Cameron, eds. *Life Underground: The Biology of Subterranean Rodents*. Chicago: University of Chicago Press, 2000. Offers considerable information on rodents and their life cycles underground. Many useful illustrations and a solid bibliography.

Stoddart, D. Michael, ed. *Ecology of Mammals*. New York: John Wiley & Sons, 1979. This classic text describes and compares many ecological aspects of mammal life, including rodents. It is well-designed, nicely illustrated, with a good bibliography.

Webster, Douglas, and Molly Webster. *Comparative Vertebrate Morphology*. New York: Academic Press, 1974. Excellent description and comparison of vertebrate anatomy, morphology, and histology, done in a clear and comprehensive fashion.

ROUNDWORMS

Type of animal science: Classification
Fields of study: Invertebrate biology, genetics, physiology

The phylum Nematoda consists of the roundworms. Nematodes are thought to be the most abundant organisms on earth, with some scientists estimating that nearly one million separate species of roundworms exist. Nematodes occupy a variety of habitats including soil, water, and even vinegar malts, and several species are significant animal and plant parasites.

Principal Terms

CUTICLE: the outer, noncellular covering of the nematode

PHARYNX: a muscular organ that is used to pump food into the digestive system of the nematode

PHASMID: sense organs located in the tail region of roundworms that are important for detecting chemical signals in the environment

PSEUDOCOEL: the unlined body cavity of the roundworm

Roundworms, also called threadworms, constitute the phylum Nematoda. Eighty thousand species of nematodes have been described, and four out of every five animals on earth are roundworms. Nematodes occupy almost every environment imaginable, ranging from soil to fresh and salt water and even vinegar and beer malts. One handful of soil generally contains thousands of free-living nematodes.

Nematodes are most notable for the huge economic impact they have as crop and animal pests. Plant parasitic nematodes cause billions of dollars in agricultural losses annually. Several species of nematodes are also human parasites. Nematode parasites and their hosts may have simple associations, in which the nematode uses the host as a means of transport from one food source to another, or more complex associations, in which the nematode is essentially a predator devouring the host from the inside.

Anatomy

Nematodes are unsegmented worms with a pseudocoelom and are round in cross section (hence the common name roundworms). The worms range in length from 0.1 millimeter to 9 meters, and most roundworms are transparent. All nematodes share a basic body plan of two concentric tubes separated by a fluid-filled space called a pseudocoelom. The outer tube is covered by a noncellular cuticle composed of the protein collagen, which is secreted by the cells immediately underneath. The inner tube is composed of the pharynx and digestive canal and includes the nervous system.

Most of the cells of the nervous system have their cell bodies clustered in a nerve ring that surrounds the pharynx. The pharynx is a muscular structure that pumps food into the worm's intestine. Roundworms have no specialized organs for circulation or excretion. Exchange of oxygen and carbon dioxide occurs across the body wall.

Roundworms move using a wavelike motion that relies on the contraction of the four sets of muscles that run the length of worm. The worms have no circular muscles, so they cannot expand and contract like segmented worms.

Reproduction

Reproduction of nematodes is sexual by internal fertilization. Most nematode species are dioecious (having both males and females), with the notable exception of the genetic model system, *Caenorhabditis elegans*. Males have specialized organs in the tail, copulatory spicules, which aid in depositing sperm into the vulva of a female. Nematode sperm do not swim using a flagellum, but instead are amoeboid and move by a crawling motion using a pseudopod.

Nematodes reproduce in massive quantities. Scientists have documented the production by the females of some species of up to twenty-seven million eggs and the laying of as many as one million fertilized eggs per day. Some parasitic species, such as the plant-root parasite, *Meloidogyne incognita*, can reproduce via parthenogenesis. Sex determination can be influenced by environmental factors in some species.

Roundworms hatch directly from eggs and undergo four molts before they become adults capable of reproducing. The life span and generation time varies; however the well-studied, free-living species, *C. elegans*, has a normal life span of two weeks and a generation time of three days.

Roundworm Facts

Classification:
Kingdom: Animalia
Subkingdom: Bilateria
Phylum: Nematoda
Classes: Adenophorea (having no phasmids, mostly aquatic, free-living species, some plant and animal parasites); Secernentea (having phasmids, mostly terrestrial, free-living species, some plant and animal parasites)
Orders: Adenophorea—Araeolaimida, Monohysterida, Demosdorida, Chromadorida, Desmoscolecida, Enoplida, Dorylaimida; Secernentea—Rhabditida (mostly terrestrial, free-living worms), Tylenchida (plant and invertebrate parasites), Aphelenchia (plant and insect parasites), Strongylida (vertebrate parasites), Ascaridida (mostly vertebrate parasites), Oxyurida (vertebrate and invertebrate parasites), Spirurida (vertebrate and invertebrate parasites)

Geographical location: Found worldwide

Habitat: Soil, freshwater, and salt water; extreme habitats such as decaying cacti and vinegar malts; several species are plant or animal parasites

Gestational period: Varies with species

Life span: Varies with species; specialized dauer larvae are dormant stages resistant to drying and can survive for months under adverse environmental conditions

Special anatomy: External cuticle made of collagen; cylindrical bilaterally symmetrical organisms with a pseudocoelom; exchange of oxygen and carbon dioxide occurs across the body wall

Parasitic Nematodes

In tropical areas, nematodes infect humans, causing filarial diseases that affect 120 million people. With the disease elephantiasis, failure of the lymphatic system results in gross swelling of the limbs and genitals. *Trichinella spiralis* is the causative agent of trichinosis, a disease acquired by eating undercooked, infected pork or other meat. The worms encyst in the meat, hatch in the gut of the consumer, and migrate to the skeletal muscle of the infected individual, resulting in muscle pain and an itching sensation for the infected person.

Nematodes are also important parasites of domestic animals. Dog and cat heartworm is caused by *Dirofilaria immitis*. Pigs are often infected by *Ascaris* species. *Haemonchus contortus* is a gut parasite of sheep that can result in anemia, weight loss, and even death of the host.

In addition to animal parasites, plant parasitic nematodes pose a serious agricultural problem. The root-knot nematodes, *Meloidogyne* species, are the most serious plant pests, infecting more than two thousand plant species. These worms hatch in soil as second stage larvae; from the soil, they invade the root tip of the host plant and feed directly from the cells of the plant.

Caenorhabditis elegans

In spite of the diverse parasites of this phylum, most roundworms are beneficial, aerating the soil and breaking down decaying matter. One free-living soil nematode that has provided much information about development and genetics is *Caenorhabditis elegans*. This millimeter-long worm is the subject of intense investigation on a cellular and molecular level. Studies in *C. elegans* have led to a greater understanding of genes mutated in Alzheimer's disease, an understanding of programmed cell death as a normal part of development in multicellular organisms, and to the development of the technologies used to sequence the human genome.

—Michele Arduengo

See also: Flatworms; Genetics; Regeneration; Reproduction; Reproductive strategies; Worms, segmented.

Bibliography

Bird, Alan F., and J. Bird. *The Structure of Nematodes*. 2d ed. San Diego, Calif.: Academic Press, 199. Provides detailed information about the anatomy and physiology of parastic and free-living nematodes.

Hartwell, L., L. Hood, M. Goldberg, A. Reynolds, L. Silver, and R. Veres. *Genetics: From Genes to Genomes*. Boston: McGraw-Hill, 2000. Chapter 20 contains a summary of *C. elegans* biology.

Mai, William F., and Peter G. Mullin. *Plant Parasitic Nematodes: A Pictorial Key to Genera*. 5th ed. Ithaca, N.Y.: Cornell University Press, 1996. Organizes the plant parasites of this phylum based on easily observable characteristics. Includes information on taxonomy, and extensive references.

Margulis, Lynne, and Karlene V. Schwartz. *Five Kingdoms: An Illustrated Guide to the Phyla of Life on Earth*. 3d ed. New York: W. H. Freeman, 1998. Contains a summary of the phylum Nematoda.

Riddle, Donald L., T. Blumenthal, B. Meyer, and J. Priess, eds. *C. elegans II*. Plainview, N.Y.: Cold Spring Harbor Laboratory Press, 1997. Includes information on the anatomy, evolution, and classification of nematodes, as well as a chapter on parasitic nematodes.

RUMINANTS

Types of animal science: Anatomy, classification
Fields of study: Anatomy, zoology

Ruminant herbivores regurgitate and chew food a second time, after swallowing it. This maximizes nutrient intake.

Principal Terms

ARTIODACTYL: a herbivore that walks on two toes, which have evolved into hoofs
CARNIVORE: an animal that eats only animal flesh
ESOPHAGUS: the tube through which food passes from mouth to stomach
GESTATION: the term of pregnancy
HERBIVORE: an animal that eats only plants
NUTRIENT: a nourishing food ingredient
OMNIVORE: an animal that eats both plants and animals

Ruminants are herbivorous animals that store their food in the first chamber of the stomach, called the rumen, when it is first swallowed, then after some digestion has taken place, regurgitate it as "cud," which is chewed again and reswallowed into another chamber of the stomach for further digestion. This maximizes the amount of nutrition the animal is able to derive from hard-to-digest plant food. Wild ruminants tend to eat very quickly, getting as much food mass into their rumens as possible, then retiring to places of safety where they can digest at their leisure. Ruminants include sheep, cows, camels, pronghorns, deer, goats, and antelope. They eat lichens, grass, leaves, and twigs.

The main ruminant suborders are the Tylopoda and Pecora. Tylopods have three-chambered stomachs. Examples are camels and llamas. Pecora are sheep, goats, antelope, deer, and cattle. Most pecorids have horns or antlers. These true ruminants have four-chambered stomachs, whose compartments are called the rumen, reticulum, psalterium, and abomasum. The abomasum is most similar to the stomachs of nonruminant mammals, while the other three compartments are developments peculiar to ruminants.

Ruminants chew or grind food between their lower molars and a hard pad in the gums of the upper jaw. The rumen collects partly chewed food when it is first swallowed. The food undergoes digestion in the rumen and passes into the reticulum. There, it is softened by further digestion intocud. Then the reticulum returns the cud to the mouth for rechewing, so that it is mixed with more saliva. Swallowing the chewed cud next sends food to the third compartment, the psalterium, for more digestion. The psalterium empties into the abomasum, where food mixes with gastric juices and digestion continues. Finally, the food enters the intestine, which absorbs nutrients that are then carried through the body by means of blood.

The stomachs of ruminants also contain numerous microorganisms. These help to break down the cellulose in plants, and also protect the ruminant from the effects of any toxins used as defense mechanisms by the plants.

Domesticated Ruminants

Cattle, sheep, goats, and reindeer are all domesticated ruminants, although there are still wild species of each. The world population of these ruminants exceeds four billion. Ruminants are useful food sources for humans because they are large mammals, providing a lot of meat, as well as milk, wool, hide, and fuel. Yet, because they eat plants, they are low on the food chain; since 90 percent of the energy from any food source is lost in digestion, domesticated ruminants are relatively efficient transmitters of food energy from plants to humans. The large size of these ruminants, however, means that their metabolisms are relatively slow, and thus they can afford the time it takes to digest grasses, leaves, and twigs through rumination, whereas smaller herbivores with higher metabolisms, such as rodents, must digest food more quickly and thus eat more nutrient-rich plant food, such as seeds.

Some Wild Ruminant Species

Wild ruminants are important to food chains because they eat plants, thereby preventing plant overgrowth. They also are eaten by carnivores and omnivores. Bactrian camels, with three-chambered, tylopod stomachs, inhabit the steppes and mountains of the Gobi Desert. These two-humped camels are domesticated as food and draft animals in Afghanistan, Iran, and China. Bactrian camels subsist on a diet of grass, leaves, herbs, twigs, and other plant parts. Their humps contain stored fat. Given the extreme aridity of their native environment, ruminant digestion allows them to derive the maximum nutrition from scarce food supplies.

Chamois goats live in the mountains of Europe and southwestern Asia. Their diet consists of grass and lichens in the summer, while in winter they eat pine needles and bark. Pronghorns live in the open plains and semideserts of the North American West. They are the only living Antilocapridae, relatives of antelope. These true ruminants eat herbs, sagebrush, and grasses in the summer, and dig under the snow for grass and twigs in the winter. The large reindeer of northern Europe and Asia inhabit forests, grasslands, and mountains. Reindeer eat grass, moss, leaves, twigs, and lichens. Sable antelope live in southeastern African woodlands and grasslands, where they eat grass and shrub leaves and twigs.

—Sanford S. Singer

Ruminant Methane Emissions

Ruminant digestion is essentially a process of fermentation, and one of its side products is huge quantities of methane gas. In fact, it is now estimated that livestock account for 15 percent of the methane in the atmosphere, actually contributing to the greenhouse effect on global warming. An adult cow will produce between thirty and fifty liters of methane per hour, while a sheep or goat will produce around five liters. This gas is released by almost continuous belching, with the gas traveling up the esophagus at the rate of 160 to 225 centimeters per second. Anything that prevents a ruminant from belching is life-threatening—the rumen blows up like a gas-filled balloon to the point that the animal asphyxiates.

—Leslie Ellen Jones

Ruminants, such as cows, have four-part stomachs where food is digested into cud, returned to the mouth for further chewing, and digested again to obtain maximum nutrition. (PhotoDisc)

See also: American pronghorns; Antelope; Camels; Cattle, buffalo, and bison; Deer; Digestion; Digestive tract; Food chains and food webs; Goats; Ingestion; Reindeer; Sheep.

Bibliography

Church, D. C., ed. *The Ruminant Animal: Digestive Physiology and Nutrition*. Prospect Heights, Ill.: Waveland Press, 1993. The standard reference for ruminant physiologists.

Constantinescu, Gheorghe M., Brian M. Frappier, and Germain Nappert. *Guide to Regional Ruminant Anatomy Based on the Dissection of the Goat*. Ames: University of Iowa Press, 2001. An introduction to ruminant anatomy, written for veterinary students.

Cronje, P. B., E. A. Boomker, and P. H. Henning, eds. *Ruminant Physiology: Digestion, Metabolism, Growth, and Reproduction*. New York: Oxford University Press, 2000. A collection of papers from a symposium on ruminant zoology.

Wilson, R. T. *Ecophysiology of the Camelidae and Desert Ruminants*. New York: Springer Verlag, 1990. Explores the adaptations of camels and other ruminants to desert life.

SALAMANDERS AND NEWTS

Types of animal science: Anatomy, classification
Fields of study: Anatomy, conservation biology, herpetology, systematics (taxonomy)

Nearly four hundred species of salamanders and newts are classified in the order Caudata. The two areas of greatest diversity of species in the world are in the southeastern United States and in Mexico and Central America.

Principal Terms

CLOACA: the common opening through which the products of the urinary, intestinal, and reproductive systems are expelled from the body

COSTAL GROOVES: parallel grooves or folds on the side of a salamander's body

EXTERNAL FERTILIZATION: the union of eggs and sperm in the environment, rather than in the female's body

GRANULAR GLANDS: one of many kinds of glands in the skin of salamanders and newts that secrete toxins for defense from predators

METAMORPHOSIS: in salamanders, the transformation of a larval salamander with gills into a juvenile salamander without gills

SPERMATOPHORE: a tiny, mushroomlike structure deposited by a male salamander for transferring sperm to a female during courtship

Salamanders and newts comprise one of three groups of amphibians living today. The other two, the caecilians (Gymnophiona) and the frogs and toads (Anura) can be readily distinguished by their body forms. Like other amphibians, salamanders and newts have glandular skin that lacks scales, feathers, or hair. Considering only species living today, salamanders and newts are a small group compared to the number of species of frogs and toads. Whereas frogs and toads are represented by about four thousand species, only about four hundred species of salamanders and newts are living today.

Systematists, biologists who study the classification of plants and animals, recognize ten families of salamanders. Newts are simply salamanders that are classified in the family Salamandridae; they can be distinguished from other salamanders by many osteological (bony) features and by their generally rough skin, compared to the smooth skin of other salamanders.

Anatomy of Salamanders and Newts

Salamanders have long, lizardlike bodies with long tails and four small legs. Many species have costal grooves along the sides of the body; the number of these grooves varies among species and can help with identification. Olfaction (sense of smell) is used during courtship, and males of many species have specialized glands on the body. Males of many species of lungless salamanders have a gland on the chin that is used to deliver hormones to the female during courtship.

Salamanders have a larval stage, but unlike frogs and toads, in which the tadpole is very different from the adult frog, larval salamanders are similar in body form to adults. Larval forms are frequently found in water and retain external gills, which are lost at metamorphosis (transformation to the adult stage). Species that breed in ponds, where oxygen levels are low, have large, bushy gills for added surface area to increase the intake of oxygen. In contrast, species that breed in streams, which have high oxygen levels, have larvae with short gills.

One of the most successful groups of salamanders are the Plethodontidae, the lungless salamanders. Most species of these salamanders live and breed on land, never entering water. They have no lungs, and oxygen uptake occurs primarily through the thin, porous skin. One requirement for this gaseous exchange is moisture, and these salamanders live primarily in damp, cool forests.

Life History of Salamanders and Newts

Primitive families of salamanders deposit eggs in water and have aquatic larvae. Other families of salamanders are unique among amphibians in producing a spermatophore for the transfer of sperm from the male to the female. The spermatophore is a gelatinous structure with a sperm cap resembling a tiny mushroom. The spermatophore is transferred from the male to the female in an elaborate courtship ritual. In some species, the male rubs secretions from a gland under his chin over the body of the female and entices her to follow him about. He then deposits a spermatophore on the substrate of the pond, which the female straddles and picks up with her cloaca. Fertilization is therefore internal in those species that produce a spermatophore.

Lungless salamanders are active on the forest floor during moist or humid periods. Males are antagonistic to one another and appear to establish small territories that they defend from other males. The territorial encounters include biting and chasing and can result in injuries, including loss of part of the tail.

Defense Against Predators

Like other amphibians, salamanders and newts have toxic skin secretions produced by skin glands that are used in various ways as defense mechanisms to repel predators. In some species of newts, glands are concentrated on the dorsum, and when disturbed by a predator, the salamander displays an "unken reflex." This display includes bending the body in a U-shape and showing bright coloration of the underbelly. At the same time, the animal becomes immobile, thus decreasing the chance that a predator will attack. Other species have glands concentrated on the tail and engage in tail lashing or tail undulation. In tail lashing, the salamander violently whips its tail toward the predator, which attacks the tail and tastes the noxious secretions. In tail undulation, the body of the salamander remains immobile while the tail is waved in a sinuous fashion above the body. The tail in these species can be autotomized, or broken from the

Salamander and Newt Facts

Classification:
Kingdom: Animalia
Phylum: Vertebrata
Class: Amphibia
Order: Caudata
Families: Ten families, including Ambystomatidae (mole salamanders), Amphiumidae (amphiumas), Cryptobranchidae (hellbender and giant salamanders), Dicamptodontidae (Pacific giant salamanders), Plethodontidae (lungless salamanders), Proteidae (waterdogs and mudpuppies), Rhyacotritonidae (torrent salamanders), Salamandridae (newts), Sirenidae (sirens)

Geographical location: Eurasia and North America, with one family extending into South America

Habitat: Many habitats, including forests, savannas, prairies, freshwater ponds and streams, and ephemeral pools

Gestational period: Varies among species; species that breed in ponds have larval stages that last two to six months, whereas the larval stage of stream-breeding species may last several years

Life span: Varies among species; larger species tend to live longer than smaller ones; large aquatic salamanders have lived for fifty to fifty-five years in captivity

Special anatomy: Long body with long tail and four legs; some species are legless

Salamanders have smooth, glandular skin with no external covering such as hair, scales, or feathers. (Rob and Ann Simpson/Photo Agora)

body by the salamander. Thus, if a predator attacks the waving tail, the salamander loses the tail but escapes with its life.

Other species actively defend themselves if attacked by a predator. Amphiumas are large, powerful salamanders that live in ponds, swamps, or marshes in the southeast United States. Adults reach one meter in length and can inflict a painful bite with their sharp teeth. In a unique manner of defense, slender salamanders can secrete copious amounts of an adhesive substance from their glands. When attacked by a garter snake, the salamander's secretions glue the snake's body to itself, and it is unable to swallow the salamander.

—Janalee P. Caldwell

See also: Amphibians; Defense mechanisms; Frogs and toads; Gas exchange; Regenerations; Skin; Tails.

Bibliography

Duellman, W. E., ed. *Patterns of Distribution of Amphibians: A Global Perspective*. Baltimore: The Johns Hopkins University Press, 1999. Covers all aspects of distribution of amphibians with respect to climate, geography, vegetation, and evolutionary history.

Duellman, W. E., and Linda Trueb. *Biology of Amphibians*. New York: McGraw Hill, 1986. A comprehensive book that covers all aspects of amphibian biology, including paleontology, morphology, reproduction, and biogeography.

Heatwole, H., and B. K. Sullivan. *Amphibian Biology*. Vol. 2 in *Social Behaviour*. Chipping Norton, New South Wales, Australia: Surrey Beatty and Sons, 1995. A thorough treatment of amphibian social behavior, including territoriality, mating behavior, and parental care.

Petranka, J. W. *Salamanders of the United States and Canada*. Washington, D.C.: Smithsonian Institution Press, 1998. An up-to-date reference work on North American salamanders. Discusses each species, with photographs of both adult and larval stages. Maps, illustrations, references, glossary.

Vitt, L. J., J. P. Caldwell, and G. Zug. *Herpetology*. 2d ed. San Diego, Calif.: Academic Press, 2001. A general introduction to the biology of frogs and toads, as well as other reptiles and amphibians. Color photographs of examples from all families. Aimed at advanced high school or beginning college students.

SALMON AND TROUT

Type of animal science: Classification
Fields of study: Anatomy, ecology, marine biology, reproduction science

Salmon and trout are salmonid food and game fishes of the Northern Hemisphere.

Principal Terms

ANADROMOUS: ocean-dwelling fish that migrate to freshwater to breed
MILT: semen that holds sperms from male fishes
SEINE: a net hanging vertically with floats at its top and weights at its bottom
TROLL: fish by dragging a baited hook through the water.

Salmon, of the family Salmonidae, are valuable Northern Hemisphere food and game fish. Most belong to five species that live in rivers and off coasts of the North Pacific, from California to Alaska and Siberia. A sixth species inhabits the North Atlantic. Salmonids also include trout, graylings, and whitefish. They eat small fish, insects, and crustaceans. Salmon and some trout are anadromous, living in oceans and migrating to freshwater to breed. The five Pacific salmon species belong to the genus *Oncorhynchus*. Different species spawn in the rivers and streams of western North America during the spring (spring run), summer (summer run), and fall (fall run).

Physical Characteristics and Habitats

Salmonids have elongated bodies, small, round, smooth-edged scales, well-developed swimming fins, and a fleshy fin between the dorsal and tail fins. There are five Pacific salmon species: chinook (king), sockeye (red), humpbacked (pink), dog (calico), and coho salmon.

Spring-run chinook salmon swim upstream to spawn. The species has red, oily flesh, and average weights of twenty-three pounds. They live from California to the Bering Strait. Sockeye (red) salmon, also spring run, weigh five to eight pounds, have deep red flesh, and their fins redden during breeding. Like chinooks, they run up to one thousand miles to spawn. Prior to spawning runs, sockeye and chinook salmon store huge quantities of oil for energy to swim upstream. During migration, nest-building, and mating they do not eat.

The other Pacific species are fall run. Spawning closer to the ocean, they do not require a huge oil energy reserve. Pink salmon weigh three to six pounds and are most abundant in Alaska. Dog salmon are not eaten much because their flesh is not tasty. Coho salmon, six to nine pounds, have pink flesh and live from San Francisco to the Puget Sound.

Atlantic (true) salmon, the largest salmonids, have orange-red flesh. Most other fish of their genus (*Salmo*) are trout. Atlantic salmon weights average twenty pounds. They migrate from ocean to freshwater to spawn in spring or summer. Nonmigrant Atlantic salmon subspecies inhabit lakes in the northern United States. These "landlocked" salmon are smaller than salmon that migrate.

Trout are also salmonids. Sea trout are anadromous, but most species inhabit only freshwater lakes and streams. They eat smaller fishes, crustaceans, and insects. The most important true trout (genus *Salmo*) is the rainbow trout of the western United States. Chars, of the trout genus *Salvelinus*, have smaller scales than true trout and inhabit colder North American, Asian, and European waters. The largest char species, lake trout, often weigh twenty-five pounds and have dark gray

Salmon and Trout Facts

Classification:
Kingdom: Animalia
Subkingdom: Bilateria
Phylum: Chordata
Subphylum: Vertebrata
Class: Osteichthyes
Order: Saliminiformes (salmonlike fishes)
Family: Salmonidae (salmon)
Genus and species: Includes *Oncorhynchus kisutch* (coho salmon), *O. gorbuscha* (pink salmon), *O. nerka* (sockeye salmon), *O. tschawytscha* (chinook salmon); *Salmo gairdneri* (brook trout), *S. salar* (Atlantic salmon), *S. trutta* (brown trout); *Salvelinus fontinalis* (speckled trout), *S. namaycush* (lake trout)

Geographical location: United States, Canada, Europe, and Asia

Habitat: Freshwater lakes, streams and rivers; oceans of the Northern Hemisphere

Gestational period: Two weeks to two months

Life span: Six to fifteen years, depending on species

Special anatomy: Bony skeletons; toothed mouths; caudal, dorsal, pectoral, and ventral fins developed for long swims; a fleshy fin midway between dorsal and tail fins

bodies with yellow-red spots. They inhabit the Great Lakes, Alaska, Labrador, New Brunswick, Vermont, and Maine.

Lives of Salmonids

Chinook salmon are typical anadromous salmonids. In autumn, adults spawn in Pacific Northwest streams, a few feet deep, where females place eggs in nests where fast-running water is rich in oxygen. Then the males squirt milt on them, and the females stir up the stream bottom so that earth and stones cover and protect the eggs, which hatch in two weeks to two months. This start of a new generation precedes the death of the adults by only a few days. After spawning, the dull-colored adults have slimy bodies and ragged fins from their treks to spawning grounds. For a few days they float downstream, tail-first and torpid, until they die. Their bodies wash ashore and become scavenger feed.

Yolk sacs nourish hatchlings for three months and they then become free-swimming. In one or two years, when they are three to five inches long, they swim to the ocean. Predators kill myriad migrants over the six-month trip, but some make it. In the ocean, salmon spend four years eating and growing. Then they swim back to their hatch sites. During salmonid runs, large numbers are caught for canning, but enough get through to restart the cycle. During their runs they leap rapids, waterfalls, and, increasingly, dams. Atlantic salmon survive spawning, return to the ocean, and later reuse the same spawning site several times.

The most widely distributed North American trout species is the brook (speckled) trout, found from Georgia to the Arctic Ocean. They spawn from September to December in holes that females scrape in gravel. Trout eggs take two to three months to hatch, and young develop as do salmon.

Catching and Disseminating Salmonids

In the United States, 4,500,000 pounds of salmon are caught yearly with nets and traps. Fishermen sell them to companies that supply consumers with fresh, frozen, smoked, or canned salmon. When salmon are caught before spawning, their eggs and milt can be removed. To maintain salmon numbers, the U.S. Fish and Wildlife Service yearly collects billions of eggs, fertilizes them and puts the eggs and young in breeding grounds. Trout caught by game and commercial fishermen are sold as food less often than salmon, though they are frequently eaten by their catchers. Their eggs are collected, fertilized, and disseminated by the Fish and Wildlife Service, too. Freshwater and ocean fish farms also produce salmonids.

—*Sanford S. Singer*

See also: Breeding programs; Eels; Fins and flippers; Fish; Lakes and streams; Lampreys and hagfish; Lungfish; Marine biology; Migration; Reproductive strategies; Sharks and rays; Whale sharks; White sharks.

A common danger for salmon is being caught by bears while swimming upstream to spawn. (Adobe)

Bibliography

Calabi, Silvio. *Trout and Salmon of the World*. Secaucus, N.J.: Wellfleet Press, 1990. A compilation of information on many types of trout, salmon, and related fishes.

Pennell, William, and Bruce A. Barton. *Principles of Salmonid Culture*. New York: Elsevier, 1996. Discusses the natural history and aquaculture of salmon, trout, and other salmonids.

Stolz, Judith, and Judith Schnell. *Trout*. Harrisburg, Pa.: Stackpole Books, 1991. Contains information on trout natural history and habitats.

Watson, Rupert. *Salmon, Trout, and Char of the World: A Fisherman's Natural History*. Shrewsbury, England: Small Hill, 1999. Describes salmonids and fishing for them.

SAUROPODS

Types of animal science: Classification, evolution
Fields of study: Anatomy, evolutionary science, paleontology, systematics (taxonomy)

Sauropods are a group of dinosaurs that include the largest land animals that ever existed. Together with the theropods, or carnivorous dinosaurs, they form a group known as the Saurischia.

Principal Terms

CRETACEOUS: a period of time that lasted from about 146 to 65 million years ago
GASTROLITHS: polished pebbles that may have facilitated dinosaur digestion
GIGANTOTHERMY: a form of metabolism in which internal temperature is maintained by the large mass of the animal
JURASSIC: a period of geological time that lasted from about 208 to 146 million years ago
THEROPODS: the carnivorous dinosaurs and the closest relatives of the sauropods
TRIASSIC: a period of geological time that lasted from about 245 to 208 million years ago

Sauropods are a suborder of the Dinosauria, a vast group of reptiles that dominated terrestrial environments during the Mesozoic (245 to 66 million years ago). Dinosaurs ("terrible reptiles") are characterized by erect limbs and a pelvis that incorporates at least five vertebrae, characteristics that are related to their active lifestyle. Dinosaurs are divided into two major groups based on the structure of the pelvis. In the Saurischia ("lizard-hipped"), the pubis points forward, while in the Ornithischia ("bird-hipped"), the pubis has swung backward to lie parallel to the ischium. The Saurischia includes the carnivorous dinosaurs, or theropods, as well as the herbivorous sauropods ("lizard-foot"), the largest land animals that have ever existed.

Systematics and Anatomy

The earliest dinosaurs are known from the late Middle Triassic (220 million years ago) of South America, but the earliest sauropods do not appear until somewhat later, in the Early Jurassic. Sauropods are most closely related to the prosauropods, a group that originated in the Late Triassic and consisted of large, heavily built herbivores that reverted to the quarupedal locomotion of their dinosaur ancestors. Although it has been suggested that prosauropods were carnivores or omnivores, their teeth are coarsely serrate and numerous and were clearly adapted for shredding coarse vegetation. They were probably adapted as the first high browsers, capable of reaching up to high foliage to feed.

The earliest sauropods occur in the Early Jurassic of Africa and Asia, and from then on are a feature of dinosaurian faunas worldwide until the end of the Cretaceous, when the dinosaurs became extinct. All sauropods were quadrupedal and most were very large. The head was small and lightly built and contained peglike or spoon-shaped teeth that were confined to the front of the mouth, and nostrils that were placed on the top of the head. The neck was long and consisted of up to fifteen vertebrae, which although large were penetrated by many openings, so that they were constructed of struts and laminae providing maximum strength for minimum weight. The tail is also very long, perhaps up to seventy feet in *Seismosaurus*, and contained up to eighty vertebrae, of which the last forty were reduced to simple rods of bone that formed a whiplash. The limbs and girdles were massively constructed to support the enormous weight of these animals and the feet were broad and elephant-like, with the animal standing on the tips of its toes like all dinosaurs.

Several main groups are recognized—the diplodocids, the camarasaurids, and the brachiosaurids—but there are other, more poorly known sauropods, particularly at the beginning of sauropod evolution, that do not fit into these groups. The earliest sauropods come from Zimbabwe and are very similar to the prosauropods. Following these are Chinese forms from the Early Jurassic, in which weight-bearing features, such as shortening of the fingers and toes, become more apparent. In addition, teeth become restricted to the front portion of the jaws only. Sauropods had become diverse and widespread by the Middle Jurassic, but it is not until the Late Jurassic that they became common in North America. At this point, the three main groups are well represented and they remain abundant into the early part of the Cretaceous.

One of the best known diplodocids is *Diplodocus*, which is represented by a number of skeletons collected from the western United States. A related form is *Apatosaurus*, which for a long time was incorrectly known as *Brontosaurus* due to confusion caused by initial descriptions based on incomplete material. These animals were characterized by a protruding muzzle and small, pencil-like teeth restricted to the front of the jaws. The tail was very long, with more than eighty vertebrae present in some species. The fore limbs are considerably shorter than the hind limbs, so that the hip was high compared to the shoulder. Brachiosaurs and camarosaurs were contemporaries of the diplodocids. *Brachiosaurus* ("arm-lizard") is particularly well known from material collected from Tendaguru in Tanzania by German expeditions between 1908 and 1912. The very fine articulated skeleton mounted in the Humboldt Museum in Berlin stands 12 meters tall and is almost 22.5 meters long, and shows the relatively short tail and very long front limbs that are characteristic of this group. In addition, the head of brachiosaurs is more compact than that of diplodocids and the jaws are lined with large, chisel-shaped teeth. The camarosaurs, such as *Camarosaurus* ("chambered reptile"), were similar in many ways to the brachiosaurs, and like them had compact skulls, with large, chisel-shaped teeth in the jaws, although the muzzle was shorter. *Camarosaurus* had a much shorter neck, however, and the front limbs were shorter than the hind limbs, as in the diplodocids, so that the hips and shoulders were at about the same level.

For unknown reasons the sauropods almost perished at the end of the Jurassic, and there are only a few species known from the Cretaceous, although they were present throughout the period. The last survivors are members of a group known as the titanosaurids, which are widespread but known only from incomplete material. One of the interesting features of this group is that they bore bony armor that consisted of large bony plates scattered across the back with masses of small nodules in the hide between them. No other sauropods are known to have possessed armor.

Sauropod Facts

Classification:
Kingdom: Animalia
Subkingdom: Bilateria
Phylum: Chordata
Subphylum: Vertebrata
Class: Reptilia
Order: Saurischia
Suborder: Sauropoda
Geographical location: Every continent except Antarctica
Habitat: Terrestrial habitats
Gestational period: Unknown
Life span: Estimated at fifty years
Special anatomy: Very large quadrupedal dinosaurs in which the head was small, the neck and tail long

Lifestyle

Sauropods were originally thought to have been aquatic animals, as it was considered that their large size would have precluded a terrestrial existence and that the long neck and nostrils on top of the head would have enabled them to reach the surface to breathe. However, it is now considered that sauropods lived a terrestrial lifestyle similar to that of modern elephants. Several features of the skeleton support this view. In particular, weight-reducing features, such as the openings developed in the vertebrae of the neck and tail, are adaptations associated with living on land where bodily support under the effects of gravity is critical in large animals. In addition, the presence of clefts in the top of neural spines on the neck vertebrae provides evidence of the presence of thick, ropelike ligaments that ran from the shoulders along the neck and allowed it to be held clear of the ground effortlessly. A similar feature is seen in the tail, indicating that it also was held horizontally, and this is supported by the evidence of tracks which only rarely show a tail-drag mark.

Trackways provide evidence that sauropods probably moved around in small groups, possibly up to twenty individuals. Although adults were probably big enough to have been immune to attack by large theropods, juveniles might have needed the protection that herding behavior would have given them. Studies of trackways have also shown that they would have moved slowly, perhaps at no more than four miles per hour, and that it would not have been possible for them to run. In all probability, they moved slowly from one stand of conifers to another, browsing high up in the trees.

As they had only small, weak teeth at the front of the mouth and no chewing teeth, it was originally thought that they fed on soft aquatic vegetation. However, stomach contents show that they fed on resistant conifers, and it is now thought that they stripped vegetation from branches and then passed it down to a gizzard, probably lined with stones, that would have processed the vegetation. The dwindling of conifers through the Cretaceous has been suggested as a cause of the decline of the sauropods during the same period.

The Size of Sauropods

Although sauropods were clearly the largest land animals that ever existed, it has been difficult to estimate the length and mass of particular species because this has to be based on reconstructions of skeletons and soft tissue estimates. The tallest sauropod seems to have been *Brachiosaurus*, as the head of the Berlin specimen is twelve meters above ground level. Other sauropods have a different build, and the neck was probably oriented more horizontally. Calculations of masses of sauropods are particularly difficult, as the amount of soft tissue probably varied for individuals depending on age, health, and even from season to season. *Apatosaurus* may have weighed between thirty and forty tons and *Brachiosaurus* thirty to ninety tons; however, some poorly known sauro-

Gastroliths

Collections of rounded pebbles are found in the rib cage of some sauropod skeletons and these have been named gastroliths, or "stomach stones." The best-documented case is the sauropod *Seismosaurus*, in which more than 240 stones were found in two groups, one in the pelvic region and another near the base of the neck. It is clear that these stones were held within the animal during life, as they are unknown in the surrounding sediment. A number of modern animals swallow stones and grit; it seems that crocodiles and turtles do this for buoyancy control, but in birds it is done to aid digestion. By analogy with birds, it has been proposed that in sauropods the gastroliths were held in a crop and a gizzard and were involved in the pulping of vegetation during digestion. In this scenario, the pencil-like teeth at the front of the mouth would have been used to strip vegetation, which would then have been passed down to the crop for initial processing before being passed on to the stomach for chemical digestion, and then to the gizzard for more mechanical breakdown before moving on to the intestine. However, a rival hypothesis is that the gastroliths were there to stir the digestive juices rather than mechanically abrade the food.

pods, such as *Supersaurus* and *Ultrasaurus*, may have been heavier. The record for the longest sauropod was long held by a specimen of *Diplodocus*, at 87 feet, but extrapolations of length from partial material suggest that *Seismosaurus* may have been as much as 150 feet long. Great size may have been a metabolic advantage for sauropods. As very large animals lose heat slowly, due to their relatively small surface area, they probably operated as gigantotherms, requiring only a small food intake to maintain their internal temperature.

—*David K. Elliott*

See also: *Allosaurus*; *Apatosaurus*; *Archaeopteryx*; Dinosaurs; Evolution: Animal life; Extinction; Fossils; Hadrosaurs; Ichthyosaurs; Paleoecology; Paleontology; Prehistoric animals; Pterosaurs; Stegosaurs; *Triceratops*; *Tyrannosaurus*; Velociraptors.

Bibliography

Alexander, R. McNeill. *Dynamics of Dinosaurs and Other Extinct Giants*. New York: Columbia University Press, 1989. Informative coverage of how large animals such as sauropods operated.

Benton, Michael J. *Vertebrate Paleontology*. 2d ed. London: Chapman and Hall, 1997. General vertebrate paleontology text that devotes one chapter to the sauropods.

Currie, Philip J., and Kevin Padian. *Encyclopedia of Dinosaurs*. San Diego, Calif.: Academic Press, 1997. Excellent coverage of all aspects of dinosaur biology.

Gillette, David D. *"Seismosaurus," the Earth Shaker*. New York: Columbia University Press, 1994. Reader-friendly discussion of the collection and description of the largest known dinosaur.

Norman, David. *The Illustrated Encyclopedia of Dinosaurs*. New York: Crescent, 1985. This older book has wonderful illustrations and an excellent text with extensive coverage of sauropods.

SAVANNAS

Types of animal science: Ecology, geography
Fields of study: Ecology, environmental science, ethology, wildlife ecology, zoology

Savannas are grasslands in tropical regions near the equator that support a larger number of plant-eating animals than any other area in the world.

Principal Terms

CARNIVORES: flesh-eating animals
HERBIVORES: plant-eating animals
MARSUPIALS: animals having a pouch on the abdomen for carrying their young
RUMINANTS: grass-eating animals that chew again food that has been swallowed
UNGULATES: hoofed animals

Savannas, or tropical grasslands, are vast open spaces on which grow a large variety of plant life. Savannas usually endure long periods of drought that are punctuated by one or two rainy seasons. When the rains come, large herds of animals, mostly ungulates, make their annual journey along centuries-old migration routes from the river valleys where they have spent the dry season, to the fresh grass on the savanna. Although the principal vegetation is grass, trees or tall bushes appear occasionally on the landscape or along streams.

The huge African savanna is ancient, having probably evolved about sixty-five million years ago. Other areas are newer, some having been created by humans when forests were cleared to accommodate farming. Savannas also exist in South America, largely in Venezuela, and in northern Australia.

Plant-Eating Animals

The African savanna varies from very dry regions to areas of swamp, lake, and woodland, and can support the largest variety of herbivores in the world. Many animals are capable of living together because most of them have their own specific feeding habits. The hippopotamus, reedbuck, and waterbuck remain near the water, while various gazelles prefer dry areas, receiving moisture from plants. While the zebra chooses open grassland, the wildebeest (gnu), giraffe, and antelope are equipped to forage in the bush and also the woodland by virtue of their long snouts for gathering leaves and stems. All parts of trees and shrubs provide food; while some animals feed on the tough outer parts of grass, others eat the tender, fresh foliage or leaves of wild flowers.

The herbivores found on the African savanna are also the world's largest land animals. African elephants live in grassland, bush, and forest, in mountainous country and near lakes. Every day, elephants eat vast quantities of grass, leaves, twigs and bark, sometimes destroying trees and helping to create savannas. Elephant herds incorporate smaller groups of four to twenty elephants, led by the older females.

The white rhinoceros, weighing more than three tons, is one of Africa's rarest animals. The herds are composed of small family groups of one male, one or two females, and several young. Its smaller relative, the black rhinoceros, exists more abundantly. Browsing on leaves and branches, the black rhinos are protected by their size and horns and tusks. Humans are their only real enemies. Many of the grass eaters on the African savanna are antelope, which belong to a suborder of animals called ruminants. Equipped with complex stomachs, ruminants eat food which passes from the mouth, through the several chambers of the stomach, and back again to the mouth. The slow process of rumination, or chewing the cud, provides the animal with more safety from predators as chewing can be accomplished later in a less dangerous place than grazing.

In the savanna of northern Australia are found marsupials—kangaroos, koalas, and wombats. Marsupials also live on the South American savanna, along with the armadillo. Other particular species found in these geographically isolated areas are long-legged, flightless birds—the ostrich, rhea, and the emu.

Predators and Scavengers

When the sun's energy is converted by plants into food for herbivores, or primary consumers, then the predators and scavengers, or animals who live by preying upon other animals, become secondary consumers in the food chain. Big cats, such as the lion, cheetah, and leopard, are powerful animals who stalk and run down their prey. Lions function in teams, with females assuming most of the work.

African wild dogs are smaller animals that have strong jaws, sharp teeth, and a keen sense of smell. They live and hunt in packs made up of six to twenty members. Other predators, such as the hyena and the jackal, kill their prey and feed at night off the carcass of the animal; powerful, far-sighted vultures and marabou stork feed by day, each eating a different part of the carcass. Smaller scavengers—crows, ravens, rats, and insects—move in, helping to dispose of dead bodies that might carry disease organisms. Smaller predators of the African savanna include the desert lynx, which pursues the smaller antelope, and the black spotted serval, which hunts ground squirrels, large rats, and guinea fowl. Other predators are the genet and the fox. Puff adders and cobras poison their victims and are themselves attacked by the mongoose and the secretary bird. The aardvark hunts ants and termites at night.

The open savannas of Africa are home to many grazing species, such as elephants, giraffes, and zebras. (PhotoDisc)

Various species of monkeys, baboons, and vervets have adapted to living on the savanna by living mostly on the ground, trying to avoid predators. In the Australian savanna, the kangaroos are preyed upon by the dingoes, or wild dogs.

—*Mary Hurd*

See also: Chaparral; Ecosystems; Fauna: Africa; Fauna: Australia; Fauna: South America; Food chains and food webs; Forests, coniferous; Forests, deciduous; Grasslands and prairies; Habitats and biomes; Tundra.

Bibliography

Alden, Peter, et al. *National Audubon Society Field Guide to African Wildife*. New York: Chanticleer Press, 1995. Visually organized mammal guide with striking photographs of birds, mammals, reptiles; contains a wealth of information on more than 850 species, including animal behavior, habitats, park, reserves.

Combes, Simon. *Great Cats: Stories and Art From a World Traveler*. Shelton, Conn.: Greenwich Workshop Press, 1998. Beautifully photographed and sketched; detailed narratives of real-life adventures with large cats. One section focuses upon lions, cheetahs, and leopards of African savanna, their habitats, behavior and feeding habits.

Miius, S. "When Elephants Can't Take It Anymore." *Science News* 155 (May 29, 1999): 341-342. Article reflects research on the decline of the elephant population in Africa. Maintains that when human population reaches a certain threshold, elephants disappear rapidly; the main threat to elephant survival is land-use clashes with humans.

Sherr, Lynn. *Tall Blondes: A Book About Giraffes*. Kansas City, Mo.: Andrews McMeel, 1997. The chapter "Where the Giraffes Are" is concerned with the exploitation and extermination of giraffes in Africa at the end of the nineteenth century and the eventual creation of wildlife laws that prevent slaughter of giraffes.

Silver, Donald M. *African Savanna*. New York: McGraw-Hill, 1997. One volume in a series of science and nature books for children called *One Small Square*. Includes stunning full-color illustrations of habitat and its creatures; also contains experiments and activities.

SCALES

Type of animal science: Anatomy
Fields of study: Anatomy, physiology

The skin of animals serves as a barrier between the exterior environment and the internal environment of the animal's body. It is, however, a barrier that is in many cases permeable and vulnerable to damage and so scales have evolved in most vertebrate groups as a protective element for the skin. Scales have the advantage of being considerably lighter than armor plating of the body with dermal bone and confer much more flexibility upon the body. Scales have evolved a number of different times in vertebrates, and variations in their form and structure reflect these separate developments of scale armor.

Principal Terms

CTENOID SCALES: scales with comblike serrations on rear edge, found on many bony fishes

DENTINE: the ivory portion of a tooth or scale; dentine or dentinelike substances such as cosmine are found in the scales of most fishes

GANOID SCALES: heavy, dense scales containing ganoine found in primitive bony fishes

KERATIN: a dense and rigid protein that makes up the scales of reptiles, birds and some mammals

PLACODERMI: extinct class of fishes characterized by dense armor plating made of dermal bone

PLACOID DENTICLES: toothlike scales found in sharks and rays

Scales have evolved in most vertebrate groups to provide a layer of protection for the integument, the outer layers of tissue that protect the internal environment of the animal from the external world. Scales are hardened plates that are made either of bonelike substance or the protein collagen, both of which are formed in the dermis (the dense, connective tissue layer of the skin) or from other rigid proteins such as keratin, which are secreted by epithelial cells in the outer layer of the skin. The protection afforded by scales does not come at the expense of flexibility, however, for most scales are attached only at one edge to folds in the skin and thus form overlapping plates which can slide over each other as the animal moves. As such, scales provide a protective armor that is lighter and more flexible than the armor formed by large plates of dermal bone such as those seen in extinct fishes such as the Placodermi.

Fish Scales

Sharks and rays possess a distinctive type of scales known as placoid denticles. These scales bear a strong resemblance to teeth; indeed, in sharks the teeth are enlarged and exaggerated placoid denticles that are replaced when lost. Each placoid denticle consists of a flat basal plate that is embedded in the dermis and a prominent, curved midline spine with a hollow interior pulp cavity. Both the basal plate and the spine are composed of dentine, the dense ivory material of teeth. The outer layer of the spine consists of a layer of vitrodentine, which is secreted by special cells (ameloblasts) in the epidermis. Within the central pulp cavity are blood vessels and nerve fibers, and fine canals (canaliculi) extend from the cavity to all parts of the dentine.

Some primitive bony fishes, such as gars and the African bichirs, possess ganoid scales. These heavy and dense scales are roughly diamond-shaped and consist principally of a dentinelike substance called cosmine. The outer surface of the cosmine layer is coated with many layers of a dense, silvery material called ganoine, which provides a lustrous, metallic sheen.

Modern bony fish, which comprise the vast majority of familiar fish species, have abandoned the heavy ganoid scales in favor of greater mobility and other protective devices. Their thinner and lighter scales contain neither cosmine nor ganoine, but rather consist of an osteoid layer (bonelike, but containing no bone cells) and an inner fibrous layer. There are two principal forms of these scales, cycloid scales, which are smoothly rounded in shape, and ctenoid scales, in which the free rear surface has comblike teeth (ctenii). Both types are fully embedded in the dermis and arranged in an overlapping fashion similar to roof shingles. The scales are covered with a thin layer of epidermis that contains single-celled mucous glands. Like the trunks of trees, these scales also exhibit "growth rings" (lamellae), the shape of which are uniquely characteristic to an individual species.

Amphibian and Reptile Scales

When amphibians became the first vertebrates to move from an aquatic to a terrestrial habitat, the weight of their scales may have become a liability in their less physically supportive environment. The earliest amphibians possessed dense and bony scales of cosmine, but these have been lost in most amphibian groups. Some of the legless modern amphibians (the caecilians) have very small dermal scales, but modern frogs, toads and salamanders lack them completely.

Scales have reappeared in reptiles; however, these scales lack any of the enamel-like dentine or cosmine components observed in aquatic vertebrates. Reptilian scales are composed primarily of a hard structural protein called keratin, which has two molecular forms, α-keratin and β-keratin. Many modifications of these scales are seen in reptiles. In turtles, the scales rest upon a layer of dermal bones (bones that arise from the layer of dermis). In snakes and lizards, the entire epidermis, including the scales, is shed periodically during growth and a new epidermis, with scales, replaces it.

Scales offer protection for an animal's skin while still allowing a great deal of flexibility. (Corbis)

Bird and Mammal Scales

Birds and mammals evolved independently from separate reptilian lineages, and birds typically retain scales on their feet, lower legs, and the base of the beak. These scales are very similar to those observed in reptiles and are also made of keratin. Most mammals have no scales whatever. However, some mammals, particularly rodents, possess scales of keratin on the epidermis of the tail. More dramatic are the epidermal scales of mammals like the armadillo, which are greatly enlarged and form an effective armor plating for these animals.

—*John G. New*

See also: Anatomy; Beaks and bills; Bone and cartilage; Brain;

Circulatory systems of invertebrates; Circulatory systems of vertebrates; Claws, nails, and hooves; Digestive tract; Ears; Endoskeletons; Exoskeletons; Eyes; Fins and flippers; Immune system; Kidneys and other excretory structures; Lungs, gills, and tracheas; Muscles in invertebrates; Muscles in vertebrates; Nervous systems of vertebrates; Noses; Physiology; Reproductive system of female mammals; Reproductive system of male mammals; Respiratory system; Sense organs; Skin; Tails; Teeth, fangs, and tusks; Tentacles; Wings.

Bibliography

Bailey, J., and L. Bailey. *Animal Life: Form and Function in the Animal Kingdom.* New York: Oxford University Press, 1994. A useful book on animal form and function targeted primarily at the young adult reader.

Moyle, P. B., and J. J. Cech. *Fishes: An Introduction to Ichthyology.* Upper Saddle River, N.J.: Prentice Hall, 1999. A thorough guide to the biology of fishes, the largest group of scaled vertebrates.

Walker, W. F., Jr., and T. S. Walker. *Functional Anatomy of the Vertebrates: An Evolutionary Perspective.* 3d ed. Fort Worth, Tex.: Saunders College Publishing, 1997. A college-level, readable text on the comparative anatomy of the vertebrate body.

Zug, G. R. *Herpetology: An Introductory Biology of Amphibians and Reptiles.* San Diego, Calif.: Academic Press, 1997. Perhaps the leading text on the biology of reptiles, generally aimed at college-level students.

SCAVENGERS

Types of animal science: Anatomy, ecology, reproduction
Fields of study: Anatomy, ecology, zoology

Scavengers live partly or entirely on carrion. They are ecologically important because they assure that carrion does not become breeding places for disease organisms. Also, they help natural selection to enhance survival abilities of species they consume.

Principal Terms

CARRION: the corpses of dead animals
CARTILAGE: the wing case of a beetle
CHITINOUS: made of fibrous chitin
ELYTRA: the wing case of a beetle

Carnivorous worms, bears, beetles, kites, vultures, gulls, and hyenas do not live only by killing prey. To varying extents they eat scraps from other predator kills and the corpses of animals that died of old age, injury, or illness. Many scavengers are poorly designed for hunting. Some have short talons, lack optimum teeth, or are awkward. Regardless, they are ecologically important. They assure that carrion does not become breeding places for disease organisms.

All scavengers have a keen sense of smell to help them to find decaying flesh. They need not worry about corpses fighting back, although they do risk dangerous diseases. However, scavengers rarely die from eating carrion, because natural selection has given them tolerance of foods that kill other animals.

Scavenger Worms and Beetles

The sea mouse is a segmented worm, related to earthworms. An undersea scavenger, it lives in shallow coastal waters worldwide. In a sea mouse, each body segment is separated from the others by chitinous tissues. On the sides of the segments are muscular protrusions having bristly hairs, used in locomotion. Finer hairs growing from the segments allow the worm to sense its surroundings. As a sea mouse crawls through the seabed, it eats carrion. The worm has a flat, three-inch wide, nine-inch long body. Scales and bristles, like gray fur, cover it and lead to the name sea mouse.

Scarab beetles are one of the twelve thousand species of colorful beetles, which grow up to six inches long. They have horns on their heads or thoraxes. Males use the horns in combat during mating. Many beetles eat carrion. Dung beetles are most often called scarabs. Some scarabs (tumble bugs) form dung into balls and roll it into their burrows. There it becomes food for them or for their larvae, hatched from eggs deposited on the pellets. Ancient Egyptians worshiped tumble bugs, viewing the dung pellets as symbolizing the world and the scarab horns as symbolizing the sun's rays. Thinking that scarabs caused good fortune and immortality, they used scarab carvings as charms and replaced the hearts of the embalmed dead with carved scarabs.

Bears, Jackals, Raccoons, and Hyenas

Many mammals, such as bears, jackals, hyenas, and raccoons, are scavengers only when the need arises. Among bears, the best known scavenger is the American black bear. It averages six feet long, approximately three feet tall at the shoulder, and weighs three hundred pounds, which is small and light for a bear. This small size may make it unable to compete for choice bear fare and explain why black bears eat carrion. Given the option, black bears prefer twigs, leaves, fruit, nuts, corn, berries, fish, insects, beehives, and honey.

Raccoons, nocturnal animals, inhabit swamps or woods near water. They eat frogs, crayfish, fish, birds, eggs, fruits, nuts, rodents, insects, and carrion. Raccoons have stout, catlike bodies, masked faces, coarse yellow-gray to brown fur, body lengths up to two feet, and ringed, bushy tails, and weigh up to fifteen pounds. They are solitary, except for mating in January and February. After two-month pregnancies, females give birth to three to seven babies, which remain with them for a year. Wild raccoons live for up to seven years.

Jackals are Old World wild dog scavengers that live on plains, deserts, and prairies. Golden jackals live from North Africa to southeast Europe and India. Black-backed jackals inhabit East and South Africa. Jackals have foxlike heads, but otherwise resemble wolves. They are nocturnal creatures, holing up in dens during the day. They eat carrion and small birds and mammals, hunting in packs. Jackals live for up to fifteen years.

African and Asian hyenas are also built something like wolves. The two main species, spotted (laughing) hyenas and striped hyenas, were once thought to eat only carrion. This notion arose from observations of their carrion eating and because hyena hind legs, shorter than their front legs, make them look awkward.

African laughing hyenas have brown fur liberally sprinkled with spots, large heads, and jaws and teeth that are able to crush bones. These hyenas, six feet long and three feet tall at the shoulder, are less clumsy than they look. It is now thought that they are the greatest killers of zebras. Nocturnal hyena packs kill prey or eat carrion. Hyenas live in groups of about sixty members. Females have one or two cubs, after four-month pregnancies. Little is known about striped hyenas, which are smaller and less aggressive. They are tan, with dark, vertical stripes and live from East Africa to Asia. Hyenas live for up to twenty-five years.

Carrion Beetles

Carrion beetles of North American forests grow 2.5 inches long. Each of their bodies, as in other beetles, has a head, thorax (midsection), and abdomen (hind section). Abdomens are covered by hard wing cases (elytra), formed from front wings. Elytra protect flying wings from harm. Beetles smell with antennae, and the smell carrion beetles seek is that of carrion from small rodents, amphibians, birds, and reptiles.

Carrion beetles eat carrion and use it in reproduction. Prior to mating, males bury carrion in holes dug in the ground. Then, after battling others for the carrion, a pair of beetles wins and mates near it. After mating, the female digs an egg chamber above the carrion, lays her eggs, climbs onto the carrion, and digs a nest in it. When her eggs hatch, young are attracted to the carrion and enter the nest. The female feeds them until they can feed themselves. Then the larvae bore chambers in the earth, spin cocoons in the chambers, and mature into adults.

Scavenger Birds

Kites, one bird scavenger group, are hawks found in warm parts of all continents. Their legs and feet are small and weak, so they eat carrion and small animals. Swallow-tailed kites are two feet long, with beautiful white bodies and black wings and tails. They inhabit the southern United States. The smaller, white-tailed kite inhabits the Americas, Europe, and Asia. Kites often hover in the air, searching for insects and small mammal carrion.

Vultures are larger carrion eaters. New World vultures are related to storks and live in the temperate and tropical regions of the Americas. The Old World vultures of Europe and Asia resemble New World vultures in eating habits, but are related to hawks and eagles. Unlike kites, vultures look like carrion eaters. Most are ugly and dark feathered. They lack plumage on their heads and necks, minimizing messiness from blood and gore. Vultures are predisposed to carrion eating because they have blunt claws, which are poor weapons for hunting. They are suited to finding carrion by their ability to sustain long flights and their sharp eyes. Vultures fly in flocks, except during mating, when pairs nest on cliffs or in caves.

Most vultures lay three eggs among bare rocks. Parents incubate the eggs and feed their young, which can fly when six months old.

There are six New World vulture species. Turkey vultures (buzzards) live in the southern United States and northern Mexico. The largest North American land bird is a vulture, the California condor, which is up to six feet long, with a wingspan up to eleven feet. Andean condors are the largest South American vultures. Condors have black body plumage. The naked heads and necks of vultures and condors vary in color and in the presence or absence of feather ruffs and wattles.

There are fourteen Old World vulture species. Most interesting are the bearded vultures called lammergeiers, four feet long and weighing up to fifteen pounds. They inhabit mountains up to fifteen thousand feet high in Europe, Africa, and Asia. Lammergeiers build nests on ledges or in mountain caves. They have tan chest and stomach plumage, white faces, and dark brown wings and tails. Masklike black feather "beards" surround their eyes and beaks. Also of interest are Egyptian vultures, two feet long, with naked yellow heads, white body feathers, and black wings. They live in Mediterranean areas and as far east as India. Many Old World vultures are not as funereal-looking as the New World breeds.

Gulls, also scavenger birds, are pigeon-sized and long-winged. Most live on the oceans and large, inland lakes. Adults have gray plumage on their wings, backs, and heads, webbed feet, and white under parts. They are graceful fliers and swimmers and nest in large colonies on rocky islands or in marshes. Gulls eat fish, other water animals, insects, carrion, garbage, and the eggs or young of other birds.

Best known are white herring gulls. Adults are two feet long and in addition to the typical gray and white gull motifs have black wing tips, yellow bills, and flesh-colored legs. They eat fish, shellfish, garbage, and carrion. Often commercial fishermen, cleaning catches, see gulls swarming behind their boats, awaiting fish offal. Herring gulls mate and lay about six eggs. Females incubate them for three weeks and feed the offspring until, at six weeks old, they strike out on their own.

Vulturelike Marabou Storks

Marabou storks (marabou) combine stork and vulture anatomy. Found throughout Africa, the adults are five feet tall. Like other storks, marabou have long legs and sharp, straight bills. Their heads and necks are bald, as in vultures, minimizing slop from blood and gore. Marabou wade in deep water, eating frogs, snakes, and small mammals. On land, they eat insects. However, most marabou food is deer, antelope, and zebra carrion.

Marabou plumage is gray-green on the back and wings. This gives way to white bellies and ruffs encircling bare, red-pink necks. Marabou heads are also red-pink. From their throats hang twelve-inch, featherless, red-pink pouches of unknown function.

Most marabous inhabit Africa's prairies, wetlands, rivers, and lakes. Each breeding pair builds a nest in a tree or on rocky terrain. Most often, three eggs are laid and incubation is shared by both parents. Chicks hatch during the dry season when carrion, from mammals who die of thirst, is plentiful. Chicks stay with parents for six months. Marabou live up to twenty-two years.

Sharks

There are over three hundred shark species. They differ from bony fish in having skeletons made of cartilage, not bone. Many sharks eat nearly all large marine animals. They vary from forty to fifty foot long, to six-inch-long species. Most abundant in tropical and subtropical waters, nearly all sharks are viewed as aggressive carnivores.

Sharks are usually gray, having leathery skins and gills behind the head. Shark tails are not symmetrical and shark skeletons end in upper tail lobes. Shark fins and tails are rigid, not erectile, as in bony fish. Sharks also have a keen sense of smell, sensing traces of blood and homing in on their sources. Despite their great strength, sharks are mostly scavengers, eating injured fish and carrion. They also eat seals, whales, and fish.

Benefits of Scavenging

Scavengers consume carrion, preventing its decay and the endangerment of the health of other animals. This is one of their main ecological functions. Species such as the sea mouse, the carrion beetle, scarabs, vultures, and condors choose to eat carrion. Others, such as American black bears, jackals, raccoons, kites, sharks, and gulls, given a choice much prefer catching live prey.

Also, some scavengers' perceived food sources may be based on incomplete data. For example, laughing hyenas were dubbed scavengers based on their awkward appearance and a few observations. More careful study showed that they are more predator than scavenger. In addition, regardless of preference in obtaining their food, scavengers such as sharks and vultures have another important ecological role in the oceans and on land, killing the injured or weak members of other species. This activity helps those species to select for strong individuals, enhancing chances of species survival.

—*Sanford S. Singer*

See also: Carnivores; Digestion; Ecological niches; Ecology; Ecosystems; Food chains and food webs; Herbivores; Hyenas; Ingestion; Omnivores; Raccoons; Vultures.

Bibliography

Ammann, Karl, and Katherine Amman. *The Hunters and the Hunted*. London: Bodley Head, 1989. This useful book, with excellent illustrations, includes an article on hyenas.

Earle, Olive L. *Scavengers*. New York: William Morrow, 1973. This juvenile book contains useful information on scavengers and puts them in perspective.

Evans, Arthur V. *An Inordinate Fondness for Beetles*. Reprint. Berkeley: University of California Press, 2000. This well-illustrated book is a good first text for readers unfamiliar with beetles.

Parker, Steve, and Jane Parker. *The Encyclopedia of Sharks*. Willowdale, Ontario: Firefly Books, 1999. This well-illustrated book holds a great deal of information on sharks.

Wilbur, Sanford R., and Jerome A. Jackson. *Vulture Biology and Management*. Berkeley: University of California Press, 1983. This professional text describes vultures.

SCORPIONS

Types of animal science: Anatomy, classification, ecology
Fields of study: Anatomy, ecology, invertebrate biology, systematics (taxonomy), zoology

Scorpions are one of the most ancient land animals, having evolved over 400 million years ago. There are over 1,300 species of scorpions occurring over every continent except Antarctica. Scorpions are known for their stinging apparatus, but relatively few species pose any danger to humans.

Principal Terms

ACULEUS: the sting, either a single or double hollow barb that delivers the venom to prey

EXOSKELETON: an external, jointed skeleton made up of chitin and protein

OVAL GLAND: the poison gland in the scorpion telson

OVARIAN DIVERTICULUM: used to house embryos and to obtain nutrients via absorptive cells among viviparous species

PECTINES: comblike structures on the ventral surface of the scorpion that are used in chemoreception

PEDIPALPS: clawlike appendages that are used to catch and hold prey

Scorpions are members of the phylum Arthropoda, and more specifically of a subphylum of that taxon known as Cheliceriformes. Scorpions are members of the subclass Arachnida, those arthropods having eight legs and chelicerae mouthparts, and lacking antennae. Scorpions are the oldest arthropod terrestrial group, whose aquatic ancestry dates back to Silurian times over 400 million years ago. Their terrestrial invasion occurred in the Devonian period.

Scorpions are cosmopolitan in distribution, occurring on all continents except Antarctica. Although most people envision scorpions as desert creatures, scorpions are found in the tropical jungles, temperate forests, and savannahs and in high elevations on mountains. They are mostly nocturnal creatures as they have little defense against the ultraviolet radiation of the sun. They are quite variable in size: Some are as small as thirteen millimeters, while others (the South African *Hadogenes troglodytes*) range up to eighteen centimeters in length.

Scorpion Physiology

Scorpions are segmental in form, the body being divided into an anterior prosoma and a segmented abdomen. A one-piece carapace covers the prosoma. The abdomen is divided into a preabdomen of seven segments and a postabdomen of five segments, ending in a stinging apparatus. The pedipalps are pincerlike and are used to capture and hold prey. The chelicerae are pincerlike as well and are used to macerate the prey. There are four pairs of walking legs, all ending in a pair of claws. All scorpions are carnivorous and are essentially liquid feeders. Copious amounts of digestive enzymes are poured over macerated areas of the prey and the liquid is then pumped into the stomach.

Scorpions are well equipped with sensory structures. They have a pair of simple eyes located in the center of the carapace. Additionally, there are from two to five pairs of eyes located along the anterior and lateral margins of the carapace. Scorpions have many setae or sensory hairs located over the dorsal surfaces of the body. These hairs function to pick up vibrations and air movement and are used to detect prey. The hairs are large on the pedipalps and are called trichobothria.

Unique to scorpions is a pair of ventrally located, comblike appendages called pectines. These structures are mainly chemoreceptors and are used to pick up pheromone trails of insects. Pectines are also used to dig burrows, although the legs mainly perform this function.

Scorpions breathe by means of book lungs that are ventrally located and open to the outside via a pair of spiracles. In this way, the book lungs are kept moist for oxygen diffusion. A circulatory system is present, with a dorsally situated heart that opens via ostia or pores into the hemocoels and book lungs. Scorpions are also well equipped to deal with excretory wastes, using Malpighian tubules. These tubules filter nitrogenous wastes from the hemocoels and deposit the waste into the gut tube for elimination.

Mating and Reproductive Strategies

There are separate sexes and the gonads are tubular in construction in both sexes. The gonopore opens on the ventral surface of the mesosoma. Males lay a spermatophore sac that is picked up by the female during a courtship dance. This dance is initiated by the male, who grasps the female's pedipalps in his and dances back and forth in a face-to-face position. When the female touches the opening lever of the spermatophore, sperm are released. Fertilization is internal, as is the development. Scorpions are either ovoviviparous or viviparous. The time from conception to birth in scorpions is quite variable. In some groups, birth requires up to five months' gestation, while in others the gestation period can last up to eighteen months. In viviparous species, the embryo is fed via a tube that extends from the digestive caeca to the embryos living in the ovarian tubes. The juvenile scorpions will exit the mother via the gonopore and climb atop her back, where they will mature and go through a molt. After this first molt, they will take up their own independent existence. Scorpions will molt from four to seven times before they reach the adult stage. As an adult no molting occurs, and limbs lost during life are not regenerated.

Scorpion Fluorescence

Looking for scorpions is made easy by the fact that scorpions will fluoresce under an ultraviolet black lamp. Scorpions appear light green under ultraviolet radiation. Fluorescence may be caused in part by chemicals known as carotinoids that are found in the epicuticle.

—*Samuel F. Tarsitano*

See also: Arachnids; Arthropods; Bioluminescence; Poisonous animals; Spiders.

Scorpion Facts

Classification:
Kingdom: Animalia
Subkingdom: Bilateria
Phylum: Arthropoda
Subphylum: Cheliceriformes
Class: Chelicerata
Subclass: Arachnida
Order: Scorpiones
Families: Bothriuridae, Buthidae, Chactidae, Chaerilidae, Diplocentridae, Euscorpiidae, Hemiscorpionidae, Heteroscoripionidae, Ischnuridae, Luridae, Microcharmidae, Pseudochactidae, Scorpionidae, Scorpiopidae, Superstitioniidae, Troglotaoysicidae, Vaejovidae, Urodacidae

Geographical location: Every continent except Antarctica

Habitat: Strictly terrestrial, found in both arid and tropical regions

Gestational period: Varies among species; lengths of between five and eighteen months have been reported

Life span: Depending on the species, the life span may be only one year, while other scorpions are known to live for twenty-five years or more

Special anatomy: Eight legs, chelicerae mouthparts and simple eyes like other arachnids; clawed pedipalps for grasping prey; telson with a sting that can deliver venom; special sensory organs, called pectines, that act in an olfactory capacity

In addition to their well-known poison glands, scorpions also have carotinoids in the epicuticle of their exoskeletons that cause them to fluoresce under ultraviolet light. (Jack Milchanowski/Photo Agora)

Bibliography

Brusca, R. C., and G. J. Brusca. *Invertebrates*. Sunderland, Mass.: Sinauer Associates, 1990. This is a textbook of invertebrates that includes a chapter on the Arthropoda. This text discusses the anatomy and physiology of scorpions including the reproductive system, locomotory and feeding structures, and sense organs. The ecology and distribution of scorpions is also discussed.

Fet, V., W. D. Sissom, G. Lowe, and M. E. Braunwalder. *The Catalog of the Scorpions of the World, 1758-1998*. New York: American Museum of Natural History, 2000. This text lists all scorpions found between the years noted.

Gaban, D. "*Androctonus australis* (L.) Fattailed Scorpion." *Forum of the American Tarantula Society* 6, no. 2 (1997): 52-53. This article discusses this species of scorpion in terms of its venom and habits.

Polis, G. A., ed. *The Biology of Scorpions*. Stanford, Calif.: Stanford University Press, 1990. Includes extensive articles on the ecology and behavior of scorpions, as well as the classification of scorpions, where they are found, and their evolutionary history. It is the authoritative work on this group of arachnids.

SEAHORSES

Types of animal science: Classification, reproduction
Fields of study: Anatomy, conservation biology, ethology, marine biology, physiology, reproduction science, wildlife ecology

At least thirty-five species of seahorses, classified in the genus of Hippocampus, *live in shallow subtropical coastal areas around the world. They swim in an upright position and have a head suggestive of a horse, a prickly looking body, and a prehensile tail. The male incubates eggs in a pouch in its abdomen and gives birth to live young.*

Principal Terms

DORSAL FIN: fin on the back of a fish
OVIPOSITOR: a tube that extends from the female's body for depositing eggs
PECTORAL FIN: one of a pair of fins just behind the head of a fish, where arms of terrestrial vertebrates are attached
PREHENSILE: adapted for seizing or grasping

Seahorses are highly unusual fish. First, they swim upright and poorly. They lack the tail fin that provides other fish with most of their swimming power; instead seahorses use a small dorsal fin to move forward, pectoral fins near the head to turn and steer, and a swim bladder to move up or down in the water. Second, their shape is unique: They have a pronounced horse-shaped head at a right angle to their rough body, and a prehensile tail. Their snout is adapted for aspirating passing crustaceans, filtering the water through their gills. Their eyes move independently, permitting them to observe prey and their environment without moving. Their body is prickly and knobby, due to bony rings perpendicular to their backbones. Their tail permits them to anchor themselves by grasping vegetation or coral. Third, as they use camouflage to escape predators, they can grow tendrils from their skin to look like sea plants and, like chameleons, can change color to match their surroundings. They can also change color in response to other seahorses, brightening in response to a mate and darkening in submission to a rival.

The thirty-five seahorse species differ in size, shape, color, and habitat. The smallest, *Hippocampus bargibanti* (called the pygmy seahorse), is a

Seahorses are unique in that the male of the species gestates the young in an abdominal pouch. (Digital Stock)

mere 1.3 centimeters (0.5 inches) from snout to tail, while the largest, *H. ingens* (the Pacific seahorse), is 35 centimeters (14 inches) long. Although seahorses generally look alike, the species differ in the number of bony rings around their bodies and tails, and one, *H. abdominalis* (the big belly seahorse), has a very pronounced abdomen. They vary in color, including pink, orange, yellow, brown, gray, and black, with the male usually the more colorful. They live in salt water at a depth of one to twenty-five meters (three to eighty feet), at a temperature of 6 to 30 degrees Celsius (43 to 86 degrees Fahrenheit), and in one of three coastal habitats: sea grass, mangroves, and coral reefs. The species are specific to different locations; for instance, *H. bargibanti* is found around the island of New Caledonia in the west Pacific, *H. ingens* inhabits the subtropical west coasts of North, Central and South America, while *H. abdominalis* exists around New Zealand, as well as on the southern and eastern coasts of Australia.

Seahorse Facts

Classification:
Kingdom: Animalia
Subkingdom: Bilateria
Phylum: Chordata
Subphylum: Vertebrata
Class: Pisces
Subclass: Teleostei
Superorder: Acanthopterygii
Order: Gasterosteiformes
Suborder: Syngnothoidei
Family: Syngnathidae
Genus and species: Hippocampus hippocampus, H. guttulatus and *H. kuda* (spotted seahorse), *H. zosterae* (dwarf seahorse)

Geographical location: Subtropical coastal regions around the world

Habitat: Shallow coastal areas that contain sea grass, mangroves, or coral reefs

Incubation period: Ten days to six weeks, depending on the species and habitat

Life span: One to four years, depending on the species and environment

Special anatomy: The male has a brood pouch on its abdomen in which the female deposits its eggs; the male fertilizes them inside the pouch and incubates them, expelling the newly hatched, fully developed, but minuscule offspring through an opening in the pouch; seahorses have a prehensile tail that permits them to attach themselves to vegetation or coral

The Life Cycle of Seahorses

Depending on the species and location, seahorses may have a breeding season (generally during the warmer time of the year) or may reproduce continuously. Most seahorses appear to form monogamous relationships, although *H. abdominalis* is promiscuous. When the male is receptive, the pair will perform a mating dance that may last for hours to a day, ending with the female depositing her eggs through her ovipositor into the brooding pouch of the male, who will fertilize them with his semen. That the male becomes pregnant is the most unusual feature of seahorses. He provides the fertilized eggs with oxygen and food through a capillary network in the pouch, which also removes waste products. The incubation period depends on the species and conditions, lasting from ten days to six weeks. When the fully formed young hatch, they are expelled from an opening in the pouch in a rhythmic process that may last up to two days. The number of offspring born typically varies from 10 for the smallest species to 200, although the record is 1,572. The male will usually become immediately pregnant again. The young must find food, such as larval crustaceans, and must avoid predation from fish, water birds, or crabs. In addition, storms may sweep them out to sea attached to seaweed, and disperse them to environments to which they may not be suited. If they find sufficient food and avoid predation, they can mature in four to six months, mate, and procreate. While most seahorses do not survive to maturity, they can live from one to four years, depending on the species.

Future of Seahorses

Overfishing is a serious threat to the future of seahorses. The demand for them is large in traditional Asian medicine, where they are touted as cures for everything from asthma and heart disease to incontinence and impotence. An estimated twenty million seahorses are so used annually. Live animals are also sold for aquariums, where maintaining them is difficult because of their nutritional preference for live crustaceans and their susceptibility to disease. In addition, drag net fishing in coastal areas harvests seahorses inadvertently. Any destruction of sea grass beds, mangroves, or coral reefs also poses a threat to these fascinating animals. Their future depends on sustainable fishing in the wild and developing seahorse aquaculture, as well as minimizing unintentional harvest and habitat destruction.

—James L. Robinson

See also: Courtship; Fish; Marine animals; Mating; Pregnancy and prenatal development; Reproduction; Reproductive strategies.

Bibliography

Schlein, Miriam. *The Dangerous Life of the Sea Horse.* New York: Atheneum, 1986. Written for children, this well-illustrated book tells the life story of seahorses and the natural risks that they face.

Vincent, Amanda. "The Improbable Seahorse." *National Geographic* 186, no. 4 (October, 1994): 126-140. Written by one of the world's acknowledged experts on seahorses, this article describes research in the wild and in the laboratory. Excellent photography.

Walker, Sally M. *Sea Horses.* Minneapolis: Carolrhoda Books, 1999. Written for adolescents, this book is comprehensive, accurate, and beautifully photographed. It also details approaches to sustainable harvesting of seahorses.

Whaley, Floyd. "Riding to the Seahorse's Rescue." *International Wildlife* 29, no. 2 (March, 1999): 22-29. This article describes Amanda Vincent's efforts to save the seahorse while preserving the livelihood of people dependent on them.

SEALS AND WALRUSES

Type of animal science: Classification
Fields of study: Anatomy, ethology, genetics, marine biology, physiology

Essentially aquatic, warm-blooded mammals, seals and walruses are widely distributed throughout the world, although most of them gravitate to the colder areas.

Principal Terms

AMULET: an ornament worn as a charm against evil
CARNIVORES: meat-eating animals
CIRCUMPOLAR REGIONS: the regions around the North and South Poles
MOLLUSKS: a phylum of aquatic invertebrates, usually shelled, such as clams, mussels, and squid
MUZZLE: the area around the nose and mouth of an animal
PINNIPEDS: web-footed animals
SCRIMSHAW: a type of art that involves etching or engraving images on pieces of ivory taken from the tusks of walruses or other animals

Seals and walruses belong to the same general family. They are pinnipeds, the term used to indicate an animal with webbed feet. Walruses are distinguished from seals by their tusks, which in mature males grow up to two feet long.

Walruses use their tusks to aid locomotion when they are on land, as a means of defense when they are physically threatened, and as shovels to plow the ocean floor to turn up the mollusks that constitute the major part of their diets. Walruses have strong, bristly hairs around their muzzles that are used to separate the meat of the mollusks from the shells.

Seal are gregarious creatures. In captivity, it is easy to train them. They bond quickly with humans. In their natural state, they tend to cluster together in groups, often with as many as a thousand of them lying in close proximity to each other on the seashore or on an ice floe.

Seal and Walrus Habitats

Seals are found on every continent, including Antarctica. Although most of them prefer the cold waters of the circumpolar regions, seals swim toward warmer waters to mate, then return to more frigid areas to give birth, often delivering their young on ice floes. Monk seals are found as close to the equator as the Galápagos Islands.

Most seals live in salt water, although Saimaa seals live in Finland's freshwater lakes. Many of them were killed off by fishermen who claimed that they were eating all the fish in the lakes. The Finnish government intervened and made killing the seals illegal in 1955, but it was not until serious conservation efforts began in the 1980's that the population began to rebound. It is also illegal to use fishing nets in those parts of the lake where the Saimaa seals live and breed, because more than half of their offspring were getting tangled in nets and drowning.

Although they are warm-blooded animals that must have air to breathe, seals and walruses spend most of their lives in the sea. They are essentially aquatic. Their webbed feet and flippers provide them with easier locomotion in the sea than on land, where most of them move quite clumsily. Seals and walruses are strictly carnivorous animals whose diet consists almost exclusively of fish and mollusks. They may ingest some seaweed, but if they do so, it is by accident. These animals can close off their nasal passages so that water does not enter them when they submerged. They can remain under water for up to thirty minutes without having to return to the surface. Seals are more streamlined than walruses. Walruses usually are found in waters that are no deeper than sixty feet, although they can dive to three hundred feet. They explore the ocean bottom, using their tusks to dig into the sand for the mollusks they live on.

Physical Characteristics of Seals and Walruses

Most seals and walruses have substantial layers of blubber, constituting almost half of the body weight of most seals. When melted down, the blubber of a mature male elephant seal can yield ninety gallons of oil. Blubber serves as the chief insulating material in the cold waters where most seals and walruses swim. The blubber also provides buoyancy. Walruses can puff up blubber-rich areas in their necks to keep their heads above water when they are bobbing about in the ocean.

There are two families of seals, the true or earless seals, family Phocidae, and the eared seals, family Otariidae. The largest species of seal is the elephant seal. It is sometimes twenty feet long and can weigh as much as four tons. Even the smallest species of seal, the ringed seal, is quite large, weighing in at about two hundred pounds. At maturity it is about five feet long. In some species, notably the monk, Weddell, leopard, and crabeater seals, the male is smaller than the female. Among Phocidae seals, the male and female are of about equal size, but among the Otariidae, the male is larger than the female.

Seals communicate at various volumes. The northern elephant seal has such a loud call that it can be heard a mile way. The Ross seal has a chirping sound that can be mistaken for bird song. The Weddell seal has a soft, gentle call.

Walruses, of the family Odobenidae, are distinguished from seals by their two upper canine teeth that grow into ivory tusks harder than those of elephants. These sometimes reach lengths exceeding two feet. Walruses have small heads and no protruding ears. The eyes are small and deepset. The body is small, but overdeveloped in the neck area. Bulls are up to twelve feet long and five feet wide. They often weigh a ton. The female is two-thirds the size of the male.

Seal and Walrus Facts

Classification:
Kingdom: Animalia
Subkingdom: Bilateria
Phylum: Craniata
Class: Mammalia
Subclass: Theria
Order: Pinnipeda
Families: Otariidae (eared seals, including fur seals and sea lions, seven genera, fourteen species); Phocidae (true, earless, or hair seals, ten genera, nineteen species); Odobenidae (walruses)

Geographical location: Seals, every continent; walruses, polar and circumpolar regions in the Northern Hemisphere

Habitat: Mostly water, but they give birth on land or on ice floes

Gestational period: Earless seals, ten to eleven months; eared seals, twelve months; walruses, fifteen to sixteen months

Life span: For males, twelve to thirty years in the wild; females may have a life span of only eight to twelve years, although some live to twenty or even thirty

Special anatomy: Strong hind flippers, strong arms; webbed digits, five toes and fingers

Mating and Reproduction

Most seals swim south to spawn, while walruses usually spawn in their northern water habitats. Male seals and walruses have multiple spouses, sometimes as many as forty at one time. The older, seasoned animals stake out the most attractive territory and attract females to it, creating substantial harems within their own venues. Older males often fight off younger males, who in the end must settle for inferior pieces of land and fewer spouses. Males mature at about five years of age and often continue to mate until past twenty.

Female walruses give birth to one baby eleven months after they mate. Female seals have similar gestation periods but sometimes deliver twins, although more typically they deliver single offspring. Walruses tend to mate again within a month of the delivery of a cub; seals in some cases mate every other year, although some deliver cubs annually.

Although some seals are solitary in winter, they gather at breeding time in rookeries on islands or ice floes, but also sometimes along stretches of beach. As few as a dozen seals may gather in a rookery, but sometimes there are as many as a million within a fifty-mile radius. The males arrive at the rookeries first to stake out their territory, the most favorable locations being directly on the water. In the northern hemisphere, females usually arrive in July.

Human Uses of Seals and Walruses

Seals and walruses are benevolent creatures that seldom attack and that are accepting of humans. They are trusting enough to be vulnerable to those who engage in mass slaughters of the animals for their fur, their blubber, and, in the case of walruses, for their ivory tusks, from which scrimshaw is made.

Inuit and other people of the Arctic depend upon seals and walruses as part of their survival strategy in hostile environments. After they kill an animal, they treat it with reverence, sometimes making a replica of it from one of its bones or from a tooth or tusk. They wear such amulets around their necks to demonstrate their gratitude to the dead animal. When these native people slaughter an animal, they waste no part of it. They eat the meat, render the blubber into oil with which they light their lamps and warm their dwellings, convert the hides into clothing and tentlike coverings, and use the tusks and bones to carve into scrimshaw.

When Europeans and Americans began to raid the seal and walrus populations, they did so indiscriminately and came close to annihilating these animals completely. Fortunately, governments stepped in to protect the endangered species and to end the brutal slaying of newborn cubs for their fur.

Kinds of Seals

ELEPHANT SEAL: The largest species of seal. It has a loud call that can be heard at great distances.

FUR SEAL: These seals were hunted nearly to the point of extinction by hunters wanting to sell their skins in the fur trade. Baby fur seals were brutally clubbed to death and skinned, sometimes when they were still conscious. Since 1974, when it became illegal to hunt them, their numbers have grown to more than five thousand.

MONK SEAL: The main varieties of the monk seal are the Caribbean monk seal, the Hawaiian monk seal, and the Mediterranean monk seal. The first two have been bordering on the edge of extinction and may already be extinct. The Mediterranean monk seal is still seen occasionally in the warm waters off Greece, Turkey, and North Africa, but its total population is estimated to be below five hundred.

ROSS SEAL: Found mostly in Antarctica, this seal has large amounts of blubber. It is characterized by the birdlike chirping sounds it makes.

SAIMAA SEAL: A freshwater seal, this variety dwells in the lakes of Finland. It was on the verge of extinction until the Finnish government, in 1955, intervened to make it illegal to kill the Saimaa and to string fishing nets near its habitat. This species is slowly rebounding, with a total population of about two hundred.

SOUTH AFRICAN FUR SEAL: This coastal seal has a limited range, seldom straying far from home. Most of its population is found within a hundred-mile range from shore.

SOUTH AMERICAN FUR SEAL: Found off the coast of South America, this species has three subspecies: sea lions, hair seals, and northern hair seals.

WEDDELL SEAL: Found mostly in the Southern Hemisphere, it prefers the frigid waters of the South Atlantic and South Pacific. It is one of the few forms of vertebrate life in Antarctica.

Walruses use their long, ivory tusks to defend themselves, to pull themselves along when they are on land, and to dig in the ocean floor for mollusks and other foods. (Digital Stock)

Those who quested after fur attacked seal cubs only days old, beating them bloody with clubs, stripping them of their fur, often while the cub was still conscious, and leaving their stripped bodies on the ice. People around the world were outraged by pictures they saw of such predation by humans upon defenseless seal cubs. They eventually called for this brutality to stop. Multinational agreements were drafted to protect these animals. The results have been encouraging as seal and walrus populations have finally begun to increase.

—*R. Baird Shuman*

See also: Dolphins, porpoises, and toothed whales; Endangered species; Fins and flippers; Marine animals; Otters; Teeth, fangs, and tusks; Whales, baleen.

Bibliography

Bonner, W. Nigel. *The Natural History of Seals*. New York: Facts on File, 1990. Excellent material on humans' relationships to seals and on seal behavior and chances for survival. Appropriate for beginners.

Fay, Francis H. *Ecology and Biology of the Pacific Walrus, "Odobenus rosmarus divergens Illiger."* Washington, D.C.: U.S. Department of the Interior, 1982. Thorough, accurate and detailed. Better for specialists than beginners.

Laws, R. M., ed. *Antarctic Seals: Research Methods and Techniques*. New York: Cambridge University Press, 1993. These fifteen essays cover a broad range. Especially relevant contributions on behavior, reproduction, and population size.

Lepthien, Emilie U. *Walruses*. New York: Children's Press, 1996. Aimed at the juvenile audience. Well researched. A sound starting point for beginners. Profusely illustrated.

U.S. Fish and Wildlife Service. *Walruses: "Odobenus rosemarus."* Washington, D.C.: U.S. Department of the Interior, 1995. Provides information about the present status of walrus populations in the United States and Canada. Somewhat specialized.

SENSE ORGANS

Type of animal science: Anatomy
Fields of study: Anatomy, behavior, ethology, neurobiology, physiology

Sense organs are responsible for detecting events that occur both outside and within the body and translating those events into the electrical and chemical language understood by the brain. As such, sense organs are the means for the perception of the surrounding world and the body's internal state, and supply the information necessary for the organization of appropriate behaviors. There are many different types of sense organs in animals, but all serve the same basic function of informing the brain of events transpiring in the internal and external environments.

Principal Terms

ADAPTATION: the decrease in the size of the response of a sense organ following continuous application of a constant stimulus

MODALITY: a specific type of sensory stimulus or perception, such as taste, vision, or hearing

PHASIC RECEPTORS: receptors that adapt quickly to a stimulus

RECEPTIVE FIELD: the area upon or surrounding the body of an animal that, when stimulated, results in the generation of a response in the sense organ

RECEPTOR CELLS: sensory cells within sense organs that are directly responsible for detecting stimuli.

RECEPTOR POTENTIAL: a change in the distribution of electric charge across the membrane of a receptor cell in response to the presentation of a stimulus

TONIC RECEPTORS: receptors that typically show little or no adaptation to a continuously applied stimulus

TRANSDUCTION: the translation of a stimulus's energy into the electrical and chemical signals that are meaningful to the nervous system

Plants are able to form complex organic compounds for nutrition from simple molecules such as carbon dioxide and water via the process of photosynthesis. Animals, on the other hand, rely upon obtaining complex organic compounds already formed by other organisms to meet their nutritional needs. Since such sources generally take the form of other organisms, these must be located and consumed by the animal. In short, animals, unlike plants, must use behavior, a set of responses to internal and external events occurring in their environment, to survive. It is the need for behavior that has formed the basis of virtually all of animal evolution.

The first and most vital element of behavior is the detection of events occurring both within the body of the animal and in the surrounding external environment. The role of the sense organs is to detect these events (which are called stimuli) and translate them into the complex series of electrical and chemical signals that is the language of the brain. It is important to understand that the stimuli (such as sound, light, or heat) which animals detect and which may possess behavioral significance are meaningless to the brain. Brains are capable of interpreting only those signals that are in its language of electrical impulses and chemical interactions. Thus sense organs have two vital functions: the detection of environmental stimuli and the translation of that stimulus into the language that is meaningful to the nerve cells of the brain. This latter process is called transduction, and it is the ultimate role of sensory organs.

Receptor Cells and Transduction

Sensory organs typically consist of several different types of cells. Receptor cells are directly responsible for transducing the stimulus into the electrical language of the nervous system. Supporting cells play a number of different roles and in some cases may themselves become receptor cells. For example, in mammalian taste buds the receptor cells routinely die after ten to fourteen days and are replaced by supporting cells that transform to become receptor cells. Accessory structures assist in the process of transduction, such as the lens of the eye. Finally, sensory nerve fibers are stimulated chemically by the receptor cells and send information concerning the presence of a stimulus into the central nervous system.

The process of transduction occurs when a stimulus interacts physically with a sense organ, causing a change in the distribution of electrical charge across the cell membrane of the receptor cell. This change in transmembrane voltage is referred to as a receptor potential, and the size of the receptor potential corresponds directly to the intensity of the stimulus applied to the receptor cell. There must therefore be a minimum intensity of the stimulus required to generate a receptor potential, referred to as the sensory threshold. In general, the sensory threshold corresponds to the smallest stimulus intensity that an animal can detect. At the same time there is a maximum receptor potential that can be generated by a receptor cell, no matter how intense the stimulus. The intensity at which this occurs is known as receptor saturation. Above this level, when the receptor is saturated, it is impossible for the animal to discriminate whether one stimulus is more intense than another. Between the upper and lower limits of threshold and saturation, a change in the intensity of the stimulus will result in a corresponding change in the magnitude of the receptor potential. This range of intensities is known as the dynamic range of the sensory organ and within this range of intensities animals can discriminate between stimuli of different intensities.

When a continual stimulus is applied, and a receptor potential is generated across the membrane of the receptor cell, the nature of the receptor potential may change with time. If the receptor potential decreases with time, even though the applied stimulus remains constant, adaptation is said to occur. If one jumps into a pool of cool water on a warm summer day, the initial sensation is that of coolness against the skin. However, this perception disappears with time as the temperature receptors in the skin adapt until the water has no perceptible temperature. There are limits to adaptation, however; immersing one's hand in very hot water does not result in an eventual disappearance of the perception of heat. Some senses exhibit sensitization, the opposite of adaptation, in which progressively less stimulus is required to elicit a sensation with increasing time. The responses to certain types of painful stimuli demonstrate sensitization.

The rates and degrees of receptor adaptation vary among different types of receptors and the specific types of information they detect. Some receptors adapt very quickly to an applied stimulus; after a short period of steadily applied stimulation the receptor potential disappears. Such receptors are said to be phasic receptors. For example, if one carefully deflects a hair on the back of one's arm with a pencil point and then holds it steadily in the deflected position, the sensation generated by the deflection quickly disappears. Other receptors show very little adaptation over time; such receptors are said to be tonic receptors. Many sorts of pain receptors are tonic, as anyone who has experienced a toothache may attest. Phasic and tonic receptors represent the extreme ends of a continuum. Most receptors can be said to be phasic-tonic receptors, which exhibit a greater or lesser degree of adaptation to continuous stimulation. Phasic and tonic receptors send different types of information concerning the nature of the stimulus to the brain. Tonic receptors send precise information concerning the duration and the intensity of a stimulus to the central nervous system. Such information may be useful, for instance, in determining the degree to which a limb is flexed. Phasic receptors, on the other hand, relay precise information about changes in the stimulus rather than its duration. Since animals live in a dynamic world, detecting small changes in the environment caused by the presence of a predator or prey may be of first importance. Both types of information are crucial to survival.

The Senses

Classically, there are considered to be five senses (vision, hearing, touch, taste, and smell), but in reality there are many different senses and these can be further divided into subsenses. For example, the sense of temperature actually consists of two different sensory systems, one that detects heat and another that detects cold stimuli. Furthermore, these are both linked, at some level, with the detection of pain (intensely hot stimuli are also painful, but mild heat is not). Neurobiologists refer to a type of sense as a modality (such as taste) and the detection of variations within that modality as a quality (such as sweet versus sour). There are numerous different modalities, and more are being discovered every year. The ability of animals to detect and use many different types of complex information from their environment is a fascinating and continually unfolding story.

Because changes in electrical and chemical activity are the only language understood by the central nervous system, information arriving in the brain from different sensory systems must be kept segregated from each other to avoid confusion. Thus, any activity arriving from the eyes via the optic nerve is interpreted by the brain as "light," whether or not light is actually present. Electrical or physical stimulation of the optic nerve in the absence of actual light will also be perceived by the animal as "light." The sensory systems within the brain are thus organized into a series of labeled lines, each dedicated to a specific sensory modality. Any electrical activity in a labeled line is interpreted by the brain as the presence of that modality.

Sensory organs can generally be broadly classified by the nature of the events that they are capable of detecting and transducing. Mechanoreceptors detect mechanical forces applied directly or indirectly to the body. The sense of touch is the most familiar sense employing mechanoreceptors, but there are others, such as hearing and balance, that are equally important. Chemoreceptors detect signals that occur when chemicals of different types come in contact with sensory organs (such as in taste or smell). Electromagnetoreceptors detect energy contained in the electromagnetic spectrum. The most familiar, and one most heavily relied on in most mammals, is vision. Visual sensory organs (eyes) detect the energy contained within a limited range of frequencies within the electromagnetic spectrum commonly referred to as visible light. Other familiar electromagnetoreceptors detect heat (infrared radiation) or its lack (cold). Other animals detect other portions of the electromagnetic spectrum and in some cases are capable of directly detecting electrical and magnetic fields. These three categories are not absolutely rigid, however. Some types of receptors such as hygroreceptors (that detect the water content of air) seem not to fall conveniently in any one category, whereas others, such as nociceptors (pain receptors) straddle several categories and may respond to mechanical, chemical, or thermal stimuli. Similarly, the sense of balance employs information from several kinds of sense organs responding to a number of discrete stimuli, such as the direction of the earth's gravitational pull, rotational acceleration of the head, and body position.

Exteroreceptors and Endoreceptors

Sensory receptors of all three types may be used to detect either stimuli originating in sources outside the body (exteroreceptors) or stimuli originating within the body itself (endoreceptors). These latter receptors play an absolutely vital role in the maintenance of a constant internal chemical environment and temperature (homeostasis). If the internal environment varies outside of a narrow set of parameters, death can quickly ensue. It is the task of endoreceptors to detect fluctuations in the internal environment and signal the body's involuntary control mechanisms (the autonomic nervous system) to effect corrections. For example, the detection of a drop in the core body temperature in mammals can result in a variety of responses, including the shunting of blood away from the skin surface (to minimize loss of heat to the environment), erection of hair or fur on the skin (what humans experience as goose bumps) to trap a layer of warmed air next to the skin, and shivering, which generates heat via muscular contractions. There are many other endoreceptors in the body, which detect stimuli such as the amount of dissolved gases in the blood (oxygen and carbon dioxide), sugars, salts, and the amount of water present in the body. Another important role of the endoreceptors is proprioception, the detection of the relative positions of the body's parts in relation to one another. It is this sense that allows an individual to touch their nose with the tip of their finger when their eyes are closed. "Muscle sense," or kinesthesia, and the detection of the amount of flexion of the joints are included as parts of the modality of proprioception.

There are very many different types of exteroreceptors, and they comprise all three general classes of receptors. All of them, however, are dedicated to detecting external events that impinge in some manner upon the body. Exteroreceptors may be scattered across the surface of the body (such as in the sense of touch) or confined within specialized structures (such as the eyes or ears). The twofold function of all of these receptors is essentially the same: Exteroreceptors relay information about the nature of a stimulus as well as its location with respect to the animal (localization). This latter purpose is crucial; it may be important to know that a given sound indicates the presence of a predator, but it is equally critical to know from whence the threat originates so that appropriate behaviors can then be generated.

All sensory organs typically possess a receptive field, that area of space on or around the body which, when stimulated, results in the generation of a response in the receptor. The size of the receptive field may vary widely among different types of receptors, and it is the receptive field size together with the density of the receptors in a given area that determines the acuity, or spatial resolution, of the sense system. For example, in humans the skin of the fingers and lips contains a very large number of tactile (touch) receptors, most of

The Electric Sense of Fishes

Many types of aquatic vertebrates, including lampreys, sharks, sturgeon, and lungfish, are able to detect the weak electrical fields produced by living organisms in the aquatic environment. This sense, known as electroreception, is a very ancient sense. Studies on the distribution of electroreception among different animal groups have demonstrated that it is as old as the other vertebrate senses, such as vision. Electroreception is useful to aquatic animals in that it can help them detect prey and potential mates by localizing the source of the bioelectric fields. It may also be a useful aid in navigation and orientation; for instance, these animals are capable of detecting the electric field produced by a large, ore-containing rock in a moving current.

Electroreception evolved early in vertebrates, but has been lost in several major groups, including most bony fishes and in terrestrial vertebrates. In the latter case, the almost infinite electrical resistance of air compared to water makes the possession of an electrosense useless. However, certain vertebrate groups have reevolved electroreception, including catfish and, in mammals, the platypus. Certain other South American and African bony fishes have also evolved a second electric sense that is designed to detect the electrical signals produced by special electric organs in these animals. They use these signals to locate objects in turbid water: The animals detect the changes in the electric field produced by objects with electrical properties different than the water. The electric sense is also used for communication among these animals, to attract mates and warn off rivals.

The Forest Fire Beetle

Animals that rely upon very specific conditions at some point in their lives frequently possess very specialized sense organs that are adapted to detect those conditions. A good example is the beetle *Melanophila acuminata*. The larvae of this beetle are unable to survive the chemical defenses of trees in the European forests in which they live, and so this species depends absolutely upon trees that have been freshly killed by forest fires. Mating takes place when the fire is still in progress and the female lays her eggs on the charred trees as soon as the flames subside. The ability to detect a forest fire from a considerable distance is therefore very important to the survival of this species. Recent studies by H. Schmitz and H. Bleckmann at the University of Bonn have demonstrated that these beetles possess pit organs on their thorax that are sensitive to the wavelengths of infrared radiation produced by forest fires, and that they can use these organs to "see" the heat signature produced by a ten-hectare fire at distances of up to twelve kilometers. Flying beetles thus localize and home in on fires where they are most likely to find potential mates and suitable locations for laying eggs, and they do so in massive swarms. Incidents have been related in which these beetles have even swarmed large outdoor sports stadiums where events were taking place. Apparently the large infrared signal produced by thousands of individuals smoking in the stands was mistaken by the beetles as a forest fire in progress.

which possess very small receptive fields. This allows individuals to discern to a very high degree precisely where on the skin a stimulus is occurring. In other areas of the body, like the back of the neck, the density of receptors in the skin is much lower and it is more difficult to localize exactly where a stimulus is being applied. The density of the receptor cells (rods and cones) in the eye decreases from the central portion of the retina toward the edges. That is why visual acuity is greatest when looking directly at an object and why it is very difficult to read using one's peripheral vision.

The localization of a given stimulus is critical to the organization of an appropriate response. The location of a given sense organ on the body corresponds to a location within the central nervous system, and adjacent receptors are represented by adjacent locations within the brain. Thus, there is a sensory map of the body within the brain, and it is via this topographic organization that information about stimulus location is maintained. There are many sensory maps within the brain and they play a prominent role in the organization of behaviors.

Responding to Sensory Data

Sense organs may be sensitive to a wide array of different qualities and provide the animal with a general sensory scene of the surrounding environment (as in mammalian vision) or may be restricted to a narrow range of stimuli that serve as channels of communication between animals of the same species. This latter case is particularly true for special chemical senses that detect chemicals that have specific behavioral meanings for members of the same species (pheromones). Very often, the sensitivity of sensory systems lies between these two extremes, with the system showing greatest sensitivity to ranges of stimuli that have greater behavioral significance to the animal. Dolphins, for example, locate objects underwater by echolocation, emitting a high-frequency call and then listening for the returning echoes. The greatest sensitivity of the dolphin auditory system is to the range of sound frequencies that are reflected back as echoes, although dolphins can hear other sounds as well.

As animals and their behaviors have evolved, so have their sense organs, providing the animals with the competitive advantages that allow them to survive and reproduce. Furthermore, natural selection frequently results in the evolution of very similar sense organs in widely divergent animals. The eyes of mammals closely resemble those of octopuses, despite the fact that these animals are not at all closely related and their common ancestor lacked complex eyes.

Such convergent evolution of sense organs has resulted from the adaptive pressures on both of these animal groups that depend strongly upon vision to organize behavior. Sometimes the sensory systems of different animal species evolve in tandem: This coevolution of sensory systems is a direct result of the interactions of the species. Bats hunt for flying moths by echolocation, emitting ultrasonic calls and homing in on the echo reflected from the moth. Many moth species, in response, have evolved "ears," located on either side of their abdomen, that are specialized to detect the calls of hunting bats. Depending upon the intensity of the detected call (and thus, the nearness of the bat) the moths display different behaviors. If the intensity is low, indicating the bat is still at a distance, the moth will fly away from the side of the body upon which the call is loudest. If the intensity of the bat's call reaches a certain level, however, the moth will execute an erratic, fluttering crash dive toward the ground in a final attempt to escape. In an additional twist, the dogbane tiger moth emits ultrasonic pulses of its own when it detects the calls of an approaching bat. Such calls may jam the bat's echolocation by interfering with the detection of the returning echoes.

A review of all of the different types of sense organs currently known in animals and the manner in which they are used to organize and shape behaviors would fill an entire volume, and more are being continually discovered. Sense organs are, in a very real sense, the keys to our individual understanding of the world. They provide the information upon which the daily understanding of reality is entirely based.

—John G. New

See also: Anatomy; Beaks and bills; Bone and cartilage; Brain; Circulatory systems of invertebrates; Circulatory systems of vertebrates; Claws, nails, and hooves; Digestive tract; Ears; Endoskeletons; Exoskeletons; Eyes; Fins and flippers; Hearing; Immune system; Kidneys and other excretory structures; Lungs, gills, and tracheas; Muscles in invertebrates; Muscles in vertebrates; Nervous systems of vertebrates; Noses; Physiology; Reproductive system of female mammals; Reproductive system of male mammals; Respiratory system; Skin; Smell; Tails; Teeth, fangs, and tusks; Tentacles; Vision; Wings.

Bibliography

Downer, J. *Supernature: The Unseen Powers of Animals*. New York: Sterling, 2000. Companion to the Discovery Channel series. A fascinating survey of life forms in the animal world and their unusual sense organs and abilities.

Gregory, R. L. *The Oxford Companion to the Mind*. Reprint. New York: Oxford University Press, 1998. A readable and useful reference for further understanding the relationships between sensory perception and behavior.

Halliday, Tim, ed. *The Senses and Communication*. New York: Springer-Verlag, 1998. A solid introduction to the physiology and anatomy of sensory systems.

McFarland, D. *The Oxford Companion to Animal Behavior*. Reprinted and corrected ed. New York: Oxford University Press, 1987. A readable and useful reference for further understanding the relationships between sensory perception and behavior.

SEX DIFFERENCES: EVOLUTIONARY ORIGIN

Type of animal science: Evolution
Fields of study: Anatomy, evolutionary science, reproduction science

In asexual reproduction, genetic material is not exchanged; offspring are genetically identical to the parent. Sexual reproduction involves the exchange of genes. Natural selection favors asexual reproduction in the exploitation of dependable resources, but selection favors sexual reproduction whenever the future is uncertain.

Principal Terms

ANISOGAMY: reproduction using gametes unequal in size or motility

ASEXUAL REPRODUCTION: reproduction in which genes are not exchanged

FEMALE: an organism that produces the larger of two different types of gametes

GONOCHORISM: sexual reproduction in which each individual is either male or female, but never both

HERMAPHRODITISM: sexual reproduction in which both male and female reproductive organs are present in the same individual, either at the same time or at different times

ISOGAMY: reproduction in which all gametes are equal in size and motility

MALE: an organism that produces the smaller of two different types of gametes

PARTHENOGENESIS: asexual reproduction from unfertilized gametes, producing female offspring only

SEXUAL DIMORPHISM: differences in morphology between males and females

SEXUAL REPRODUCTION: reproduction in which genes are exchanged between individuals

SEXUAL SELECTION: selection for reproductive success brought about by the behavioral responses of the opposite sex

The evolutionary origin of sex differences can be understood only by examining the relative benefits of sexual as compared to asexual reproduction. Those forms of reproduction in which genes are not exchanged are considered asexual. Asexual reproduction may take place from already developed body parts (vegetative reproduction) or from special reproductive tissue. In either case, however, asexual reproduction results in the rapid production of numerous individuals genetically identical to their parents. Because asexual reproduction allows numerous offspring to be produced in a short time, it is favored in situations in which a species can gain an advantage by exploiting an abundant but temporary resource, such as a newly discovered cache of food. There is also a further advantage: The individual that finds a resource that it can effectively exploit, if it can reproduce asexually, is assured that all its offspring will possess the same genotype as itself, and will thus be equally able to exploit the same resource for as long as it lasts. Despite these advantages, asexual reproduction is much less common than sexual reproduction among animals. It is a temporary stage in many species, alternating with sexual reproduction. Asexual reproduction is far more common among microorganisms such as bacteria.

Forms of Sexual Reproduction

Sexual reproduction may take many forms, but all of them involve the exchange of genes. Some algae and protozoans exchange chromosomes without gametes in a process called conjugation. Most other forms of sexual reproduction use special sex cells called gametes, which exist in different "mating types." Two gametes can combine only if their mating types are different. Some simple organisms, such as the one-celled green alga *Chlamydomonas*, have gametes that are indistinguishable in size or appearance, a condition known as isogamy. Most other organisms have gametes of unequal sizes, a condition called anisogamy. Selection often intensifies the differences between gametes, producing a small, motile sperm and a much larger, immobile egg, laden with stored food (yolk).

Some sexually reproducing organisms have separate sexes, a condition called gonochorism. Individuals producing eggs are called female, while individuals producing sperm are called male. Since sperm are generally small and can be produced in great numbers, males tend to leave more offspring if they reproduce prolifically, indiscriminately, and often. Females, on the other hand, have fewer eggs to offer, and in many species they must also invest nutritional and behavioral energy in the laying of eggs and the care of the resultant offspring. Selection in these species favors females who choose their mates more carefully and take better care of their offspring.

The differing selective forces operating on the two sexes often give rise to sexual dimorphism, or differences in morphology between the sexes. Sexual dimorphism can also be reinforced by competition for reproductive success, a phenomenon first studied by Charles Darwin. Darwin called this type of competition sexual selection. It takes two basic forms—direct competition between members of the same sex, and mate choices made by members of the opposite sex.

Direct male-male competition often takes such spectacular forms as rams or stags fighting in head-to-head combat. Similar fights also occur in many other species, including a variety of turtles, birds, mammals, fishes and invertebrates. Many more species, however, engage in ritual fighting in which gestures and displays substitute for actual combat. Male baboons, for example, threaten each other in a variety of ways, including staring at each other, slapping the ground, jerking the head, or simply walking toward a rival.

Although male-male rivalry has attracted more attention in the past, female-female competition also occurs in many species. Now that more ethologists and sociobiologists are looking for evidence of such direct competition among females, it is being discovered that it is a fairly widespread occurrence which had previously escaped notice only because so few scientists suspected its existence or were interested in looking for it. Female-female competition has been found among langur monkeys, golden lion-marmosets, ichneumon wasps, and several other species.

Sexual Selection

Sexual selection in mating is selection in which reproductive success is determined at least in part by mate choice. No matter what form sexual selection may take, it results in greater reproductive success for those individuals chosen as mates, while those not chosen must try again and again if they are ever to succeed in leaving any offspring at all.

Sexual selection of this kind occurs in nearly all gonochoristic species. In some species, males will attract females by means of a visual display or by various sounds (also called calls or vocalizations). Females in such species will exercise choice by selecting among the available males. For example, male peacocks, lyre-birds, and birds of paradise will court females by showing off their elaborate tail feathers in bright gaudy displays. In other species, the females perform the display and the males do the selecting.

Sometimes, the display will include an object such as a nest constructed by one partner as an attraction to its mate. Bowerbirds, for example, construct elaborate nuptial bowers as a means of attracting their mates. These bowers, which contain a nest in the center, are sometimes adorned with attractive stones, flowers, and other brightly colored objects. In some species of animals, males and females will respond to one another by performing alternating steps; in this manner, each sex selects members of the other.

Many sexually reproducing organisms have both male organs which produce sperm and female organs which produce eggs, a condition known as hermaphroditism. Earthworms and many snails are simultaneous hermaphrodites, meaning that both male and female organs are present at the same time. Hermaphrodites often have their parts so arranged that self-fertilization is difficult or impossible. One system that guarantees cross-fertilization is serial hermaphroditism. In this system, each individual develops the organs of one sex first, then changes into the opposite sex as it matures further.

Some sexually reproducing organisms have become secondarily asexual through a process called parthenogenesis, in which gametes (eggs) develop into new individuals without fertilization. In bees and wasps, males develop parthenogenetically from unfertilized eggs, while females (with twice the chromosome number) develop from fertilized eggs.

The Cost of Sexual Reproduction

Sexually reproducing organisms experience a cost associated with the energy devoted to courtship behavior and to the growing of sexual parts. In addition, the act of courtship usually exposes an individual to a greater risk of predation, and the distractions of mating further increase this risk. In view of these costs, many evolutionists have wondered how sex ever evolved in the first place, or why it is so widespread. Any adaptation so complex and so costly would long ago have disappeared if the organisms possessing it were at a selective disadvantage. The widespread occurrence of sex, and of numerous sexual systems, shows that there must be some advantage to all the various forms of sexual reproduction, and that this advantage is sufficient to overcome the recognized advantages of asexual reproduction in terms of rapid proliferation with relatively low investment of energy.

The answer to this puzzle is based on the fact that asexually produced offspring are all genetically similar to the parent, while sexually produced offspring differ considerably from one another. Organisms exploiting a dependable habitat or food supply often leave more offspring if they produce numerous genetically similar offspring rapidly and asexually. On the other hand, organisms facing uncertain future conditions have a better chance of leaving more offspring if they reproduce sexually and therefore produce a more varied assortment of offspring, at least some of which might have the adaptations needed to survive in the uncertain future. Examination of those species that are capable of reproducing either way confirms this hypothesis: Whenever favorable conditions are likely to persist, they reproduce rapidly and asexually. Faced with conditions of adversity or future uncertainty, however, these same species reproduce sexually. In species that alternate between sexually produced and asexually produced generations, the asexual phases typically occur during the seasons of assured abundance, while the sexual phases are more likely to occur at the onset of harsh or uncertain conditions. Sex, in other words, is a hedge against adversity and against an uncertain future.

Studying Sexual Reproduction

Most biologists who study reproduction are either ecologists, ethologists, or geneticists. Their methods include counting various kinds of offspring and measuring their genetic variability. Reproductive ecologists and ethologists also measure parental investment, or the amount of energy used by individuals of each type (and each sex) in the courting of their mates, in the production of gametes, and in caring for their young. Energy costs of this kind are generally measured by comparing the food consumption of individuals engaged in various types of activity using statistical methods of comparison among large numbers of observations.

The morphology of sex organs in various species is also studied by comparative anatomists and by specialists on particular taxonomic groups such as entomologists (who study insects), helminthologists (who study worms), malacologists (who study snails and other mollusks), and ich-

Lions exhibit marked sexual dimorphism, with males not only being much larger than females but also developing distinctive manes when they reach maturity. (Corbis)

thyologists (who study fishes). In most hermaphroditic species, for example, the organs are so arranged as to make cross-fertilization easier and self-fertilization more difficult.

The above explanation of sexual reproduction as resulting from the greater variability among offspring facing an uncertain future is partially confirmed by studying species that can reproduce either sexually or asexually. Among these species, asexual reproduction is always favored in situations in which an individual discovers a resource (such as a habitat or a food source) too large to exploit by itself. These conditions favor individuals that can reproduce rapidly and asexually produce numerous individuals genetically similar to themselves, who then proceed to exploit the resource. Aphids, for example, produce one or several asexual generations during the spring and early summer, when plant food is abundant. In seasons or situations of great risk or uncertainty, however, the same species often reproduce sexually at somewhat greater energetic cost, leaving a wider variety of offspring but a smaller total number. Under unpredictable conditions (such as those associated with wintering in a cold, temperate climate), the greater energetic costs of reproducing sexually are more than made up by the greater genetic and ecological variability among the offspring. Sexually reproducing individuals leave more offspring (on the average) than asexual individuals under these conditions. Similarly, among hermaphroditic species, cross-fertilization results in more varied offspring than self-fertilization, and is therefore favored under such conditions.

Testing Theories

The several reproductive methods studied by biologists provide a natural laboratory for the testing of several theories. Among these are theories concerned with genetic variability, natural selection, the evolution of sex, and the allocation of resources, including the theory of parental investment in the care of their offspring.

In terms of the two most general types of reproductive strategies, those species using a system called the r strategy (reproducing prolifically at small body size) may be either sexual or asexual, or may alternate between these two methods of reproduction. On the other hand, species following the K strategy (reproducing in smaller numbers at larger body size and investing time and energy in parental care) are invariably sexually reproducing and most often gonochoristic as well.

In addition to the theoretical considerations mentioned above, the study of alternative methods of reproduction gives us important insights into the reasons that our species, like other K strategists, is sexually reproducing and gonochoristic. In most species, sexual behavior is largely controlled by instincts, but learned behavior plays a major role among higher primates. Beyond what is necessary in copulation and childbirth, much of sex-specific behavior in humans is culturally defined and may differ from one society to another. This includes the norms of what behavior is appropriate (or inappropriate) for each sex and what personal qualities are considered masculine or feminine. All attempts to redefine sex roles will lead nowhere, unless one is aware of both the biological and the social underpinnings of these roles.

—*Eli C. Minkoff*

See also: Asexual reproduction; Copulation; Development: Evolutionary perspective; Gametogenesis; Gene flow; Genetics; Isolating mechanisms in evolution; Natural selection; Nonrandom mating, genetic drift, and mutation; Parthenogenesis; Reproductive systems of female mammals; Reproductive systems of male mammals; Sexual development.

Bibliography

Alcock, John. *Animal Behavior: An Evolutionary Approach.* 7th ed. Sunderland, Mass.: Sinauer Associates, 2001. Perhaps the best overall textbook on the subject of animal behavior, this book takes an evolutionary approach in that it attempts to examine the adaptive reasons behind each behavior pattern. The book is also good in its coverage of a wide variety of organisms, including insects, aquatic invertebrates, fishes, amphibians, reptiles, birds, and mammals. The book has many good black-and-white illustrations and a good, lengthy bibliography.

Brown, J. L. *The Evolution of Behavior.* New York: W. W. Norton, 1975. Another good review of animal behavior, including an entire chapter on mating systems and sexual selection. A lengthy bibliography is included.

Campbell, Bernard, ed. *Sexual Selection and the Descent of Man, 1871-1971.* Chicago: Aldine, 1972. A series of eleven articles, each written by a different contributor, outlining the theory of sexual selection as applied to a variety of species.

Campbell, Neil A. *Biology: Concepts and Connections.* 3d ed. Menlo Park, Calif: Benjamin/Cummings, 2000. An innovative college textbook for students with some biology background.

Clutton-Brock T. H., ed. *Reproductive Success: Studies of Individual Variation in Contrasting Breeding Systems.* Chicago: University of Chicago Press, 1988. A series of twenty-nine individual studies by fifty-three contributors, this book holds a treasure of data on a variety of mating and reproductive systems in both insects and vertebrates (mostly birds and mammals). Most of the articles deal with lifetime measures of reproductive success. The book contains many tables of data, a moderate number of illustrations, and an extensive bibliography.

Daly, Martin, and Margo Wilson. *Sex, Evolution, and Behavior.* 2d ed. Belmont, Calif.: Wadsworth, 1983. Perhaps the best summary of the issues related to the advantages and costs of various reproductive strategies, including those related to sex.

McGill, T. E., D. A. Dewsbury, and B. D. Sachs. *Sex and Behavior: Status and Prospectus.* New York: Plenum, 1978. A series of sixteen chapters, each by a different author or authors, on various topics related to sex-related behavior. A large portion of the book is devoted to studies of sex differences in humans. There is a bibliography at the end of each chapter. There are very few illustrations, mostly in the form of graphs.

Maynard-Smith, John. *Evolution and the Theory of Games.* New York: Cambridge University Press, 1982. This short book presents a more theoretical approach to the problem of evolutionary strategies in general. Sexual reproduction is discussed as a reproductive strategy as is parthenogenesis. There are also discussions of sex ratios and parental care. The book contains a few graphs and many equations. There is a bibliography, but most of the works listed are articles in technical journals.

Rosenblatt, J. S., and B. R. Komisaruk, eds. *Reproductive Behavior and Evolution.* New York: Plenum, 1977. This book has seven chapters, each by a different author. Included are good chapters on the genetic control of reproductive behavior and reproductive isolation, on mating and child-rearing systems, and on parental care by both mothers and fathers. Most of the discussion centers on mammals. There are black-and-white illustrations of several types, and a bibliography is included at the end of each chapter.

SEXUAL DEVELOPMENT

Type of animal science: Development
Fields of study: Anatomy, embryology, genetics, physiology, reproduction science

The development of an organism into a male or female involves a complex series of interactions, including differential growth, influence of the external or internal environment, and genetic factors.

Principal Terms

ANDROGENS: the general term for a variety of male sex hormones, such as testosterone and dihydrotestosterone

GENITAL TUBERCLE: a small swelling or protuberance toward the front of an embryo's genital area; it is destined to become the penis tip or clitoris

GENITALIA: the external sex structures

GONAD: the structure that produces eggs or sperm cells and sex hormones; the ovary or the testis

HERMAPHRODITE: a single organism that produces both eggs and sperm

LABIAL FOLDS: the paired ridges of tissue on either side of the embryo's genital area, which become penis and scrotum in males and labia in females

MÜLLERIAN DUCTS: the embryonic ducts that will become the female oviducts or Fallopian tubes, uterus, and vagina

PARTHENOGENESIS: the development of an unfertilized egg

UROGENITAL GROOVE: a slitlike opening behind the genital tubercle that will become enclosed in the penis but remain open in females

WOLFFIAN DUCTS: an embryonic duct system that becomes the internal accessory male structures that carry the sperm

While some lower forms of life with no recognizably different sexes exchange genetic material in a form of sexual reproduction, sexual reproduction in most organisms involves individuals with some obviously different physical and behavioral features. Biologically, the real difference between males and females is the type of sex cells they produce—whether large eggs specialized to support embryonic development or tiny sperm specialized for moving to the egg. Eggs and sperm are produced in gonads—the ovaries of females and the testes of males. The gonads of higher animals also produce sex hormones, chemical messengers that affect both embryonic and adult sexual development.

Even these basic sex distinctions are rather flexible in some organisms. Sometimes, sex is determined entirely by the environment. One kind of marine worm becomes a female unless it attaches as a larva to an adult female, whereupon it becomes a male—probably because of hormones secreted by the female. Temperature can control sex development in some animals, such as mosquitoes and amphibians. Sex may also be determined by size. Since it takes more energy to produce eggs than sperm, when food is scarce it may be more adaptive to be male. The European oyster begins adult life as a male, changes to a female as it grows larger, and reverts to being a male after shedding eggs.

In territorial animals, being a large male may be an advantage. A tropical wrasse, or "cleaner fish," travels with a harem of smaller females. If he is removed, the largest female becomes a male within a few days. Many organisms, including earthworms, snails, and some fish, are hermaphrodites—functional males/females that can fertilize themselves or exchange sperm with others. Some insects, worms, crustaceans, goldfish, whip-tailed lizards, and even turkeys lay eggs that can develop without fertilization, a process called parthenogenesis. This strategy is not sound in the long run, since it does not promote genetic diversity. It is an advantage for an organism living under good conditions, however, where an all-female population can exploit the ideal environment most efficiently.

The Origin of Sex Differentiation

Sex differentiation probably originated as differential growth of either the ovary or the testis, mediated in various ways by hormones or other environmental factors. Later in evolution it came under genetic control, which made the process more independent of environment and made possible the development of more complex reproductive structures and behavior.

The genetic sex of an animal is determined by the father at fertilization. In most species, females have two matching X chromosomes, males have an unmatched X and a smaller Y chromosome. If a normal egg with one X chromosome is fertilized by an X-bearing sperm, the XX embryo is genetically female. A Y-bearing sperm will produce a genetic male, XY. In butterflies, fishes, and birds, however, females have XY chromosomes and males have XX. Initially, XX and XY embryos look identical and in a sense are still sexually bipotential. Their gonads are "indifferent," that is, able to form either an ovary or a testis. Each has two sets of undifferentiated sex ducts. One set, the Wolffian ducts, will become the sperm ducts and other male structures. The other set, the Müllerian ducts, form the female oviducts, uterus, and vagina.

Soon, however, genes on the Y chromosome direct the inner part of the indifferent gonad to become a testis, which then produces the male sex hormones (androgens) and Müllerian-inhibiting substance (MIS), which control further events in male development. An androgen called testosterone causes the male duct system to persist and develop, and MIS makes the female duct system degenerate. Testosterone has other developmental effects, as indicated by the fact that in monkeys, male behavior is linked to the length of embryonic exposure to testosterone.

Without the influence of the Y chromosome an XX gonad begins to develop into an ovary. The role of female sex hormones in development is unclear, since in mammals female embryonic development can occur in the absence of female hormones. The mammalian embryo has a tendency to develop in the female direction unless specific influences prevent it. The Wolffian ducts are actually remnants of a drainage system from a temporary embryonic kidney that disappears before birth. Only the presence of male sex hormones will keep these tubes from disintegrating. The Müllerian ducts, on the other hand, tend to persist unless acted upon by the anti-Müllerian substance. In birds, the embryonic ovary is the dominant gonad, and it actively feminizes the reproductive tract. It has been suggested that the early male development in mammals is necessary to allow male differentiation in the female-hormone-rich uterine environment.

Until differentiation begins, both sexes also have the same vaguely female-looking external sex structures or genitalia. In both sexes, a small protuberance called the genital tubercle is found toward the front or belly side of the embryo. Behind the genital tubercle is a slitlike opening, the urogenital groove; it is flanked by two sets of paired folds or swellings, like a river valley paralleled by two sets of ridges on either side. In the female, the genital tubercle will form a small structure called the clitoris. The urogenital groove will remain open, forming a vestibule into which the vagina and the urethra open, which empties the urinary bladder. The folds on either side of the groove will remain relatively unchanged to form labial folds.

In the male, the genital tubercle will become the tip of the penis, and the innermost urogenital folds will fuse together to form the body of the penis; the "scar" of this joining may be seen on the underside of the penis. This fusion closes off the urogenital groove and encloses the male's urethra within the tubelike penis. The outer pair of ridges will fuse to form the scrotum, the sac that encloses the two testes, which descend into the scrotal sac before birth. Another androgen, dihydrotestosterone, may be responsible for the development of these external male structures.

Studying Sexual Development

Many sex-determining mechanisms can be studied by simple modification of the environment. For example, by varying temperature, hormone level, social-group composition, or other environmental factors it is actually possible to reverse the sexes of some invertebrates, fish, and amphibians. Castration experiments are commonly used to study the effects of hormones on the sexual development of birds and mammals. For example, castrated mammals of either sex develop in the female pattern. Since in birds only the left ovary develops, castration of hens may result in the transformation of the right gonad into a testis, with complete functional sex reversal.

Sex development can be studied with naturally occurring hormone imbalances, as in freemartin calves—sterile, masculinized females whose male twin exposed them before birth to male sex hormone. The same masculinizing effect on female fetuses can be achieved by injecting a pregnant mammal with androgens or even by growing an embryonic ovary and testis together in an organ culture outside the body. In each case, the female structures are masculinized by the male hormones.

Sex chromosome mutants in animals as diverse as fruit flies, mice, and birds can be used to study chromosomal influences in sex determination. To use a human example, there are sterile XX men with a tiny piece of the short arm of the Y chromosome attached to one X. There are also XY females who show a deletion of the same short arm of the Y. These observations have led geneticists to think that the testis-determining genes are on the short arm of the Y, since without it an individual—even one with a Y—is female.

The Evolutionary Advantage of Sexual Reproduction

In spite of its great biological costs, sexual reproduction is practiced by almost every kind of living thing. It confers tremendous evolutionary advantage on a species by producing a new individual with the genetic characteristics of two parents but with unique combinations of features that may make the offspring more successful than either parent. The advantage of having separate sexes for reproduction is that it permits the development of extremely specialized reproductive organs for the very different requirements of sperm or egg production, and, when needed, intrauterine support for the embryo. Though some organisms show great sexual flexibility, higher animals and plants have tended toward sexual stability, probably because of the high cost of sex reversal for organisms with highly specialized sexual structures.

The study of sex differentiation helps advance scientific knowledge in many areas. Modes of sex determination often provide clues to evolutionary relationships among groups of organisms. In addition, the development of sex differences makes a good model system for the study of more general questions. For example, one might use the control of sexual size differences to attack the broader question of what makes mice smaller than elephants. The control of sex differentiation by environmental factors, to use another example, might provide geneticists with a way to study how genes are turned on by hormones, temperature, or other external influences.

For embryologists, the stepwise determination and differentiation of the mammalian reproductive system is an excellent general development model; it involves a genetically controlled sequence of events that includes both the preservation of one embyronic structure (the Wolffian duct) and the removal of another (the Müllerian duct). Hormonally controlled events include a wide variety of developmental sequences, from the externally visible large-scale changes involved in the shaping of the external genitalia to the biochemical differentiation that programs the brain hypothalamus for its complex control of the menstrual cycle.

—Michele Morek

See also: Determination and differentiation; Embryology; Endocrine systems in invertebrates; Endocrine systems in vertebrates; Gametogenesis; Hormones and behavior; Hormones in mammals; Reproduction; Reproductive systems of female mammals; Reproductive systems of male mammals.

Bibliography

Carlson, Bruce M. *Patten's Foundations of Embryology*. 6th ed. New York: McGraw-Hill, 1996. Detailed treatment of general anatomy of the reproductive organs and their embryonic development. Though this is an upper-level college embryology textbook and too technical for many general readers, it has unsurpassed illustrations and an extensive bibliography.

Hickman, Cleveland, et al. *Biology of Animals*. 7th ed. St. Louis: Times Mirror/Mosby, 1998. This freshman college zoology text has a special chapter on reproduction and others on major animal groups. It is beautifully illustrated and has a good glossary, index, and bibliographies.

Hopf, Alice. *Strange Sex Lives in the Animal Kingdom*. New York: McGraw-Hill, 1981. This short book, intended for the general reader, gives a very readable account of the importance of sexual reproduction, unusual types of sex determination, sex switching, hermaphroditism, and parthenogenesis. The bibliography and index are brief.

Naftolin, Federick, et al. *Science* 211 (March 20, 1981). The entire issue of this magazine is devoted to sexual differences and how they develop. Though difficult for the casual reader, most of the articles were written for a general scientific audience and can be understood by someone with a strong interest in the subject. Each article has a good bibliography and deals with the genetic or hormonal mechanisms that govern sex organ development.

Rothwell, Norman V. *Understanding Genetics*. 4th ed. New York: Oxford University Press, 1988. This college-level genetics textbook requires little more than a basic familiarity with genetic terminology. Contains a very thorough treatment of sex and inheritance, including information on sex chromosomes and their role in sexual differentiation, and related topics such as sex-linked, sex-influenced, and sex-limited genetic traits in various animals. Good bibliography and glossary.

Wrangham, Richard W., W. C. McGrew, Frans B. M. De Waal, and Paul G. Heltne, eds. *Chimpanzee Cultures*. Cambridge, Mass.: Harvard University Press, 1996. Essays by authorities on chimpanzee biology and behavior. Discussions of social relations include sexual development and behavior.

SHARKS AND RAYS

Type of animal science: Classification
Fields of study: Anatomy, systematics (taxonomy), zoology

Cartilaginous fishes constitute one of the seven living classes of vertebrates. There are about seven hundred living species of sharks, rays, skates, and ratfishes in this vertebrate group that has ancient origins.

Principal Terms

CARTILAGE: a gristlelike supporting connective tissue that forms the entire skeleton of cartilaginous fishes

CHONDRICHTHYES: the scientific name for the taxonomic group, or class, in which the jaw-bearing cartilaginous fishes are placed; it includes sharks, rays, skates, and ratfishes

DEVONIAN: a geological period from about 400 million years ago to about 350 million years ago; during this period, ancestral sharks were abundant and diverse

PLACOID SCALES: hard, toothlike scales, sometimes called denticles, that are embedded in the skin of most sharks, rays, and skates

SILURIAN: a geological period from about 440 million years ago to about 400 million years ago; the first jawed fishes appeared during this period

VERTEBRATE: a member of the chordate subphylum Vertebrata, characterized by the possession of a vertebral column made of cartilage or bone

The Chondrichthyes, or cartilaginous fish, constitute a large group of cold-blooded fishlike vertebrates. Like other fish they are characterized by gills, fins, and a dependence on water as a medium in which to live. Unlike the more primitive jawless fishes (class Agnatha), they possess well-developed jaws; unlike the bony fish (class Osteichthyes), they possess a skeleton composed entirely of cartilage, although this is often highly calcified (strengthened and hardened by calcium compounds).

Additional features characteristic of cartilaginous fish include teeth, lack of a swim bladder (the buoyancy organ present in most bony fishes), a spiral valve intestine, internal fertilization, and osmoregulation (salt balance) by means of the nitrogen compound urea. The fossil record for the Chondrichthyes begins in the late Silurian period, over 400 million years ago. They were abundant throughout the next period, the Devonian. Within the class Chondrichthyes are two distinct subclasses, the Elasmobranchii and the Holocephali; this separation dates back to the Devonian period.

Most living cartilaginous fishes are members of the Elasmobranchii. This group includes at least six hundred, perhaps seven to eight hundred, living species of sharks (order Selachii) and rays (order Batoidea). The elasmobranchs are characterized by five to seven separate gill openings on each side of the head region; members of this group also usually have placoid scales. Sharks and rays are almost exclusively marine, although there are about ten ray species (the river stingrays) that inhabit only freshwater environments and several other rays and at least one shark species (the bull shark) commonly found in rivers.

Sharks

Sharks are a diverse group of carnivorous species, ranging in size from the tiny dwarf shark (*Squaliolus laticaudus*), which matures at less than fifteen centimeters in length, to the enormous whale shark (*Rhincodon typus*), which reaches fifteen meters or more in length and represents the largest fish species of any kind. Curiously, the whale shark and the nearly-as-large basking shark (*Cetorhinus maximus*) are plankton feeders. They capture their tiny food organisms by swimming open-mouthed through the water and straining out the plankton with fine comblike structures in their gills, called gill rakers.

Most sharks, however, have sharp, bladelike teeth, suitable for attacking and feeding on more active prey. The white shark (*Carcharodon carcharias*) is a voracious roving predator that may grow to twelve meters in length. It has been implicated in more fatal shark attacks than any other species.

In addition to well-developed eyes, inner ears, and olfactory (smell) organs, sharks possess a lateral line system, as do most bony fishes. This is a sense organ consisting of a canal beneath the skin, on each side of the body, connected to the surface by numerous pores. It is sensitive to vibrations in the water, giving sharks a sense of "distant touch" that enables them to navigate and hunt their prey in murky water. Another sensory feature of sharks and other elasmobranchs is an electroreception system, consisting of receptors, called ampullae of Lorenzini, on the surface of the snout. Apparently, this system is useful in hunting, since it allows the weak electric fields produced by the muscle contractions of prey species to be detected. It may also function in intraspecific communication (communication with others of the same species), since many elasmobranchs possess electric organs.

Sharks are typically torpedo-shaped and slightly depressed in form—that is, flattened from top to bottom. They swim by means of rhythmic undulations of the body, which are produced by sequential contraction of the myomeres (body-muscle segments). The tilt of the shark's pectoral fins (the paired fins toward the front of the body) and heterocercal tail (the upper lobe of the tail fin being larger than the lower lobe) enable it to maintain its relative depth position as it swims forward, despite the fact that the shark lacks a swim bladder. Also improving the buoyancy of sharks are their cartilage skeletons, which are lighter than bone, and their large, oily livers. Some shark livers contain a unique low-density oil called squalene.

Sharks and other cartilaginous marine fishes regulate the concentration of solutes (dissolved substances) in the body in a manner very different from that of the bony fishes, which either retain salt (freshwater bony fishes) or secrete salt (marine bony fishes). Sharks maintain a concentration close to or higher than that of seawater by retaining urea and trimethylamine oxide, two relatively nontoxic nitrogenous waste products.

Reproduction in the sharks and other cartilaginous fishes is characterized by internal fertilization. A pair of intromittent, or copulatory, organs called claspers are located on the pelvic fins (the paired fins nearer the tail region) of the male. These are used to transfer sperm to the female genital opening. Embryos remain in the body or are released in egg cases, for a long gestation, or development, period. A small number of young either are born alive or hatch from an egg case in active, well-developed form.

Common Shark Species

Among the more familiar shark species are members of the family Lamnidae. This family includes the dreaded white shark and other "mackerel sharks," such as the shortfin mako (*Isurus oxyrinchus*)—a popular game fish and food fish, but a dangerous species as well. Another family, the Carcharhinidae (requiem sharks), with dozens of species, includes two man-eaters, the tiger shark (*Galeocerdo cuvieri*) and the bull shark (*Carcharhinus leucas*). Bull sharks have been found in rivers and lakes in Central and South America; they have penetrated the Amazon River as far as Peru.

Yet another group of dangerous sharks is the family Sphyrnidae, the hammerheads. These species are distinguished by a laterally expanded head, having the eyes and nasal openings at the ends of the hammerlike extensions. The function of this arrangement is unclear, but it probably aids in detecting and homing in on prey organisms. To students of comparative anatomy, the spiny dogfish (*Squalus acanthias*) is perhaps the most famil-

The unusual flat body shape of rays is an adaptation to living on the bottom of the ocean. (Digital Stock)

iar shark, since it is often dissected in the classroom as a typical representative of the lower vertebrates. This worldwide species, inhabiting temperate coastal areas, is also an important food fish in many parts of the world. It commonly appears in England, for example, in fish and chips.

Rays and Skates

Most of the more than three hundred living species of the order Batoidea, the rays and skates, are adapted for living on the bottom. In body form they are strongly depressed (flattened), with enlarged pectoral fins extending forward to the head region. Their teeth are usually pavementlike, for crushing their hard-shelled invertebrate prey. Most species give birth to live young, except the skates (family Rajidae), in which the eggs develop in a leathery egg case (the "mermaid's purse" that beach visitors often find in the sand).

Several ray families include members with a venomous spine on the tail, including the Dasyatidae (stingrays), the Potamotrygonidae (river stingrays), and the Myliobatidae (eagle rays). The largest species among rays and skates is the giant manta ray or devilfish (*Manta birostis*), which may attain a width of over six meters between the tips of its pectoral fins and a weight in excess of 1,300 kilograms. Like two other cartilaginous fish giants mentioned earlier (the whale shark and basking shark), the giant manta ray is a plankton feeder. It directs plankton into its mouth as it swims by means of large scooplike extensions on its head—the "horns" responsible for the name "devilfish." It then filters out the plankton with its comblike gill rakers.

Rays and skates swim by means of flapping movements of their winglike pectoral fins. Some species, including eagle rays and manta rays, can make spectacular leaps from the water. Among the more remarkable rays are the electric rays (family Torpedinidae). These sluggish fishes use electrical discharges of up to two hundred volts, produced by a pair of disk-shaped electric organs on the sides of the head, to stun their prey and perhaps to repel predators. Another specialized group among the rays is the sawfish family (Pristidae). A sawfish resembles a somewhat flattened shark in body form but has a long, flat, toothed extension (the "saw") on the end of its snout. This is used to slash through a school of prey fish.

Ratfish

The other major subgroup of cartilaginous fishes, the subclass Holocephali, comprises about twenty-five or thirty living marine species, most or all of which are placed in a single family, the Chimaeridae. They have a single gill opening on each side, like the bony fishes, but unlike them have a soft (rather than bony) gill cover. These fishes, commonly called chimaeras or ratfish (because of their long, slender tails), live and feed on the ocean bottom, usually in deep water. They have pavementlike teeth for crushing their mollusk and crustacean food, and they have a venomous spine on the leading edge of the first dorsal fin (the forwardmost of the unpaired fins on the upper surface of the body) for defensive purposes. Male ratfishes have a fingerlike barbed clasper, of unknown function, on the top of the head, and two pairs of claspers on the ventral (belly) side of the body. At least one pair of these ventral claspers is involved in mating. The female lays eggs in leathery capsules somewhat like skate egg cases.

Ichthyology

The study of sharks, rays, and related species is part of the larger discipline known as fish biology, or ichthyology. This science has its origins in the writings of Aristotle more than twenty-three centuries ago. He was the first to report, for example, that the sex of sharks can be determined by the structure of the pelvic fins, that is, by the presence of claspers in the male. Aristotle also contrived some rather fanciful interpretations of shark anatomy and behavior, as in his explanation for the fact that the shark mouth is on the under side of the head, far back from the tip of the snout, unlike the mouths of most other fish. In his view, this made it difficult for the shark to feed on its prey, requiring it to turn on its back, and thus nature allowed some chance for the poor animals to escape the jaws of this ravenous predator.

Modern study of cartilaginous fishes, like fish biology in general, involves several disciplines. Ichthyology, or systematic ichthyology, is particularly concerned with the naming and classifying of species and higher taxa (taxonomic categories) and determining their interrelationships. Living cartilaginous fishes are probably better known (that is, more of the extant species have been discovered) than living bony fishes, simply because they tend to be larger, more conspicuous, and less secretive. Yet it was not until 1976 that one of the largest shark species, a deep-water filter-feeding species called the megamouth shark (*Megachasma pelagios*), was discovered near Hawaii. There may exist many additional Chondrichthyes species in deep ocean waters and remote coral reef areas.

Chondrichthyes systematics (that is, the classification of the fishes) has undergone many changes and revisions as more has become known about fossil representatives and about the characteristics of the anatomy, biochemistry, and the like, of the living species. Studies of fossil cartilaginous fishes are limited almost entirely to samples of teeth, since these are virtually the only body parts durable enough to be preserved in the fossil record. Nevertheless, there is enough information in the characteristics of the teeth so that knowledge of the interrelationships of fossil species, both among one another and with living species, is quite advanced. It is known, for example, that the enormous hand-sized fossil teeth of *Carcharodon megalodon*, which lived about twenty million years ago, are so similar to the smaller teeth of the white shark (*Carcharodon carcharias*) that they both belong in the genus *Carcharodon*.

Systematic study of modern species requires collection of specimens, generally by means of nets, traps, hooks, and lines, or spearing. Specimens are then preserved in some way and maintained in a museum collection. Entire specimens, if they are relatively small, can be maintained in diluted alcohol after fixation in formalin (formaldehyde solution). Other specimens, especially large ones, are dissected, and only certain parts are preserved, particularly the head skeleton with jaws and teeth. Certain new techniques provide taxonomic information from samples of living tissue. Karyotyping (analysis of the chromosomes), protein analysis (determination of the amino acid sequence), and DNA hybridization (estimation of genetic similarities) are all techniques that can elucidate interrelationships among the Chondrichthyes. Other disciplines concerned with the study of cartilaginous fishes include fisheries biology (the science of management and exploitation of commercially important fish species) and comparative physiology.

The Place of Sharks and Rays in Evolution

Cartilaginous fish represent an early line in the evolution of vertebrates. Understanding their interrelationships is crucial to an understanding of the ancestry of other fishes and of tetrapods (amphibians, reptiles, birds, and mammals) and thus, ultimately, of humans. Even though humankind's ancestors split from the ancestors of sharks and rays more than 400 million years ago, many anatomical and physiological features are shared. A prime example is the eye, which is extraordinarily similar in all vertebrates. The same system of eye movement, involving six muscles innervated by the same three cranial nerves, has remained unchanged throughout vertebrate evolutionary history. Thus, the study of shark eyes, or any other aspect of shark biology, deepens the understanding of the evolution of higher animals.

Sharks and their relatives are important and interesting in other ways as well. Many species have importance as food, especially in Asia and the South Pacific. Other products derived from sharks include shark liver oil (which was an important vitamin A source before the development of synthetic vitamin A), shark skin (for leather products), and shark cartilage derivatives (used in medicine).

Even though the real risk of shark attack anywhere in the world is statistically very small, sharks have been known to be such brutal killers that interest in preventing shark attacks is widespread. Various chemical shark repellants such as "shark chaser" have been tried. This water-soluble mixture of dye and copper acetate was given to U.S. military personnel during World War II for use if they were stranded in the sea after their ships were sunk or planes downed. It was, however, later shown to have little or no effect on sharks. Other techniques have included the cartridge-loaded "bang-stick," which is probably more dangerous to the untrained user than to a shark. A more promising device is the "shark screen," a floating plastic bag that can be filled with water and entered—masking the odors, sounds, and movements that might attract sharks.

Much remains to be learned about sharks, rays, and other cartilaginous fishes. Studying their ecology, behavior, and evolutionary relationships is important for further understanding of their basic biological nature. It is also essential for maximizing the benefit of commercially important species and minimizing the risk posed by dangerous species.

—George Dale

See also: Amphibians; Evolution: Historical perspective; Fish; Invertebrates; Lungs, gills, and tracheas; Marine animals; Marine biology; Systematics; Whale sharks; White sharks.

Bibliography

Budker, Paul. *The Life of Sharks*. Translated by Peter J. Whitehead. New York: Columbia University Press, 1971. A medium-sized, semipopular account of shark biology. The taxonomy and bibliography are somewhat dated, but in other respects it is highly interesting and informative. Has drawings and a few black-and-white photographs. Widely considered a classic.

Dingerkus, Guido. *The Shark Watcher's Guide*. Englewood Cliffs, N.J.: Julian Messner, 1985. This short book is intended as a popular introduction to the diversity of shark species, but it is scientifically rigorous and very well illustrated. Very useful for fishermen who might have occasion to seek information about what they have caught.

Ellis, Richard. *The Book of Sharks*. New York: Grosset & Dunlap, 1975. This large-format book is illustrated with the magnificent paintings of Richard Ellis, widely considered the world's premier painter of sharks and whales. The paintings are matched with a concise, highly readable introduction to sharks and shark biology. An unusual feature is a section of brief biographies, with photographs, of the world's leading shark investigators—a who's who of shark research.

Lineaweaver, T. H., and R. H. Backus. *The Natural History of Sharks*. Reprint. New York: J. B. Lippincott, 1997. This classic account is one of the best popular introductions to sharks. It weaves a story like an extended magazine article, full of anecdotes and historical perspectives, on shark biology and shark-human interactions. Illustrated with drawings and black-and-white photographs. Includes an unusual glossary, giving the origins and meanings of shark scientific names.

McCormick, H. W., T. Allen, and W. Young. *Shadows in the Sea: Sharks, Skates, and Rays*. Rev. and updated ed. Baltimore: Johns Hopkins University Press, 1996. This highly readable popular book reads like a history of all man's interactions with sharks. Packed with anecdotes, diary excerpts, and newspaper accounts, it is an encyclopedia of shark information, although the taxonomy and other biological aspects are somewhat out of date.

Moss, S. A. *Sharks: An Introduction for the Amateur Naturalist*. Englewood Cliffs, N.J.: Prentice-Hall, 1984. This introduction to selachians, their diversity, and all aspects of their biology, is a fine popular book on the subject. It is well illustrated and aimed at the general reader.

Moyle, P. B., and J. J. Cech. *Fishes: An Introduction to Ichthyology*. 4th ed. Englewood Cliffs, N.J.: Prentice-Hall, 2000. This college-level text is one of the best such volumes on general ichthyology available. Its exhaustive bibliography runs to fifty-three pages. Its thirty-four chapters deal with all aspects of fish biology, but chapter 15 describes, in detail, the sharks and their relatives. Other aspects of cartilaginous fish biology are scattered throughout the book.

Pough, F. H., J. B. Heiser, and W. N. McFarland. *Vertebrate Life*. 5th ed. New York: Macmillan, 1999. This is a thoroughly revised edition of one of the best undergraduate texts in vertebrate biology, reflecting the extraordinary advances in philosophy and methodology of vertebrate taxonomy. It is a rather large volume, but three of its twenty-four chapters (6, 7, and 8) are a particularly useful account of the origins of cartilaginous and other fishes.

Rosenzweig, L. J. *Anatomy of the Shark*. Dubuque, Iowa: Wm. C. Brown, 1988. This clearly written and well-illustrated text and dissection guide is intended for use in college comparative anatomy classes. It describes shark anatomy, system by system, using the spiny dogfish *Squalus acanthias* as the representative species. Not intended for general reading; it would, however, be useful as a guide to the anatomy of any shark species.

Springer, V. G. and J. P. Gold. *Sharks in Question*. Washington, D.C.: Smithsonian Institution Press, 1989. Written in question-answer format by a leading ichthyologist and a technical writer, this attractive volume is an excellent popular introduction to sharks. It is well illustrated and includes many fine color plates. It has several appendices, a glossary, and a concise and useful bibliography.

SHEEP

Type of animal science: Classification
Fields of study: Anatomy, zoology

Over nine hundred breeds of sheep have been identified worldwide, and four hundred have significant research data available to allow for distinctive, productive evaluation.

Principal Terms

BLOOD SYSTEM: grading for fleece, as in fine, ½, ¾, ¼, low, and braid
CLIP: the fleece that is removed from a single sheep
EWE: female sheep
FLEECE: wool fiber, which comes in various colors
FLOCK: group of sheep
LAMB: baby sheep, either ram or ewe, under one year of age
RAM: male sheep, also called a buck
WOOL: outer covering of sheep, similar to hair but with a crimp

Sheep are estimated to have evolved over two million years ago. They were the first agricultural animal to be domesticated, about twelve thousand years ago, just after the end of the last Ice Age. It is estimated that sheep domestication occurred in the Middle East, with the nomadic tribesmen that lived near the current Iran and Iraq. The sheep herding practice spread west to the Mediterranean Sea area to Africa and Spain. By 4000 B.C.E., sheep could be found in northern Europe, India, and China. The domesticated sheep evolved from either the European mouflon or the Asiatic mouflon, which can still be found in the wild. These sheep were very large in comparison to current sheep, and weighed about the same as some breeds of cattle. They had huge curved horns, and had thick woolen coats covered with long guard hairs that helped repel water and snow. Modern sheep resemble the sheep of old, but only weigh about one third as much. The shepherds strove to improve their flocks, and culled out the sheep that did not improve the breeding.

In general, sheep are classified as either carcass animals, developed for meat consumption, or fiber animals, which were used primarily for wool clothing. If a sheep is used for food consumption and is under one year of age, its meat is called lamb, but if it is a year old or older it is referred to as mutton.

Description

Sheep come in all shapes and sizes, with many different variations. Many have horns that are large and curving, others are polled and have no horns at all. Depending on the breed, some sheep have horns in the male, where the female has none; in other breeds, both sexes have horns. Some sheep are covered with wool and need to be sheered on a regular basis; others have no wool and are hair sheep, raised for either meat or hides. Most hair sheep were developed in the tropics and are relatively immune to parasites. Sheep are excellent grazers and prefer a varied diet of green soft plants of almost every description. Sheep prefer to eat as they walk up hills rather than grazing when walking down. They can be very selective in choosing what they eat, as their mouths are flexible due to a cleft in their lips that serves them like fingers. Sheep are agile and can move very quickly on cloven hooves. Sheep are ruminants and have a four-compartment stomach that allows them to regurgitate the herbage they have eaten and then

Sheep Facts

Classification:
Kingdom: Animalia
Phylum: Chordata
Subphylum: Vertebrata
Class: Mammalia
Order: Ungulates
Suborder: Artiodactyla
Family: Bovidae
Genus and species: Ovis aries (domestic sheep)
Geographical location: Every continent except Antarctica
Habitat: Land, including mountainous regions, desert and dry land, forested areas, tropics
Gestational period: Approximately five months, depending on breed
Life span: Up to ten years, although some have survived up to twenty years
Special anatomy: Narrow muzzle divided by a center cleft; very flexible lips, which allow them to graze close to the ground and to be very selective as to the plants and food eaten; wool

chew cud. Sheep do not have teeth on the upper front of the jaw but have a hard pad, where food is pressed between the pad on top and the teeth on the bottom. The sheep uses a jerking, upward motion to pull up its food, then swallows it.

Most sheep have wool as an outer cover, although several of the more primitive breeds are covered by hair. The color white is the most dominant of colors, but other colors and spots are present in several breeds. A sheep's skin is very thin, less than 0.1 inches in thickness. The skin also has sweat glands, which are missing in most other domesticated animals. The number of fibers that grow out of the skin of a sheep varies from breed to breed. The carcass-bred sheep have the least, whereas the fine wool sheep the most. The fiber number can be as low as 4,500 follicles, to as much as 80,000 follicles per square inch.

Sheep Life

Sheep in general live for about eight to ten years. Depending on the breed, they will mature in one or two years. The ram is able to reproduce earlier than the ewe, and can be put into service before the female can. Sheep tend to breed during the late autumn or early winter. The gestation period for a ewe is five months.

When the lambs are born, the average weight is about nine pounds, about the same size as the average domestic cat. Most are single births, but twins are frequently born. Lambs nurse for about eight weeks.

Sheep are followers, and they will follow a leader in groups. At the first sign of danger, they run. They would rather live in totally dry conditions, and do not do well where it is moist and muddy, even though they have cloven feet. Sheep also have the ability to do without water

Bighorn sheep retain the large, curved horns possessed by the moufflon ancestors of today's domesticated sheep. (Digital Stock)

for extended periods of time, but prefer to have it available.

—*Earl R. Andresen*

See also: Cattle, buffalo, and bison; Domestication; Fur and hair; Goats; Herds; Horns and antlers; Ruminants.

Bibliography

Briggs, Hilton M., and Dinus M. Briggs. *Modern Breeds of Livestock*. 4th ed. New York: Macmillan, 1980. Reviews the various breeds of modern sheep and the history of each major breed.

Parker, Ron. *The Sheep Book*. Reprint. Athens: Ohio University Press, 2001. Discusses the basic sheep raising formula for a small farmer or the new person interested in sheep production.

Ross, C. V. *Sheep Production and Management*. Englewood Cliffs, N.J.: Prentice Hall, 1989. Explores in detail the history, genetics, and feed requirements of sheep production.

Simmons, Paula. *Raising Sheep the Modern Way*. Rev. ed. Pownal, Vt.: Storey, 1989. A thorough review of sheep farming on a small scale. Reviews various breeds, medical problems, and covers solutions to many sheep problems.

SHELLS

Types of animal science: Anatomy, evolution, physiology
Fields of study: Anatomy, biochemistry, evolutionary science, invertebrate biology, physiology, zoology

External shells have been produced by many phyla, including both plants and animals. The majority of external shells are composed of calcium carbonate, but there are also examples of silica shells or tests.

Principal Terms

CARAPACE: the exoskeleton of arthropods

CARBONIC ANHYDRASE: an enzyme used in the mineralization process to convert carbon dioxide to bicarbonate

LAMINATE STRUCTURE: having a layered shell, as in the exoskeletons of crustaceans and the valves of clams

MATRIX: composed of proteins or protein-chitin polymers that act as nucleation sites for mineralization

NACREOUS LAYER: the pearl-like inner layer of molluscan shells

PORE CANALS: sites that house the cytoplasmic extensions of the crustacean hypodermis

PRISMATIC LAYER: the outer crystalline layer of the molluscan shell

Shells are external coverings that are produced by an organism. As such, they require a process by which the constituents of the shell are deposited in a site-directed fashion. What this means is that calcium and carbonate cannot come in contact with one another when their concentrations exceed their solubility product; otherwise, they will precipitate (form a solid). Thus, these ions must be directed to the area where the preferred precipitation is to take place. For this reason, a matrix is needed to provide a negative attractive force for calcium or other bivalent ions, such as magnesium. Other parts of the matrix may be composed of or house the enzyme carbonic anhydrase for the conversion of carbon dioxide reversibly to bicarbonate. The bicarbonate will degrade to carbonate ion that can then react with the positively charged calcium ion. Calcium can be taken out of the seawater or diet and concentrated; likewise, bicarbonate ions can be formed in the gills of some of these organisms and transported to deposition sites. However, having a matrix containing the enzyme carbonic anhydrase ensures that the calcium will only precipitate at that matrix site and nowhere else. In this way, the organism can control the shape of the shell by laying down a fiber matrix, usually composed of protein, as in mollusks, or a protein-chitin mixture, as is found in arthropods. The dumping of bicarbonate outside of a tissue where calcium is present will cause precipitation on the tissue membranes, a consequence with detrimental effects on the cells of the tissue. How calcium and carbonate ions are brought together varies from phylum to phylum.

The Protozoa

Among protists, mineralization is usually accomplished within a membrane-bound vacuole. The precipitated calcium carbonate is then exported by exocytosis to reside on the external surface of the cell. A matrix is secreted in the vacuole to facilitate the mineralization process. External calcium carbonate shells are found among the order Sarcodina of the phylum Sarcomastigophora. The shells of these amoebas become chambered with growth and are perforated by small openings, through which protrude slender pseudopodia that may be interconnected to form reticulopodia. Other sarcodines, the radiolaria and heliozoa, use silica as their shell material. In this case, too, the shell or test is perforated with puncta, or small openings, through which extend slender axopodia that will collect food.

Shells of Arthropods

The mineralized shell makes up the exoskeletons of the subphylum Crustacea and the subphylum Merostomata, which includes the horseshoe crabs. The mineralization process in the Crustacea is more complex than that of mollusks because the exoskeleton must be periodically shed and a new exoskeleton must be formed in the growth process. A percentage of the old exoskeleton is reabsorbed prior to molting and then redeposited in the new exoskeleton. Like mollusks, an epithelial, sheetlike tissue, the hypodermis, is responsible for the formation of the fiber matrix upon which is deposited calcium and carbonate. Unlike mollusks, the hypodermis forms cellular extensions that elongate as the layered exoskeleton is constructed. These cellular processes come to lie in pore canals as mineralization proceeds. The first layers laid down are the first to be mineralized, so new matrix layers are being formed while the mineralization of earlier deposited layers is proceeding.

Other differences found between the mollusks and crustaceans is the amount of protein-chitin matrix of Crustacea and the mainly protein matrix of the Mollusca. The fiber matrix of crustaceans averages about 40 percent by volume, a volume as high as the protein matrix of bone. Because the enzyme becomes entombed in the matrix during the mineralization process, new matrix must be laid down in stepwise fashion. This gives the exoskeleton a laminate structure that is divided up into an outer epicuticle, a middle exocuticle, and an inner endocuticle. A noncalcified membrane lies between the hypodermis and endocuticle.

Shells of arthropods are quite variable, but most have a carapace covering the dorsal part of the body (the head and the thorax). The jointed plates of the abdomen are also calcified, as are the legs and antennae. Ostracod arthropods have a clamlike shell that is hinged dorsally. Barnacles are the other crustacean group that has a different sort of exoskeleton. Their exoskeleton is composed of plates called parietes that form a wall around the organism.

Shells of Mollusks

The shells of mollusks are usually arranged into two major layers: an outer prismatic layer and an inner nacreous or pearly layer. The prismatic layer is composed of vertically oriented prisms that are bounded by matrix to separate each prism from one another. The prisms are elongated and extend upward to the organic layer of the shell, termed the periostracum. The nacreous layer is composed of cross-laminated lamellae. The lamellae are oriented in different directions, much like plywood layering. This serves to increase the strength of the shell in its resistance to cracking. The amount of matrix in the shells of mollusks is quite low and may be less than 1 percent of the shell volume. The mantle tissue that lines the shell is responsible for

Mineralization Debate

There is currently a debate concerning the mineralization process of the Crustacea and Merostomata. While it is clear that these hypodermal extensions are supplying the calcium ions for mineralization, it is still unclear as to the origin of the carbonate ions. Presently there are two views. First, bicarbonate ions are supplied by these cellular extensions to the mineralizing tissues. Second, the large quantities of carbonic anhydrase deposited into the external matrix of the shell function to maintain the bicarbonate concentrations in a reaction that is favored at a pH above 7.0 found in seawater. It is likely that both hypotheses are valid. Carbonic anhydrase has been extracted from formed shells and has been found to be active. This strongly implies that carbonic anhydrase is indeed active in the mineralization process.

Bivalve shells, such as those of mussels, are hinged at the back by an elastic ligament. (Joerg Boetel/Photo Agora)

exporting the matrix, calcium, bicarbonate, and carbonic anhydrase for shell mineralization. A space, the extrapallial space, lies between the mantle and the shell and is filled with fluid to facilitate the transfer of ions to the shell. Whereas shell is deposited along the area of the mantle, new shell is deposited along the edge of the mantle.

The shells of mollusks are found in many members of this taxon. The class Monoplacophora have a single cap-shaped shell covering a muscular foot. The class Polyplacophora have an elongated body and eight shell plates covering their dorsal surface. More familiar shells of mollusks are found among the classes Gastropoda and Bivalvia. Gastropods have a single coiled shell that may be twisted, with an opening for the protrusion of the foot and head. The opening may be covered by an operculum that, in some groups, is calcified. The coiling of the shell may keep the weight of the shell over the foot for balance. The shell is used for protection against predation and to prevent desiccation in land forms. The animal can retreat into the shell when frightened or if poor water conditions occur. Many groups lose the shell in both terrestrial and marine forms and are sluglike.

Bivalves, as the name implies, have two shells or valves that enclose the animal and are hinged dorsally with an elastic ligament. Some species of both gastropods and bivalves grow spines by the evagination and elongation of the mantle to export a matrix for mineralization. This is carried out along the edge of the mantle; a ventral slit in the spine often remains to show where the mantle was during spine formation. Spine formation appears to be a device used by gastropods and bivalves to thwart predation, although this idea has not been fully tested. Spines may make these animals difficult to grasp, for example, by the claws of crabs or lobsters. In addition, the spines may make the animal difficult to swallow or crush in the mouths of fishes. Another strategy seen by these mollusks is the thickening of the shell, especially around its lip or edge. This makes it difficult for a crab to start breaking the shell from the edge, which is usually the thinnest part of the shell as this area represents new shell deposition.

Other molluscan groups with external shells are the nautiloids and extinct ammonites. These cephalopods often have coiled shells that resemble those shells of gastropods in their external appearance. Their shells are chambered, with the animal living in the endmost chamber. Gas can be secreted into the old chambers to act as a buoyancy device. The chamber partitions add strength to the shell to withstand ocean pressure at great depths. Nautiloid shells lack a periostracum. Ammonites were a similar cephalopod that occurred in great numbers during the Mesozoic era. Their shells were coiled or straight and developed complex sutures between the separate chambers. Rib patterns developed as well.

The class Scaphopoda has tusklike shells that are open at both ends. The animal extends out the lower opening and the upper opening extends above the sediment in this burrowing form for water flow into and out of the mantle cavity.

—*Samuel F. Tarsitano*

See also: Anatomy; Beaks and bills; Bone and cartilage; Brain; Claws, nails, and hooves; Defense mechanisms; Endoskeletons; Exoskeletons; Muscles in invertebrates; Muscles in vertebrates; Physiology; Predation; Skin; Teeth, fangs, and tusks.

Bibliography

Carter, J. G., ed. *Skeletal Biomineralization: Patterns, Processes, and Evolutionary Trends*. Vol. 1. New York: Van Nostrand Reinhold, 1990. This book contains a series of articles on the construction and mineralization process of both invertebrates and vertebrates.

Compére, P., and G. Geoffinet. "Elaboration and Ultrastructural Changes in the Pore Canal Systems of the Mineralized Cuticle of *Carcinus maenas* During the Molt Cycle." *Tissue and Cell* 19 (1987): 859-875. This paper describes in detail the pore canal system of crustacean exoskeletons and their role in the mineralization process.

Kamat, S., X. Su, R. Ballarini, and A. H. Heuer. "Structural Basis for the Fracture Toughness of the Shell of the Conch, *Strombus gigas*." *Nature* 405 (June, 2000): 1036-1040. These authors describe the constructional principles of the molluscan shell and the biomechanical properties that processes such as cross laminate shell construction impart to the shell. Construction is discussed in terms of stress resistance.

Roer, R., and R. Dillaman. "The Structure and Calcification of the Crustacean Cuticle." *American Zoologist* 24 (1984): 893-909. This paper describes the construction of the arthropod cuticle in terms of its morphology and mineralization.

Simkiss, K., and K. M. Wilbur. *Biomineralization*. New York: Academic Press, 1989. This is the best review of the mineralization process of all phyla to date. Each phylum of both plant and animal is examined and the shell construction and mineralization process is described.

Vermeij, G. J. *Evolution and Escalation*. Princeton, N.J.: Princeton University Press, 1987. This book is an extensive review of how shell construction influences the ecology of mollusks and other organisms as well as how specialization of the shells is used to protect animals from predation.

SHREWS

Types of animal science: Anatomy, behavior, classification, reproduction
Fields of study: Anatomy, classification, zoology

Small, ferocious shrews sometimes have venomous saliva. They are ecologically useful because they eat carrion and harmful insects.

Principal Terms

CARRION: dead, decomposing animal bodies
GESTATION: the time developing mammalian offspring are in the uterus
VENOM: a poison made by an animal

Shrews are usually very small; the Savi's pygmy shrew is probably the smallest known mammal on earth, less than 1.5 inches long and weighing 0.07 ounces. The shrew family, Soricidae, is the largest family of the order Insectivora, and shrews inhabit all major land masses except polar regions, Australia, New Zealand, and Greenland. Shrews are useful to gardeners because they eat many insects. They also inhabit fields, marshes, and woodlands. Some shrews are semiaquatic.

Shrews are often mistaken for mice because of their small size and vaguely similar bodies. Their habits and bodies are actually more like those of moles. Shrews have long, mobile snouts, tiny ears and eyes, and bodies covered with short, thick hair colored gray, brown, or black. Smaller shrews are under two inches long, while the largest are one foot long. They eat insects, worms, small fish, and plants. Weasels, foxes and owls eat shrews. However, the shrew's unpleasant odor and taste protects it from excessive predation.

Physical Characteristics of Shrews

Shrews have sharp teeth and are both vicious and ferocious. When attacked or disturbed, they fight wildly. They live alone, stake out territories, and hold them against invaders. As added protection, some shrews have venomous saliva, which they use to poison prey. Shrews also have whiskers that aid detection of prey at night. Shrews live in gardens, forests, woodlands, and grasslands. Most dig burrows, though some species do not burrow and others are semiaquatic.

The largest shrew is an elephant shrew, also called the golden-rumped shrew. These "giant" shrews have maximum body lengths of one foot, tails ten inches long, and weigh one pound. Elephant shrews are so named for their long snouts. Their tails are like rat tails and their hind legs are designed for hopping. A small elephant shrew, the short-eared shrew, has maximum body length of four inches, a tail of five inches, and a weight of 1.5 ounces. Elephant shrews inhabit South Africa and eat insects, snails, and plant roots, fruits, and seeds.

Shrews forage day and night and must eat every two to three hours to survive. This great need for food leads shrews to eat anything available. Depending on habitat and species, they eat insects, worms, carrion, seeds, nuts, plants, shellfish, frogs, and fish. Shrews hear and smell well. This helps them find prey and avoid predators. Some creatures kill shrews, but most will not eat them because they smell and taste bad. The smell and taste is due to secretions from skin glands most plentiful at knee and elbow.

The Life Cycles of Shrews

Shrews live alone except when mating. This occurs year round, except in cold climates, where mating is from March to September. Gestation, two to three weeks, takes places in nests that females dig. A female can have ten litters a year.

Each litter contains three to ten furless, blind young. An exception to standard litter size occurs in elephant shrews, who have one or two offspring. Offspring are full grown four weeks after birth.

Because of long mating seasons, some mothers nurse a litter while pregnant. Young leave nests after two to three weeks. Females mature in 1.5 months and males in 3 months. Courtship is short since shrews fight, even with mates. The life spans of shrews range from one to five years, depending on species.

The American Short-Tailed Shrew

American short-tailed shrews inhabit southeastern Canada, the northeastern United States, Texas, and Louisiana. They have thick, gray-black fur, are four to five inches long, have one-inch tails, and weigh one ounce. Like other shrews, American short-tailed shrews constantly seek food, eating insects, earthworms, snails, small vertebrates, centipedes, spiders, mice, frogs, and plants. They have venomous saliva, used to stun and kill prey, which is also painful to humans and large animal predators. Predators of the shrews are owls and other raptors.

American short-tailed shrews live alone except when mating, which happens from late winter to the next September. Courtship is short because shrews are harsh, even to mates. Females give birth after two to three weeks to litters of four to nine young. Young leave after a month. As with other shrew species, males are mature in 1.5 months, females in three. The life span of these shrews is two to three years.

Shrew Facts

Classification:
Kingdom: Animalia
Subkingdom: Bilateria
Phylum: Chordata
Subphylum: Vertebrata
Class: Mammalia
Order: Insectivora
Family: Soricidae (shrews), with twenty-two genera and 246 species

Geographical location: Worldwide except for poles, Australia, New Zealand, and Greenland

Habitat: Most in gardens, fields, marshes, and woodlands; some are semiaquatic and spend much time in freshwater

Gestational period: Thirteen to twenty-four days

Life span: One to five years in the wild

Special anatomy: Mobile snout, venomous saliva, glands that make offensive excretions to dissuade predators

Most shrews are harmless to humans, though bites of some species cause severe pain. They are useful in gardens and farms, where their consumption of insects and grubs cuts down harmful insect populations. Shrews are also useful ecologically because they eat carrion, preventing its decay and resultant endangerment of humans and other animals.

—*Sanford S. Singer*

See also: Gophers; Mice and rats; Moles; Poisonous animals; Predation; Rodents.

Bibliography

Bailey, Jill. *Discovering Shrews, Moles, and Voles*. New York: Bookwright Press, 1989. This brief book covers shrew characteristics, diet, habitat, and behavior.

Churchfield, Sara. *The Natural History of Shrews*. Ithaca, N.Y.: Comstock Publishing Associates, 1990. This book contains many maps and other illustrations, and gives a valuable exposition of shrew natural history.

Crowcroft, Peter. *The Life of the Shrew*. London: M. Reinhardt, 1957. This solid illustrated book covers life cycle, biology, habitat, diet, and characteristics of shrews.

Merritt, Joseph F., Gordon L. Kirkland, Jr., and Robert K. Rose. *Advances in the Biology of Shrews*. Pittsburgh: Carnegie Museum of Natural History, 1994. A collection of detailed colloquium presentations holding much information on shrew biology.

SKIN

Types of animal science: Anatomy, physiology
Fields of study: Anatomy, cell biology, histology, invertebrate biology, physiology

Skin is the organ that covers the body surface of an animal, and it is composed of cells. The specialized structures and cells associated with the skin are involved in a variety of physiological functions, including protection, communication, regulation of body heat, and respiration.

Principal Terms

CHROMATOPHORES: pigment-producing cells
DERMIS: layer beneath the epidermis, primarily connective tissue but also containing nerves and blood vessels
EPIDERMIS: surface layer of epithelial cells
INVERTEBRATE: animal without a backbone
MITOTIC CELLS: cells capable of dividing and forming new cells
VERTEBRATE: animal with a backbone made up of individual bones called vertebrae

Survival in animals requires that the internal body components be separated and protected from the external environment. Most single-cell organisms are separated from the environment only by the plasma membrane (cellular membrane). In multicellular organisms the body surface is covered by a tissue consisting of epithelial cells and connective tissue. The covering is commonly referred to as skin, but the skins of invertebrates and vertebrates have distinct differences. Invertebrates often have a single layer of surface epithelial cells, which is generally referred to as an integument. However some invertebrates, specifically flukes and tapeworms, have a unique, living surface covering called a tegument. In this situation, the epithelial cells have fused and formed a single bag of cellular components called a syncytial epidermis. Thus, the word "skin" is often reserved specifically to describe the surface covering in vertebrates, but the word "integument" is also used.

The Structure and Physical Properties of Skin

The general structure of the skin is similar in all vertebrates. There are two primary regions, an epidermis and a dermis. The upper region, the epidermis, is made up of multiple layers of epithelial cells. All vertebrate skin has a basal layer (the stratum germinativum) consisting of mitotic cells. These mitotic cells divide to replace the cells closer to the surface as they are worn away, and to heal skin wounds. In most vertebrate species, the outermost layer of cells (the stratum corneum) is dead. The dead cells are filled with a waterproofing protein called keratin, which is produced by keratinocytes, the major type of cells forming the epithelium. An exception can be noted in many species of fish, where the epidermis is composed entirely of living cells and the stratum corneum is absent. The epidermis will differ the most between aquatic (water dwelling) and terrestrial (land dwelling) organisms. When compared to mammals and reptiles, the epidermis of amphibians, birds, and fish is thinner and the stratum corneum may only be one or two cells thick. The epidermis lacks its own blood supply (is avascular). Nourishment reaches the living cells by diffusion from the underlying dermal blood supply. No nerves are present in the epidermis.

The underlying region, the dermis, is primarily composed of connective tissue. Although epidermal and dermal thicknesses vary between groups of animals and thickness may vary along an individual's body surface, the dermal layer is always thicker than the epidermis. The hypodermis, a layer of subcutaneous tissue immediately beneath the dermis, connects the skin with underlying tissues such as muscles and bone. In birds and mammals in particular, this layer often contains a significant amount of fat, which provides insulation and a reserve source of energy.

The protective aspect of skin does not mean that it is entirely impenetrable or that the body is completely isolated from the environment. Materials that are fat-soluble or that disrupt cellular membranes can be absorbed across the skin surface. In some cases, beneficial chemicals cross skin. For example, frogs, which are amphibians, actively take up oxygen and expel carbon dioxide across the skin as well as the surface of the lungs. Amphibian skin is also permeable to water and, in fact, some species absorb amounts comparable to that obtained by drinking in other organisms. In other cases, detrimental materials such as solvents and potential environmental pollutants cross the skin. Acetone present in nail polish remover, methanol which is sometimes used to remove old finish on furniture, and salts of heavy metals such as mercury or lead are some examples.

Skin, like muscle, has the properties of extensibility (stretch) and elasticity (ability to return to the original shape after being stretched). These properties are made possible by the presence of collagen and elastic fibers as major components of the tissue comprising the dermis. When the elastic properties of the skin have been exceeded, white lines known as stretch marks appear.

Specialized Secretions and Structures Associated with Skin

Some epithelial cells produce and release protective secretions onto the external surface. Vertebrates and invertebrates both have mucous-secreting cells. On internal surfaces, such as the digestive tract, mucus protects cells from being broken down along with the food. On the external surface, mucus may trap bacteria or, as in earthworms, prevent death from desiccation (drying out). Another example of an invertebrate secretion is a covering called a cuticle. In insects this cuticle includes a mixture of proteins that eventually harden and form the exoskeleton.

Vertebrates, including fish, birds, and humans, as well as invertebrates, insects, secrete a group of antimicrobial (bacteria-killing) proteins called defensins (originally called magainins). Species of poison dart frogs have another type of protective secretion which is toxic to potential predators. Some of these secretions are used on poison dart arrows.

Changing Protective Layers

Snakes and other reptiles shed their skin, arthropods shed their exoskeleton, and some worms shed their cuticle in a process called ecdysis. In snakes, the mitotic cells of the basal layer replicate to form a new layer of epidermal cells that will be beneath the cells that are breaking down. There is a layer of cells in a region between the mitotic cells and the outer epidermal layers that is not keratinized and begins to break down. A separation or fission zone occurs at the level where the cells are breaking down. This results in the old epidermal layers breaking away above the newly formed layers. The old skin can now be shed and the new skin will be exposed. The surface of the new skin will be keratinized and scaly just like the old. Most snakes shed their skin in one piece.

Arthropods, which includes the insects, shed their exoskeleton many times during the growth stage. The epidermis detaches from the old exoskeleton and secretes a new epicuticle that will be the new exoskeleton. Molting fluid, which contains enzymes, is secreted into the region between the old and what will be the new exoskeleton. The enzymes degrade the inner region of the old exoskeleton. The old skeleton now splits and the animal emerges with a new exoskeleton. However, this skeleton will remain soft for some time and during that time it is able to stretch to accommodate the larger body size of the arthropod.

Structures with quite varied functions are derived from skin cells, particularly the epithelial cells. The feathers of birds function in flight but they also provide insulation. A bird's beaks and claws provide a method of defense and a way to secure food. Mammalian hair is an epithelial derivative. Body hair provides protection from abrasion and sunlight and has some insulation value. In animals that have them, sweat, oil, and mammary glands are groups of specialized epithelial cells. Reptiles, for example, lack sweat glands. Light organs of deep-water fish are modified epithelial glands. The scales of reptiles, the rattles of snakes, and the claws of turtles are other examples of epithelial derivatives. Geckos are able to walk up walls because they have modified epidermal scales on the tips of digits which serve as suction cups.

Cells of the dermis also are the origin for specialized structures in some organisms. Although there are fewer examples, dermal derivatives include shark teeth, fish scales, and the protective armor plates of an armadillo.

Skin and Temperature Regulation

The skin is a major organ in controlling body temperature. Mammals and birds are animals that generate internal body heat (warm-blooded or endothermic). Species of reptiles, fish, and amphibians, which are often called cold-blooded (ectothermic), are unable to control their body temperature through internal regulators in the same way that warm-blooded animals can. Both groups of animals depend upon the rich supply of blood vessels in the dermis as one mechanism for maintaining a safe body temperature. When internal body temperatures rise in endotherms, an increase in blood flow carries internal body heat to the surface, where it is lost to the environment. A similar increase in blood flow in ectotherms carries heat from the environment into the body and helps to warm internal organs and tissues. A decrease in blood flow will work in the opposite direction in both groups. In animals with sweat glands, the evaporation of sweat secreted onto the body surface also helps to lower body temperature. One unique feature of birds is a specialized region of skin, the brood patch, located on the ventral (stomach) surface. This area is rich in blood vessels (is highly vascularized) and is used to transmit heat from the female to the eggs or the hatchlings.

Skin Coloration

Vertebrate skin contains pigment-producing cells called chromatophores. Pigment production pro-

Hair-Raising Response

Originating in the superficial dermis and attaching to the hair shaft (long part of the hair) near its base is a small muscle, the arrector pili muscle. This is a smooth muscle, which means that its contraction and relaxation are involuntary responses. The muscle is innervated by a part of the nervous system (autonomic nervous system) that is under subconscious control. One part of the autonomic nervous system regulates specific responses during periods of stress. These responses are generally referred to as "fight-or-flight" responses.

When the muscle attached to a hair is stimulated to contract by the nervous system, the hair will stand up. In many mammals, dogs for example, this phenomenon occurs when a nonfriendly situation is detected. The raised hairs indicate alarm. In other cases, erected body hair serves as a protective method of defense since it will make animals appear larger than normal. In many mammals, elevated body hair traps air and assists in maintaining body warmth when exposed to cold temperatures, or helps in cooling when the core body temperature gets too warm. Hair normally emerges from the body surface at a slight angle. "Goose bumps," noticeable in humans who are frightened or experiencing some type of stress, are the result of a slight elevation of the skin when the hair shaft moves into a vertical position.

vides many benefits to animals. Skin pigments help to limit the amount of damaging ultraviolet light or irradiation to which the deoxyribonucleic acid (DNA) in the mitotic cells and the underlying tissues are exposed. Melanocytes located in the epithelium of mammals produce a brown-black pigment called melanin. They are the only pigment-producing cells in most mammals. In addition to epithelial melanocytes, amphibians, fish, reptiles, and birds have other types of pigment-producing cells which are located in the dermis. Examples are lipophores, which use carotene, a naturally occurring pigment in food, to synthesize yellow, orange, and red pigments, and iridophores, which use molecules called purines to synthesize pigments that are iridescent. The amount of pigment produced, the final location of the pigment in the cells, and the combination of cells producing it result in a range of body and feather coloration. Chromatophores account for the changes in body color that allow chameleons, flounders, and octopuses to easily blend in with different surroundings. Changes in body color also provide a means of communication between individuals.

Skin color, particularly in organisms with a thin, fair surface covering such as humans, is also influenced by the amount of oxygen bound to hemoglobin in the blood. Fully oxygenated hemoglobin is red and it gives a pink coloration to the skin. Hemoglobin that is not fully oxygenated can cause the skin to appear blue or take on a purplish hue, a condition called cyanosis.

—*Robert W. Yost*

See also: Anatomy; Beaks and bills; Bone and cartilage; Brain; Circulatory systems of invertebrates; Circulatory systems of vertebrates; Claws, nails, and hooves; Digestive tract; Ears; Endoskeletons; Exoskeletons; Eyes; Fins and flippers; Immune system; Kidneys and other excretory structures; Lungs, gills, and tracheas; Muscles in invertebrates; Muscles in vertebrates; Nervous systems of vertebrates; Noses; Physiology; Reproductive system of female mammals; Reproductive system of male mammals; Respiratory system; Sense organs; Shells; Tails; Teeth, fangs, and tusks; Tentacles; Wings.

Bibliography

Hickman, C. P., Jr., L. S. Larsen, and A. Larsen. *Biology of Animals*. 7th ed. Boston: WCB/McGraw-Hill, 1998. General biology text with sections on invertebrate and vertebrate systems.

Linzey, Donald. *Vertebrate Biology*. Boston: McGraw-Hill, 2001. A good midlevel text outlining structure and function in vertebrates. Excellent use of descriptive features and terminology as part of the text. Good list of related texts in preface.

Miller, S. A., and J. P. Harley. *Zoology*. 4th ed. Boston: McGraw-Hill, 1999. A lower-level text with introductory sections on animal systems.

Solomon, E. P., L. R. Berg, and D. W. Martin. *Biology*. 5th ed. Philadelphia: Saunders College Publishing, 1998. Fairly comprehensive general biology text with sections on invertebrate and vertebrate systems.

Walker, F. W., Jr., and K. F. Liem. "The Integument." In *Functional Anatomy of the Vertebrates: An Evolutionary Perspective*. 4th ed. Philadelphia: Saunders College Publishing, 2001. An upper-level comprehensive text on vertebrate systems. Excellent chapter on comparative anatomy and physiology of vertebrate skin.

SKUNKS

Types of animal science: Anatomy, classification, reproduction
Fields of study: Anatomy, zoology

Skunks, related to weasels, are known for the vile odor of their musk. They are helpful to farmers, eating animals that prey on agricultural products. Their fur is used for garments and some people keep them as pets.

Principal Terms

GESTATION: the term of pregnancy
MUSK: vile-smelling liquids that skunks use for self-defense
NOCTURNAL: occurring at night
PERINEAL: located between scrotum and anus in males or the equivalent region in females

Skunks belong to the weasel family (Mustelidae), which also contains badgers, weasels, and otters. They form three mustelid genera found in southern Canada, throughout the United States, in Mexico, and in Central America. The average skunk is sturdily built and cat-sized. It has a long, pointed nose or a hoglike nose, an arched back, and short legs. Skunk fur is long, soft, shiny, and black with wide, white stripes down the back. Stripe patterns differ among skunk species. Many skunks also have white forehead patches. Skunk tails are long, bushy, and usually black on top and white underneath.

Skunks live in hollow trees, burrows, or under sheds when dwelling among humans. They eat insects, mice, gophers, reptiles, squirrels, birds, and eggs. They help farmers by killing other animals that eat or prey upon agricultural products. Skunks are best known for their vile-smelling musk, originating in perineal glands on either side of the anus. When frightened, a skunk squirts out this fluid with considerable force. The musk's vile odor usually keeps enemies away. A human or animal sprayed with the fluid smells bad for weeks. For this reason, most people and animals learn not to attack skunks. One predator of skunks is the great horned owl, unaffected by the musk.

Physical Characteristics of Skunks

Striped or common skunks live in small groups in underground dens in pastures, meadows, and fields. They eat insects, gophers, reptiles, squirrels, birds, grubs, and eggs. These skunks are noc-

Striped or common skunks are the largest skunk species and can spray their foul-smelling musk distances of over six feet. (Rob and Ann Simpson/Photo Agora)

Skunk Facts

Classification:
Kingdom: Animalia
Subkingdom: Bilateria
Phylum: Chordata
Subphylum: Vertebrata
Class: Mammalia
Order: Carnivora
Family: Mustelidae
Subfamily: Mephitinae
Genus and species: Three genera and thirteen species, including *Mephitis mephitis* (striped skunk); *Spilogale angustifrons* (southern spotted skunk), *S. gracilis* (western spotted skunk), *S. putorius* (eastern spotted skunk), *S. pygmaea* (pygmy spotted skunk)
Geographical location: The United States, Mexico, southern Canada, and Central America
Habitat: Rocky crevices or hollow trees; may live in the suburbs, where they make dens in burrows or under buildings or sheds
Gestational period: Two to seven months
Life span: Up to seven years in the wild, ten years in captivity
Special anatomy: Musk-secreting perineal glands; long, very bushy tails

turnal hunters. They either dig their own dens or use those vacated by other animals. Striped skunks are the largest skunks, reaching body-to-tail lengths of 3.5 feet and weights of five pounds. Their coats are glossy and black, with two wide, white stripes running from head top to tail tip. They spray musk in self-defense up to 6.5 feet. The musk hurts the eyes of predators and its vile odor lingers for many days.

Another type of skunk is the hog-nosed skunk. There are seven hog-nosed skunk species. They differ in habitat from striped skunks, living in rocky areas and inhabiting rocky crevices. They have sharp claws for digging hard, rocky soil. Hog-nosed skunks are two feet long from nose to tail tip and weigh about 3.5 pounds. Their glossy black coats have a white head-to-tail stripe. Unlike other skunks, they lack white stripes down the middle of the face and their tails are all white. Their long, bare snouts look like pig snouts, hence the name. Hog-nosed skunks are nocturnal and eat the same food as striped skunks.

Spotted skunks differ from striped and hog-nosed varieties in having four to six broken stripes or spots in different patterns on body and tail. Some dig burrows; others live in rock crevices. They are much smaller than other skunks, being only 1.25 feet long from nose to tail tip and weighing only one pound. Spotted skunks are like other skunks in nocturnal predation and diet.

The Life Cycle of Skunks

Striped and hog-nosed skunks mate during February and March in dens lined with grass and leaves. Males do not help raise young and live alone during the summer. Gestation lasts 2 to 3.5 months and a female gives birth to between two and ten babies, depending on species. The babies initially weigh about an ounce and can spray musk before they can walk. They are nursed for 1.5 months and then follow their mother around, learning to hunt. At six months old, they strike off on their own. These skunks can mate when they are eleven months old, and can live for six to seven years in the wild and ten years in captivity.

Mating of eastern, southern, and pygmy spotted skunks occurs in the usual February to March period, but gestation is only five weeks. In contrast, western spotted skunks mate in the late summer. Females of all four spotted skunk species give birth to two to six babies. Nursing and life spans are the same as for hog-nosed and striped skunks.

Skunk Fur and Skunk Pets

Wild skunks produce a valuable fur, but skunk farming is not profitable because of the low prices paid for each small pelt. However, skunk fur is beautiful. Coats made of this fur were once sold as "black marten," but are now sold under their real name due to Federal regulations. Buyers like skunk fur for its appearance and durability compared to most other furs. Skunks, often "destunk" by surgical removal of perineal glands, have some popularity as pets because they are attractive, friendly, and cat-sized.

—*Sanford S. Singer*

See also: Defense mechanisms; Fauna: North America; Otters; Weasels and related mammals.

Bibliography

Blassingame, Wyatt. *Skunks*. New York: Putnam, 1981. This brief book holds basic information on characteristics of skunk species and skunks as pets.

Lepthein, Emilie U. *Skunks*. Chicago: Children's Press, 1993. This brief, basic book, aimed at juveniles, provides descriptions of physical characteristics and habits of skunks.

Verts, B. J. *The Biology of the Striped Skunk*. Urbana: University of Illinois Press, 1967. This book describes striped skunk biology, behavior, distribution, and reproduction.

Wilson, Don, and Sue Raff, eds. *The Smithsonian Book of North American Mammals*. Washington, D.C.: Smithsonian Institution Press, 1999. Extensive coverage of all North American mammals, including habitat, ecology, behavior, diet, reproduction, and evolution. Distribution maps and photographs.

SLEEP

Type of animal science: Behavior
Fields of study: Ecology, ethology, neurobiology

Almost all animals exhibit behavioral sleep, a form of rest, but only birds and mammals have the same kinds of brain states as humans when they sleep.

Principal Terms

CIRCADIAN RHYTHM: a physiological or behavioral cycle that occurs in a twenty-four hour pattern
DIURNAL: habitually active during the day
ELECTROENCEPHALOGRAM (EEG): a chart of brain wave activity as measured by electrodes glued to the surface of the skull
NOCTURNAL: habitually active during the night
NONRAPID EYE MOVEMENT (NREM) SLEEP: sleep characterized by relaxed muscles and slow brain waves
RAPID EYE MOVEMENT (REM) SLEEP: sleep characterized by fast brain waves, during which dreaming typically occurs

Behaviorally, sleep can be recognized by four basic features. It generally consists of 1) a prolonged period of physical immobility during which there is 2) reduced sensitivity to environmental stimuli, and which 3) typically occurs in specific sites and postures and 4) in a twenty-four hour (circadian) pattern. Using this broad definition of sleep, almost all animals can be said to sleep.

The Ecology of Sleep

Since animals are more vulnerable to danger when sleeping than when awake, most animals sleep in sites and postures that help to maximize their safety. For an insect or a small lizard this might mean wedging into a crack in tree bark or burying themselves under leaf litter. For a snake, small bird, or mammal it might mean sleeping in a nest, burrow, or tree hollow. Some animals can adopt a particular sleep posture that helps them to blend into the background to avoid detection. Animals that cannot hide or camouflage themselves might try sleeping while semiprotected in the center of a family or larger group.

Animals can also modify their sleep sites and postures to help regulate their temperature. In cold temperatures, sleep sites and postures can be chosen so as to cover exposed skin on the face or feet; birds fluff their feathers and mammals fluff their fur to trap air like a blanket; small animals huddle together to keep warm. In particularly hot temperatures, well-chosen sleeping sites may protect an animal from direct exposure to the sun, and specific postures can be adopted to facilitate heat loss.

Sleep periods also tend to be taken at times that are most safe. Diurnal species are those that are typically active in the day and do most of their sleeping at night; nocturnal species those that are more active at night and do most of their sleeping during the day. In general, birds, reptiles, and shallow-water species tend to be active during the day while mammals and deep-water species tend to be active at night, but there are many exceptions to this generalization. Whether a particular species is primarily diurnal or primarily nocturnal depends upon many aspects of its ecology and physiology, but most exhibit some kind of circadian pattern of rest and activity.

The Physiology of Sleep

Neurophysiologically, "sleep" can be distinguished from "rest" in vertebrate animals only by

Do Animals Dream?

While we can never ask animals whether they dream, there is every indication that some of them do. During sleep, the brain activity of birds and mammals cycles between slow and fast waves just like that of humans going in and out of the dream state. During fast-wave sleep, birds and mammals also have rapid eye movements which, in humans, are associated with the visual imagery of dreams. Because the neurophysiology of rapid eye movement (REM) sleep in birds and mammals is so similar to that of humans, it is reasonable to think that they, too, are dreaming.

In mammals, a specialized part of the brain suppresses most muscle activity during REM sleep. Experiments show that if this area is destroyed, an animal will move around just as if it is awake, presumably because it is acting out its dreams. In an undamaged animal, small body movements and twitches during REM sleep are often visible. Many people know when their dog is dreaming because they see it making tiny running motions or hear it making muffled barking sounds. Cats often twitch their limbs and whiskers as they go through a period of REM sleep.

In birds, REM sleep takes place in very brief bursts of only a few seconds or a minute at a time, so the dreams of birds, while probably visual, could never follow much of a story line. Part of the reason for this is related to body size: The larger an animal, the longer its REM sleep periods. Since most birds are small, their REM sleep cycles tend to be short.

measuring changes in brain state. Fish, amphibians, reptiles, birds, and mammals all show changes in brain waves that accompany the progressive muscle relaxation that characterizes deeper and deeper states of sleep.

During their sleep periods, fish, amphibians, and reptiles slowly progress into more and more relaxed stages of sleep, then remain in their deepest state for a prolonged period of time, eventually returning slowly back to the waking state. Birds and mammals, on the other hand, show a pattern of alternating states of sleep within each sleep period. The first state is called NREM sleep (for "nonrapid eye movement sleep"). NREM sleep is characterized by relaxation of the muscles, slowed breathing and heart rate, and slow waves in the EEG (a measure of brain activity). Alternating with periods of NREM sleep are periods of REM sleep. REM sleep is characterized by rapid eye movements, irregular heart rate and breathing, and fast waves in the EEG that look identical to brain activity while awake. Although the brain is very active during REM sleep, most muscles are deactivated, leading some people to refer to REM sleep as "paradoxical sleep." In humans, it is during REM sleep that dreaming typically occurs.

Amazingly, some birds and marine mammals can sleep on one side of their brain and body while the other side remains awake. In marine mammals, it is thought that one-sided sleeping may enable an animal to keep swimming and stay near the surface in order to breathe. In birds, one-sided sleeping is thought to be a wayn for a particularly vulnerable animal simultaneously to get some rest and still remain alert for predators.

Across-Species and Developmental Patterns of Sleep

Large animals have longer sleep periods than small animals. Large animals also sleep more deeply (are more relaxed and have fewer arousals) than small ones and, among birds and mammals, have a greater proportion of REM sleep. According to the vigilance model of sleep, this is because large animals are less vulnerable than small animals, and so can afford to be less alert. Supporting this idea is the fact that for a given size animal, species that are predators typically sleep more and sleep more deeply than animals that are prey. Cougars, for example, sleep more and sleep more deeply than the deer they hunt, while falcons sleep more and sleep more deeply than pigeons and ducks.

Although large animals tend to have longer sleep periods than small animals, small animals generally have more total sleep time than large animals because they sleep more often. It is not known whether this pattern results because small animals need more sleep or because what sleep they do get is shallower and more disrupted.

Consistent with the fact that small animals sleep more than large animals is the fact that in any particular species, young animals sleep more than adults. Not only does total daily sleep time drop as an animal ages, so does the relative percentage of time spent in REM sleep. Human babies have more total sleep time and a greater percentage of REM sleep than adults, and young adults have more sleep time and a greater percentage of REM than elderly adults. The same pattern seems to hold true for other species as well.

Possible Functions of Sleep

Besides the vigilance model, there are three other models which try to explain across-species and across-age differences in sleep. One of these suggests that sleep is necessary for learning. Since large animals generally live longer than small animals, they typically have a greater capacity for learning. Likewise, for a given size animal, predatory species typically rely more on learning, while prey species rely more on instinct. (A prey animal who makes a mistake is dead, whereas a predatory animal who makes a mistake can always try again.) According to this model, larger animals and predatory animals not only can afford to sleep long and deeply, they actually need more sleep in order to process and encode information. This model also accounts for the facts that young animals sleep more than older ones (they have more to learn), and that after accounting for body size and predator/prey status, mammals sleep more than birds.

A second model suggests that sleep is necessary for visual-motor coordination. This model was originally formulated to try to explain why birds and mammals have REMs during sleep but fish, reptiles, and amphibians do not. Birds and mammals have a much more complex visual system than other vertebrates. This model also attempts to explain why young animals sleep more

It is theorized that large animals sleep more than small animals because they are less vulnerable to predation. (Corbis)

Do Animals Sleepwalk?

In humans, sleepwalking occurs during NREM sleep when muscles are relaxed but not deactivated. Birds that sleep on the water are known to go through automated swimming movements while they are in NREM sleep—what might be called "sleepswimming." It is also thought, but not yet proven, that some birds can "sleepfly." Albatrosses, for example, are known to fly for extended periods without landing; it is possible that they fly on a kind of autopilot while they sleep.

When humans sleepwalk, they certainly autopilot. Although sleepwalkers are not conscious of any visual input, lower brain centers use information from the open eyes to navigate around objects. Presumably birds that are sleepswimming or sleepflying also have their eyes open in order to navigate, so it would be difficult to tell whether they were awake or asleep unless one had a very long range radio transmitter measuring their EEG.

In fact, birds sleep with their eyes open much more often than humans or other mammals. They also open their eyes frequently during sleep quickly to assess the status of the environment. This "peeking" is particularly common in birds that are especially vulnerable. Brightly colored breeding male mallard ducks, for example, peek more often than their camouflaged mates. Birds sleeping alone are also more vulnerable than birds sleeping in pairs or groups and, not surprisingly, they sleep less, are more easily aroused, and are more likely to peek than are birds sleeping in groups.

than older animals—their visual and motor systems are not yet fully developed—and why young animals of altricial species (those born or hatched relatively helpless) sleep more than young animals of precocial species (those born or hatched at an advanced stage of development).

A third model suggests that sleep functions as a mechanism for thermoregulation. Among warm-blooded species, small animals, having a greater surface-to-volume ratio, both lose heat and overheat more rapidly than large ones; thus, they would need to rest more frequently but for shorter periods. Young animals, according to this model, need to sleep more than older animals because their thermoregulatory abilities are not yet fully developed. Likewise, altricial animals sleep more than precocial animals because their thermoregulatory mechanisms are less well developed.

All or none of these models may be correct; although virtually all animals sleep, no one yet really knows why.

—Linda Mealey

See also: Aging; Brain; Camouflage; Defense mechanisms; Groups; Invertebrates; Learning; Predation; Thermoregulation; Vertebrates.

Bibliography

Campbell, S. S., and I. Tobler. "Animal Sleep: A Review of Sleep Duration Across Phylogeny." *Neuroscience and Biobehavioral Review* 8 (1984): 269-300. Discusses differences in sleep time in different animal phyla.

Hobson, J. Allan. *Sleep*. New York: Scientific American Library, 1995. This is the best general book on the topic of sleep. Like other books in the Scientific American series, it is written by one of the world's foremost experts, yet is quite readable and abundantly illustrated.

Kryger, M. H., T. Roth, and W. C. Dement, eds. *Principles and Practice of Sleep Medicine*. Philadelphia: W. B. Saunders, 1989. Although this is a technical volume likely to be available only at research libraries, it includes three excellent and comprehensive chapters on animal sleep: chapter 4 (by Harold Zepelin) on sleep in mammals; chapter 5 (by C. J. Amlaner and N. J. Ball) on sleep in birds; and chapter 6 (by Kristyna Hartse) on sleep in insects, fish, amphibians, and reptiles.

Meddis, Ray. "On the Function of Sleep." *Animal Behaviour* 23 (1975): 676-690. Still the best user-friendly summary article of sleep across the animal kingdom.

SMELL

Types of animal science: Anatomy, behavior, development, evolution
Fields of study: Anatomy, biochemistry, biophysics, cell biology, developmental biology, genetics, herpetology, histology, human origins, immunology, marine biology, neurobiology, ornithology, pathology, physiology, reproduction science, zoology

For most animal species, smell is the main sense that is used in locating food and in detecting harmful agents and predators. Smell is used to recognize or exclude members within a social group, such as a pack, herd, or flock, and to find appropriate home sites. In some species, smell is also used to attract and identify mates.

Principal Terms

ANOSMIA: the clinical term for the inability to detect odors

CHEMOTAXIS: an oriented response toward or away from chemicals

OLFACTION: the sense of smell

OLFACTORY RECEPTORS: receptor organs which have very high sensitivity and specificity and which are "distance" chemical receptors

PHEROMONES: species-specific compounds (odors) which, acting as chemical stimuli at a distance, have a profound effect on an animal's behavior

Responses to chemicals are fundamental at all stages in biological organization. Chemotaxis, an oriented response toward or away from chemicals, has been observed in species ranging from single-cell animals such as bacteria and protozoa to very complex multicellular animals including humans. An attraction to a chemical is referred to as positive chemotaxis whereas a rejection or repulsion is called negative chemotaxis. The development of sensitivity (both positive and negative) to particular chemicals is the dominant sense in most animals. In general, receptor organs which have very high sensitivity and specificity, and which are distance chemical receptors, are called olfactory; the receptors of moderate sensitivity, usually found in the mouth, which are associated with feeding and are stimulated by dilute solutions are called taste receptors. Smells can be delivered to the olfactory receptors through air, as is the case with terrestrial animals ranging from insects to humans. On the other hand, smells can be delivered to the olfactory receptors through water, as is the case with aquatic animals such as insects and fish.

Smell in Insects

In insects, the olfactory (smell) receptors are located on the antennae. Because of the superficial location of their receptors and, more especially, because of their suitability for electrophysiological studies, insects have contributed much basic information about the mechanisms of olfaction. The olfactory receptors of most insects are highly specialized and can detect very trace amounts of compounds that are biologically important to the animal.

Olfaction is an important sensory modality for insects, particularly in mating, egg laying, and food selection. Numerous male insects, such as moths and cockroaches, are attracted by species-specific compounds called pheromones that act as chemical stimuli at a distance. Pheromones can be thought of as a language based on the sense of smell. Pheromones are often divided into two categories. Releaser pheromones initiate specific patterns of behavior. For example, they serve as powerful sex attractants, identify territories or trails, signal danger, and bring about swarming or similar types of grouping behavior. Primer pheromones trigger physiological changes in metabolism related to sexual development, growth, or metamorphosis. These changes are usually mediated through the endocrine system.

Male silk moths and gypsy moths may be attracted from a distance of a mile or two by a releaser pheromone from the scent glands of the females. Males will attempt to mate with any object that has touched the female scent gland; however, males deprived of their antennae do not even orient toward the female. Synthetic releaser pheromones are now being used in traps to attract pest insects such as the gypsy moth and the Japanese beetle.

Chemical communication in social insects is used for alarm, attraction, recruitment, and recognition of nest mates and of castes. Ants give off alarm releaser pheromones from mandibular glands and so are able to warn other ants of impending danger. Army ants deposit releaser pheromones on trails to food sources or to nest sites. Primer pheromones secreted by the queen bee cause the worker bees to cluster and swarm, and they suppress the rearing of other queens in the hive.

Mosquitoes are attracted chemically to warm-blooded animals and are sensitive to several chemicals. Carbon dioxide (the metabolic waste product excreted through the lung) attracts them and they are able to orient themselves and fly to the source of this compound. They also react positively to other mammalian body products. Most common insect repellants work by interfering with the olfactory ability of the mosquito, so that the insect can no longer follow an odor toward its source.

Smell in Fish

The olfactory receptors in most fish are located in olfactory sacs in a pit on the head. Chemicals are brought to the receptors while swimming or during respiratory movements.

Odors and the olfactory sense play a major role in the life of many species of fish. For example, homing in salmon is controlled mostly by the "smell" of the water in which the fish was born. By following the smell trail composed of the minerals found in the water, salmon are able to return to breed in the same stream in which they were born.

Fish can also become rapidly conditioned to odors. For example, once a pike has attacked a school of minnows, the odor of other pike in the water becomes associated with an alarm response in the minnows.

Smell in Terrestrial Vertebrates

In vertebrates, the olfactory receptor cells are located in the nose along the respiratory airflow path. As a result, when air is brought into the nose either during breathing or sniffing, odorant molecules are delivered to the headspace above the mucus-coated olfactory receptors. The odor molecules then bind to hairlike cilia on the olfactory receptors, producing a signal that is transmitted to the central nervous system. Because they stimulate different receptors, different smells produce different patterns of electrical activity. These odorant-specific patterns are used by the brain in smell identification.

Olfactory receptor cells are primary receptors, with axons running directly to the brain. This makes olfactory receptor cells unique, since most other sensory cells send their signals through processing centers (called synapses) before the message is carried to the brain. In the case of the olfactory receptor cells, all the information recorded by the cell is transmitted to the central nervous system. Once in the brain, the output of the olfactory receptors is sent to the limbic system (a portion of the brain involved with memory), the endocrine system, and throughout the rest of the central nervous system. The connections to the limbic system result in the very strong association that odors have in memory recognition. In humans, smells can often trigger very vivid memories. The rest of the brain also sends messages back to the bulbs, amending the pleasure of a food aroma when the stomach is full. Unlike other neurons, olfactory receptor cells constantly replicate. As a result, after a life span of about thirty days, olfactory receptor cells are replaced.

Odors help bond mothers to their newborn babies. A mother cuddling her infant will invariably brush her nose in the baby's hair to inhale its sweet aroma. She can identify her baby by its smell as much as by its cry. Additionally, one-day-old infants of many species have been shown to be able to recognize the smell of their mothers. A mother rat licks her nipples so that her blind pups can follow the scent of her saliva to the milk. Likewise, a mother kangaroo produces a saliva trail so the newly born and blind babies can follow the trail from the uterus to the mother's pouch. Wash the nipples and eliminate the saliva trail, and the pups are lost.

Female rodents who periodically smell male urine will move more quickly into puberty than females that do not. If a pregnant female mouse smells the urine of a male of another colony, she will immediately terminate her pregnancy. Also, if the olfactory nerves of a newborn rat pup are cut, the rats will never develop sexually.

A diminished sense of smell is termed hyposmia. Hyposmia can occur following a cold or after head trauma, and humans experience some reduction in the sense of smell with age. Also, most conditions that reduce the flow of air through the nose will reduce olfactory acuity. For example, a stuffy nose as a result of an allergy, a cold, or a nasal polyp often creates hyposmia. Anosmia is the complete loss of the ability to detect airborne odorants. Head trauma and severe nasal obstructions can produce anosmia. If the cause of hyposmia or anosmia is related to a blocking in the nasal airflow passageways, then treatment with steroids and/or surgery often can restore the olfactory loss.

Human experience seems to draw a sharp contrast between taste and smell. Taste is the chemical sense related to sampling compounds that come in directed contact with the inside of the mouth whereas smell is the ability of the nose to monitor airborne chemicals, often from distant sources. However, the sensations of taste and smell are not completely independent, since smell can influence taste and vice versa. For example, a lemon smell in the nose can make distilled water appear to "taste" bitter, and a sugar solution in the mouth can affect the perception of a fruit smell such as cherry. Much of what is usually perceived of as being a taste is really a smell. For example, with the nose blocked, it is difficult to tell coffee from bitter water or an onion from a potato. As humans chew, volatile compounds in the food are released into the air in the back of the throat. These compounds then make their way up the back of the nasal cavity, where they stimulate the olfactory receptors, producing a smell sensation that dramatically enriches the perception of the taste. This combination of smell and taste is referred to as flavor. What is often thought of as "taste" is actually a combination of smells and tastes, with additional contributions to the flavor coming from temperature and pain receptors in the nose and mouth.

—David E. Hornung

See also: Brain; Communication; Hearing; Insect societies; Nervous systems in vertebrates; Noses; Pheromones; Vision.

Bibliography

Association for Chemoreception Sciences. www.achems.org. This Web site contains a description of current work in the field and well as a discussion of smell disorders.

Getchell, T. V., R. L. Doty, L. M. Bartoshuck, and J. B. Snow, eds. *Smell and Taste in Health and Disease*. New York: Raven Press, 1991. Discussion of the clinical aspects of smell problems.

Gibbons, Byron. "The Intimate Sense of Smell." *National Geographic* 170, no. 3 (1986): 321-361. An excellent overview of the anatomy, physiology and psychology of the sense of smell. The making of perfumes, use of dogs for tracking, and the history of smell are all well covered.

Vroon, Piet. *Smell: The Secret Seducer*. Translated by Paul Vincent. New York: Farrar, Straus and Giroux, 1997. A cultural history and compendium of odd facts and a tribute to the sense of smell.

SNAILS

Types of animal science: Anatomy, classification, reproduction
Fields of study: Anatomy, invertebrate biology, zoology

Snails are marine, freshwater, and land mollusks belonging to the class Gastropoda. Some are used for food; others are ecologically useful.

Principal Terms

FOOT: muscular bottom portion of snails, on which they walk

HERMAPHRODITE: an organism having both male and female reproductive systems in the same individual

RADULA: a tonguelike, toothed organ used to grind food

The first gastropods appeared 600 million years ago. Among these mollusks are snails and slugs, related animals which have a single shell and an asymmetric body. There are thirty-five thousand species, varying hugely in structure and lifestyle. The smallest snails are barely visible, while the largest, sea slugs, weigh up to thirty pounds. They occur in marine, freshwater, and land habitats.

The first gastropods were bilaterally symmetrical. Later, many evolved into asymmetric snails, with gills and anus above the head and coiled shells. In land snails, the gill-holding cavity became a lung. The snail head has eyes and tentacles, enabling good sight and smell. The mouth has a rasplike radula, used to harvest food. Digestive, nervous, circulatory, and reproductive systems are also well developed. Some snails have sexes and lay eggs in water, where they hatch, settle down, and mature. In many snails, fertilization is internal and hermaphroditic, and they can mate with any mature animal of the same species.

Physical Characteristics of Snails

Most snails have hard outer shells and slimy bodies. Slugs have no shells or internalized shells. On their heads, snails and slugs have two pairs of antennae. One pair holds the eyes. The other antennae sense the environment. The bodies of

Most snails have hard, spiral shells that protect their soft, invertebrate bodies. (Adobe)

these gastropods grow from 0.1 inch to several feet in length, depending on species. The shell of a snail may be rounded, long and pointy, or flattened. Shells are homes and protection. When afraid, snails close them up via plates under their bodies.

Snails and slugs eat algae, leaves, lichens, small insects, and small marine organisms. As they also eat decaying plant and animal matter, the gastropods are ecologically important. Snails are also important to food chains, serving as food sources for fish and birds. A snail or slug eats with a radula, a ribbonlike tongue having thousands of tiny teeth. Radulas are drawn along rocks, leaves, or plants to scrape off food. Carnivorous snails have radulas that bore holes through shells of other mollusks to eat their flesh. Snails move by wavelike muscle contraction along the bottom of their muscular feet. This motion is aided by cilia in aquatic snails or slugs and by a slime track on land.

Types of Snails

Abalone, marine snails of the family Haliotidae, live on underwater rocks near shores of warm ocean regions. Their flat, ovoid shells, often a foot long, are nice souvenirs. Their thick feet are tasty.

The common garden snail, *Helix aspersa*, a land snail, inhabits Great Britain and continental Europe. It lives in moist, shady places, not restricted to gardens, is nocturnal, and can be four inches long.

Conches, marine snails of the order Mesogastropoda, close their shells with their digging claws. The largest conches (queen conches) inhabit warm Atlantic regions. Their three-foot-long shells make beautiful ornaments. Humans eat conch feet.

Great pond snails, order Bassommatophora, occur in ponds in Europe, Asia, and Africa. Their shells grow two inches long and one inch wide. Air-breathers, with lungs, they frequently come to pond surfaces for air.

Limpets, marine snails of the order Archaeogastropoda, are found worldwide. Limpets have arched, nonspiral shells and are found clinging to rocks. They scrape out rock areas the size of their shells, returning to their safety nightly. The New England species is 1.5 inches long.

The Life Cycle of Snails

Most snails are hermaphrodites, having both male and female sex organs. However, they usually mate with another individual of the same species, passing sperm to its partner and getting sperm from the partner. Fertilized eggs hatch in two weeks to two months. They are usually laid on marine or land plants, depending on the species involved.

The offspring pass through complex developmental cycles before hatching. Often, they hatch as miniature replicas of parents. In some cases hatchling snails have weak shells and very fragile bodies. Such snails become stronger and obtain hard shells, as they grow. Snails of most types are sexually mature at one to two years of age and live for up to ten years.

Abundant snails and slugs are important to food chains. They are herbivores, carnivores, and omnivores. As they eat decaying flesh and plants,

Snail Facts

Classification:
Kingdom: Animalia
Phylum: Mollusca
Class: Gastropoda
Orders: Include Archaeogastropoda (limpets), Bassomatophora (land snails), Mesogastropoda (conches)

Geographical location: Europe, Asia, Africa, the Americas, and Australia

Habitat: Oceans; freshwater lakes, streams, ponds; moist land environments

Gestational period: Eggs hatch in two weeks to two months, yielding miniature versions of parents

Life span: Two to ten years, depending on species

Special anatomy: Antennae, radula, muscular foot for locomotion, univalve shells, eyes on stalks in head, hermaphroditic reproductive system these gastropods have another ecological function, environmental cleanup. They are also important human foods. Escargot, in French cuisine, are cultivated land snails. Other edible snails are abalone, periwinkles, and queen conches. A few of these gastropods are harmful. For example, snails and slugs damage crops and gardens. Others are pests in oyster beds.

—*Sanford S. Singer*

See also: Hermaphrodites; Home building; Marine animals; Mollusks; Shells.

Bibliography

Buholzer, Theres. *Life of the Snail*. Minneapolis: Carolrhoda Books, 1985. This book describes the physical characteristics, behavior, and environments of snails.

Fisher, Enid. *Snails*. Milwaukee: Gareth Stevens, 1996. The physical appearance, behavior, and habitats of snails are described briefly but clearly.

Hughes, Roger N. *A Functional Biology of Marine Gastropods*. Baltimore: The Johns Hopkins University Press, 1986. The text is clear and contains a lot of information on gastropods, including snails.

Jacobson, Morris H., and David R. Franz. *Wonders of Snails and Slugs*. New York: Dodd, Mead, 1980. This illustrated book describes characteristics and behavior of a variety of snails and slugs and their uses by humans.

Simon, Hilda. *Snails of Land and Sea*. New York: Vanguard Press, 1976. This illustrated book clearly describes habitats, appearance and natural history of marine, freshwater, and land snails.

SNAKES

Types of animal science: Classification, evolution
Fields of study: Evolutionary science, herpetology, systematics (taxonomy)

Snakes are squamate reptiles with slender, elongate bodies that lack limbs. In spite (or because) of these features, snakes have successfully exploited many habitats and lifestyles.

Principal Terms

CLADE: a group of animals and their common ancestor

CONSTRICTION: a method of killing prey using increasingly tight coils around the body to trigger stress-induced cardiac arrest

FANGS: enlarged teeth that are hollow like a hypodermic needle or grooved to facilitate the injection of venom

KINETIC SKULL: a highly moveable arrangement of bones that allows independent action of the snout and jaws on both sides

VENOM: a toxic substance that must be injected (instead of ingested) to immobilize or kill prey

Humans are fascinated by snakes, arguably more so than any other kind of animal. This fascination may be based on fear (some snakes are undeniably dangerous), religious fervor (snakes figure, for good or ill, in nearly all world mythologies and religions), or curiosity (how does an animal move without limbs?). Regardless, this fascination has led to myths and an inclination to attribute mystical powers or malevolent intentions to these bizarre (by human standards) creatures.

Snakes are fascinating for purely scientific reasons as well. Based on diversity and total numbers, they are very successful. They have adapted to amazingly varied lifestyles in spite (or maybe because) of many unique features, the most obvious of which is the combination of extreme body elongation and lack of limbs. These, in turn, help explain other features. For example, long, slender bodies limit the size of the body cavity, requiring that paired internal organs are offset (placed one in front of the other) or that one organ is disproportionately elongated and its counterpart reduced or absent. Similarly, all snakes are obligate carnivores (meat-eaters; that is, they eat other animals). Plants are difficult to digest, and the simple, straight digestive tracts that can be accommodated in a snake's body are not long enough to provide enough time in passage for vegetable matter to be adequately broken down. However, these characteristics that appear to define snakes are shared with many other squamate reptiles, whereas those that are unique to snakes (number of body vertebrae; modification, reduction, or absence of some skeletal features; location of the ophthalmic nerve; eyes without ciliary bodies to facilitate focusing) are comparable to differences between other squamate reptiles.

Relationships Among Squamate Reptiles

For many years, squamate reptiles were divided into three suborders, Sauria (lizards), Serpentes (snakes), and Amphisbaenia (odd, mostly limbless burrowers), but recent studies of relationships indicate that snakes and amphisbaenians are cladistically nested within lizards; in other words, instead of being "cousins" of lizards, they are siblings. Consequently, the traditional suborders no longer reflect true relationships. They are retained for the sake of convenience by some authors, but only because they reflect clades within Squamata. As suborders, they have no formal taxonomic status.

Several distinct groups of snakes are recognized. The Microstomata (small-mouthed snakes) include small burrowing forms with blunt heads and tails. Among other snakes, two primitive families (uropeltids and aniliids) are distinct from macrostomatan (large-mouthed) forms, but many relationships within the latter remain unresolved.

Body Forms and Habitats

Evidence suggests that snakes evolved as burrowers, but snakes in many families have subsequently and successfully exploited various habitats. Fossorial snakes (burrowers) are small, have conical heads (sometimes with modified snouts that help them dig), reduced scales that often overlap and reinforce sutures between skull bones, tiny eyes, and mouths located underneath the snout to avoid ingesting soil while burrowing. Many have spines on their tails for use as anchors when using the head to dig.

Arboreal snakes live in trees, and often are very slender, although the center of gravity may have shifted posteriorly so the rear of the body or tail can be used as an anchor while extending the head and anterior body over open spaces. To facilitate extension, many have laterally compressed bodies and enlarged scales along the back and belly; this allows them to form a structure much like an I-beam for support. Tails are long and may be prehensile (capable of grasping branches). Eyes are often large, as vision is more important than chemical cues when searching for prey in trees.

Snakes that inhabit loose ground litter in prairies or forests are small and often have enlarged scales on their snouts, with which they root for food. Like burrowers, body scales are usually smooth to reduce friction and ease passage through tight spaces.

Aquatic snakes often have eyes and nostrils on the tops of their heads, and nostrils may be equipped with valves. Many are heavy-bodied and laterally compressed to increase the surface

Snakes have remarkably flexible spines that allow them to coil tightly. (Corbis)

area with which they push against water while swimming. Scales are generally rough to increase friction. Some sea snakes have fins supported by extensions of their vertebrae. Most aquatic snakes exhibit countershading (dark above, light underneath), reducing their visibility to both prey and potential predators.

Locomotion

Most snakes move by means of rhythmic waves of muscle contractions that cause the body to undulate laterally. Very heavy-bodied snakes often rely on rectilinear propulsion. Instead of muscles on the sides of the body working alternately, they work in concert, contracting and relaxing while drawing the body forward in a straight line. Enlarged belly scales or friction with the surrounding substrate (water in aquatic forms) prevents backsliding. Concertina locomotion and thrust creeping are used by burrowers. In the former, snakes first wedge the anterior part of the body and draw the posterior portion forward, then wedge the posterior region while extending the head and anterior body. Thrust creeping is very similar, except the spiked tail is used as a posterior anchor instead of a body loop. The most unusual form of snake locomotion is sidewinding, which is employed on loose or slick substrates. Loops of the body are elevated and thrown laterally to serve as the contact points while the intervening areas are raised. The advantage of this method is that force on the substrate is directly downward, which reduces slippage. Regardless of method, snakes are not very fast. The world's fastest snakes, large, slender mambas (*Dendroaspis*) can achieve top speeds on ideal substrates of about 12 kilometers per hour (about 7.5 miles per hour). Humans, in contrast, can sprint at speeds up to about thirty-five kilometers per hour (over twenty miles per hour).

Sensory Systems

Except for many arboreal and some actively foraging terrestrial forms, most snakes rely primarily on chemical cues to monitor their environment for food, prospective mates, or danger. Limbless forms are close to the ground and have a limited scope of vision (in fact, most are nearsighted). Snakes also lack external and middle ears, and are essentially

Snake Facts

Classification:
Kingdom: Animalia
Subkingdom: Bilateria
Phylum: Chordata
Subphylum: Vertebrata
Class: Reptilia
Order: Squamata
Superfamilies: Typhlopoidea, Booidea, Colubroidea
Families: Fifteen recognized, including Typhlopidae (blind snakes or worm snakes); Leptotyphlopidae (slender blind snakes); Aniliidae; Xenopeltidae (sunbeam snakes); Uropeltidae (shieldtail snakes); Boidae (pythons, boas, and wood snakes); Acrochordidae (wart snakes); Colubridae, with subfamilies Dasypeltinae (egg-eating snakes), Pareinae, Dipsadinae, Xenodontinae, Homalopsinae, and Colubrinae; Viperidae (vipers, rattlesnakes, moccasins, and relatives), with subfamilies Viperinae (Old World vipers) and Crotalinae (pit vipers); Elapidae (cobras, mambas, coral snakes, and relatives); Hydrophiidae (sea snakes)
Genus and species: 450 genera, nearly three thousand species
Geographical location: Worldwide
Habitat: Terrestrial, aquatic, and marine habitats, except at very high latitudes
Gestational period: Some snakes lay eggs, whereas others bear live young; clutch sizes range from one or two (tiny threadsnakes and some sea snakes) to one hundred or more (large boas and pythons), and the incubation period varies according to temperature
Life span: Varies by species
Special anatomy: No legs

deaf to airborne sounds, although they are very sensitive to vibrations of the substrate. Thus, the importance of smell and taste is not surprising. Both smell and taste are very well developed in almost all forms, and are further enhanced by vomeronasal systems. The moist, forked tongue collects chemical molecules by flicking in the air or licking a substrate, and delivers these to paired sensory pits in the roof of the mouth. The importance of chemical senses is reinforced by the role of pheromones in snake reproductive ecology and observations that many snakes will follow a convoluted scent trail rather than rely on vision to directly approach a mate or food.

Specialized infrared (heat) receptors occur in several boid snakes and in pit vipers (Viperidae). In boids, receptors are in pits on the lips or snout, whereas pit vipers have a single pit between the nostril and eye on each side of the head. These sensors apparently are able to generate infrared images that are superimposed on visual images. Although obviously valuable to species that consume mammals or birds, which generate body heat, infrared imaging works equally well for species that eat frogs or salamanders. Amphibians have moist skins and lose evaporative heat to the environment, appearing as distinct "cold" spots in thermal images.

Some Types of Snakes

THREAD SNAKES (Leptotyphlopidae) include the smallest snakes, extremely slender forms less than ten centimeters (about four inches) long. Most are burrowers, but some have been found in bird nests. These were presumably brought to the nests by birds as food for their young, but escaped and subsist on the insects that also inhabit the nests.

BOAS and PYTHONS (Boidae) include the world's largest snakes. Although mythical accounts abound, individuals of some species (invariably females) may exceed twenty feet, and a very few may approach thirty. Tales of larger snakes are probably fanciful. Large snakes can exploit large prey, and can swallow deer, pigs, and even humans.

SEA SNAKES (subfamily Hydrophiinae) are the most aquatic snakes. Some species never leave water and give live birth. All are venomous and most feed on fish.

KING COBRAS (*Ophiophagus hannah*) are the longest venomous species, sometimes exceeding five meters (over sixteen feet) in length. King cobras feed largely on other snakes and are one of the few species that builds nests and cares for its eggs.

Killing and Consuming Prey

All snakes eat animal food. Some are know to scavenge or eat dead animals, but most hunt and kill live prey. Venom, often prey specific, is used by some to immobilize and kill especially large and dangerous prey. Constrictors use teeth to grab prey, then rapidly loop one or more coils of their body around it, tightening with each breath until the animal dies of stress-induced cardiac arrest and suffocation (constrictors do not crush their food). A few species may use loops of their bodies to press prey against the substrate. Some venomous forms constrict or compress prey while envenomating it; others quickly inject venom and then follow the prey's scent trail, ingesting it only after it dies. Other snakes eat animals with limited defensive capabilities, and prey is swallowed alive.

Snake teeth vary in number and size, but species that hunt prey that is hard to hold (slimy fish or snails that must be pulled from their shells) generally have the longest teeth. Bird-eaters often have long teeth to grasp the body of their prey and avoid a mouth full of feathers while the prospective meal escapes. Some snakes have upper jaw teeth of approximately equal size, rear-fanged species have enlarged posterior teeth that may be grooved, and others (cobras and their kin) have enlarged anterior fangs. Vipers and pit vipers have very large fangs on small maxillary bones that are rotated to erect the fang.

Most snakes can consume food much larger in diameter than their bodies by using a kinetic skull, a protrusible glottis (opening to the lungs) that is used like a snorkel when the mouth and throat are filled with food, and an elastic body wall not enclosed by ribs. Using teeth as anchors, upper and lower jaws on alternating sides separately "walk" over the prey until it is ingested. Rhythmic muscular contractions of the body wall then move the meal through the digestive tract. The ability to consume large prey reduces the need for frequent meals, allowing snakes time to rest and conserve energy before eating again.

Many snakes forage actively for food, rooting through debris or scanning their environment with one or more sensory systems. Some species, however, lie motionless, often well camouflaged, along a game trail, patiently waiting to ambush a meal. Because this sit-and-wait strategy does not require constant movement (other than the strike, which may be lightning quick), snakes that employ this method are frequently heavy-bodied and capable of eating the largest prey items relative to their body sizes.

Snakes in Need of Conservation

Snakes, because of the fear and revulsion they engender in many people, are subject to greater persecution than almost any other kind of animal. They often are actively hunted and, in many parts of the world, quickly killed whenever encountered. Nevertheless, habitat destruction and alteration are responsible for more declining snake populations than any other single factor. Many species are exploited as food, especially in parts of eastern Asia, and others are killed for body parts thought to have medicinal or aphrodisiac qualities. Introduced exotic species exact a toll. Many island populations of diurnally active terrestrial snakes have been extirpated by the mongoose, an effective predator introduced to control introduced rat populations. However, rats are nocturnal and mongooses active by day; consequently they have a much greater impact on ground-dwelling snakes than on rats. The leather industry is responsible for the deaths of hundreds of thousands of giant snakes each year. In addition, many wild snakes die each year as a consequence of being caught for the pet trade, much of it illegal. Many species become roadkills as they migrate. Only a few species are formally protected in at least some parts of their ranges, and many others may be nearing extinction in spite of increasingly frequent efforts to conserve them and their habitats.

—*Robert Powell*

See also: Camouflage; Digestive tract; Lizards; Poisonous animals; Predation; Reptiles; Scales; Smell; Vertebrates.

Bibliography

Ernst, Carl H., and George R. Zug. *Snakes in Question: The Smithsonian Answer Book.* Washington, D.C.: Smithsonian Institution Press, 1996. A highly informative volume organized around commonly asked questions about snakes.

Greene, Harry W. *Snakes: The Evolution of Mystery in Nature.* Berkeley: University of California Press, 1997. This remarkable work discusses in eminently readable fashion the ecology, diversity, and evolution of snakes.

Murphy, John C., and Robert W. Henderson. *Tales of Giant Snakes: A Historical Natural History of Anacondas and Pythons.* Malabar, Fla.: Krieger, 1997. This engaging book examines historical accounts, many fanciful, and realities of the four species of snakes known to exceed twenty feet in length.

Pough, F. Harvey, Robin M. Andrews, John E. Cadle, Martha L. Crump, Alan H. Savitsky, and Kentwood D. Wells. *Herpetology.* 2d ed. Upper Saddle River, N.J.: Prentice Hall, 2001. This ambitious textbook attempts to address all pertinent aspects of herpetology.

Shine, Richard. *Australian Snakes: A Natural History.* Ithaca, N.Y.: Cornell University Press, 1991. A delightful book that uses the author's experiences with Australian snakes to teach more universal lessons.

SPARROWS AND FINCHES

Type of animal science: Classification
Fields of study: Anatomy, ornithology, systematics (taxonomy)

The finches and sparrows are some of the most widespread and successful of all songbirds. Of relatively recent origin and uncertain taxonomic status, they are united in possessing several basic characteristics: All are small, seed-eating songbirds that have a bill adapted for gathering, holding, and, in some cases, crushing seeds.

Principal Terms

CONVERGENT EVOLUTION: unrelated animals that have evolved similar features which enable them to exploit a habitat in a similar fashion, and have the same way of life

PASSERIFORMES: the largest order of birds, consisting mostly of perching songbirds

PRIMARY FEATHERS: feathers of the hand that provide most of the lift and thrust for bird flight

The name "finch" at one time was applied to ten different subfamilies of songbirds such as the chaffinches, goldfinchlike birds, buntings, grosbeaks, tanagers, weaver-birds, and sparrows. All of these birds have large jaw muscles that power cone-shaped bills. They differ, however, in skull structure and in the ways they open seeds. These differences suggest that at least some "finches" may have evolved independently and share characteristics as a result of convergent evolution. In view of these differences, most modern taxonomists agree that the name "finch" should be limited to the family Fringillidae in the order Passeriformes.

The 112 species of fringillids are among the most successful of the seed-eating passerines. All have conical bills or crossed bills adapted for eating seeds. Several fringillid subfamilies are recognized. The subfamily Carduelinae includes the New World rosy finches, purple finches, crossbills, redpolls, siskins, some grosbeaks, and saltators. Old World chaffinches and bramblings are placed in the subfamily Fringillinae. Two other New World fringillid subfamilies include the Hawaiian honeycreepers (subfamily Drepanidinae), which are restricted to the Hawaiian Islands, and Darwin's finches (subfamily Geospizinae), which are found only on the Galápagos Islands.

Several finchlike birds have recently been placed in a separate family called the Emberizidae. This family is further subdivided into the subfamily Emberizinae, which includes the New World sparrows, juncos, longspurs, and tropical brush finches, and the subfamily Cardinalinae, which includes the forty species of cardinals and their allies.

Old World sparrows, which include the familiar house sparrow (*Passer domesticus*), are more distantly related and placed in a separate family called the Passeridae. They are widespread seed eaters that occur in a wide variety of habitats throughout most of Eurasia and Africa. Some, such as the house sparrow, have been deliberately and successfully introduced in many other areas of the world.

The Ecology of the Finches

Most finches are forest-dwelling, seed-eating songbirds that have nine instead of ten primary feathers in the wing and twelve tail feathers. The outermost part of ten primary feathers is usually small and hidden.

Most species have sweet, melodious songs and often sing in winter, which is why they were

Finches have beaks that are specialized to eat seeds and caterpillars and other larvae. (Kenneth Layman/Photo Agora)

named finch (from Latin *frigus*, "cold," because finches sing in the cold of winter). The female builds an open, cup-shaped nest with her tail feathers and also uses the feathers to incubate the eggs. Incubation and the fledgling period usually last between eleven and fourteen days. Members of the subfamily Fringillinae feed insects to their young and inhabit large territories while breeding. There are usually about three or four eggs and they are blue-gray with purple-brown spots. In the summer, the birds eat caterpillars from trees, and in winter seeds from farmland, including spilled grain and weed seeds. Over most of their territory they are migratory, but females tend to move farther away from their territory than males.

The Carduelinae form the largest branch of the finch family, with about 122 species. These birds are more specialized seed-eaters than Fringillinae and they feed their young mostly seed, sometimes augmented with insects. They nest either alone or in loose colonies and feed away from the nest in packs. Many feed directly on plants and are adept at clinging to stems or hanging on twigs. They demonstrate a wide range in bill shape and adaptation for extracting the seeds from different types of seed pods. The fringillids range in size from the *Mycerobas* grosbeaks of the Himalayas, which reach eight inches in length and 3.5 ounces in weight, to the relatively tiny Lawrence's goldfinch (*Spinus lawrencei*) of eastern North America, which just reaches four inches in length and weighs no more than 0.3 ounce.

Most fringillids are found in temperate regions, with fewer in the Arctic, deserts, tropics and subtropics. About sixty-eight species occur in Eurasia, thirty-six in Africa, and twenty-five in the New World. Fringillids are absent only from Madagascar and the islands of the South Pacific. Some species have been introduced into Australia and New Zealand.

Sparrows

The twenty-seven species of Old World sparrows are closely related to weaver finches and were once grouped with them in the family Ploceidae before being placed in their own family. They are all small, seed-eating birds that occupy a wide variety of habitats throughout Eurasia and Africa. Over half of the true sparrows (*Passer*) coexist well in human habitations and at least one, the house sparrow, has for so long adapted to exist in human-modified landscapes that it is no longer found naturally in the wild. House sparrows coexist with humans in habitats and climates ranging from tundra to tropics. Other *Passer* species occur in habitats as diverse as desert, woodlands, or tropical rain forests. Dry brush country and mountain habitats are occupied by species of rock sparrows (*Petronia*) which may range upward of fifteen thousand feet in the Himalayas.

Males claim nest sites and sing to attract females and deter other males. Following pair formation both sexes help build the nest. Many *Passer* species, such as tree sparrows (*Passer montanus*), are solitary nesters, usually nesting in tree cavities, but some are at least partly social and nest in loose aggregations. Several species also build a ball-shaped nest with a feather-lined interior and an entrance toward the side. Depending on species, from two to eight whitish or mottled eggs are laid in a clutch, generally fewer in the tropics and more in temperate zones. Incubation is brief, generally eleven or twelve days, and the young remain in the nest for another two weeks (up to seventeen days). The young are fed on a variety of foods, mostly seeds and insects, but discarded bits of human food, grains, and other discarded edible substances are often readily used to feed the young.

Tropical sparrows generally raise one brood a year, but temperate species may raise two or three broods or occasionally even four broods in good weather conditions and availability of sufficient food. Following breeding, the sparrows are gregarious, gathering in loose aggregations of foraging flocks that may include other species. In many locales their numbers may cause appreciable destruction of seed crops.

Some Typical Finches and Sparrows

The northern cardinal (*Cardinalis cardinalis*) belongs to the subfamily Carduinalleae. Males are bright "cardinal" red with black about the bill. Females are a duller brown-red. Both have a distinctive crest. The species is widespread throughout eastern North America, in western North America from the Great Plains south and west into California, Arizona, New Mexico, Texas, and into Mexico. It is essentially nonmigratory and winters in the same general area. The cardinal's range has expanded northward in recent years, especially into New England, and it has been successfully introduced in Hawaii.

Cardinals dwell in mostly edge habitats, especially the interface of woodland and meadow, and edges of swamps of shrubby wetlands, especially areas that offer shrubby evergreens for nesting. They have adapted well to human-modified landscapes and commonly nest in landscaped yards in which the mix of ornamentals and grass essentially mimics its natural habitats.

One of the earlier nesting species, cardinals may claim their territories in late February. The female typically constructs a nest of leaves, weeds, grasses, and thin bark strips in dense shrubbery, often in either evergreens or vine tangles from one to twenty feet

Sparrow and Finch Facts

Classification:
Kingdom: Animalia
Phylum: Chordata
Subphylum: Vertebrata
Class: Aves
Order: Passeriformes
Families: Fringillidae (finches and allies, New World sparrows); Emberizidae; Passaridae (Old World sparrows)
Subfamilies: Carduelinae (New World rosy finches, purple finches, crossbills, redpolls, siskins, some grosbeaks, saltators); Fringillinae (Old World chaffinches, bramblings), Drepanidae (Hawaiian honeycreepers); Geospizinae (Darwin's finches); Emberizinae (New World sparrows, juncos, longspurs, tropical bush finches); Cardinalinae (forty species of cardinals and allies)

Geographical distribution: All continents, except Madagascar and South Pacific islands

Habitat: Prefer temperate regions, although some species are found in Arctic, desert, tropic, and subtropical regions

Gestational period: Varies by species, but most eggs are incubated for eleven to fourteen days

Life span: Varies; three to ten years in the wild, five to eight years in captivity

Special anatomy: Beaks adapted for gathering, holding, crushing, and eating seeds; finches have nine primary wing feathers and twelve tail feathers; melodious song

high, but generally lower. Clutch size is generally three or four spotted white eggs which hatch in about twelve to fourteen days. Three broods may be raised in a season in southern states, about two broods in more northerly states. The young are fed seeds and small insects. Adults divide roles and responsibilities to raise a second and sometimes a third family; the male cares for the first brood while the female incubates the second brood. The young and adults also form family groups that may remain together in fall. Groups often form loose flocks of up to seventy birds in winter. Cardinals are common birds at feeders.

The American goldfinch (*Carduelis tristis*) belongs to the family Fringillidae. It is widespread in North America, breeding from southern Canada southward to Gulf Coast states, west to California. At least partially migratory, it winters along the Gulf Coast and Florida but may winter anywhere in its breeding range in mild winters.

The American goldfinch is a bird of fields and open woodlands. It is a late-nesting species, often not beginning until August in its northern range, late May or June in western and more southern states. Nests are constructed in branch forks, often from one to thirty feet high. The highly compact, cup-shaped nests are lined with thistle or thistledown and are so tightly constructed that they hold water—young sometimes drown following heavy rainstorms when the nest is flooded. Generally four to six (usually five) bluish eggs are laid and incubated by the female for about two weeks. The male attends the incubating female at the nest, bringing her food.

Young are fed by regurgitation as the parents first fill their crops with dandelion, burdock, thistle, or chicory seeds, berries, and insects, then regurgitate or cough them up to feed each of the young. Seeds are usually augmented by insects, especially caterpillars, grasshoppers, aphids, and plant lice.

The house sparrow (*Passer domesticus*) belongs to the family Passeridae. It is the most widely distributed of Old World sparrows, ranging across Eurasia and Africa and successfully introduced in South America and Australia. Repeated introductions of birds in North America between 1850 and 1867 resulted in the establishment of this aggressive songbird. At the turn of the last century (1900), it was probably the most abundant bird in North America but has declined with the switch to combustion engine vehicles, for it fed on the scattered feed grain fed to horses. The house sparrow gathers in large winter roosts in urban and suburban areas or evergreen plantations.

House sparrows are aggressive cavity nesters. Their success has at least partly been at the cost of other cavity nesters such as bluebirds and swallows. Nesting sites are selected in cultivated areas, especially buildings, outbuildings, farms, and edge habitat between urban and suburban landscapes and natural landscapes. Artificial cavities, including nest boxes, are also readily appropriated by this adaptable species. The nest of small twigs and leaves is lined with grasses, feathers, hair, and bits of paper and other discarded materials. The four to six white or greenish eggs are incubated for about fourteen days. Young are brooded by the female and fed mostly insects and spiders, along with seeds and blossoms.

—*Dwight G. Smith*

See also: Beaks and bills; Birds; Domestication; Feathers; Flight; Molting and shedding; Nesting; Respiration in birds; Wildlife management; Wings.

Bibliography

Austin, O. L., Jr. *Life Histories of North American Cardinals, Grosbeaks, Buntings, Towhees, Finches, Sparrows, and Allies*. 3 vols. 1968. A comprehensive compilation of ecological and behavioral information about North American finches and sparrows, along with other passerines. Highly recommended.

Byers, C., J. Curson, and U. Olsson. *Sparrows and Buntings: A Guide to the Sparrows of the World*. Boston: Houghton Mifflin, 1995. This well-illustrated book is the only guide to offer complete coverage of the world's sparrows and buntings.

Clement, Peter. *Finches and Sparrows*. Princeton, N.J.: Princeton University Press, 1999. A well-illustrated, comprehensive guide to sparrows and finches.

Rising, J. D. *Sparrows of the United States and Canada*. New York: Academic Press, 1996. Thorough and scientific treatment. Contains a great deal of ecological information about North American sparrows.

SPIDERS

Types of animal science: Classification, ecology, evolution, geography
Fields of study: Ecology, evolutionary science, systematics (taxonomy)

Spiders are members of phylum Arthropoda, the dominant animals on earth in terms of both number of species and number of individuals. Of an estimated ten million species of arthropods, only 10 percent have been identified. A close examination of spiders helps to understand other arthropods.

Principal Terms

BOOK LUNGS: specialized respiratory structures of arachnids, such as spiders

CLASS: the taxonomic category composed of related genera; closely related classes form a phylum or division

COMPOUND EYES: eyes that are made up of multiple lenses or light detectors

EXOSKELETON: an external skeleton that encloses the body like a suit of armor

MOLTING: in animals with an exoskeleton, the process of shedding the old shell to allow the increase in body size due to growth

PHYLUM (pl. PHYLA): the taxonomic category of animals and animal-like protists that is contained within a kingdom and consists of related classes

TRACHEA: a network of narrow, branching respiratory tubes

The major classes in the phylum Arthropoda include the insects (class Insecta), the spiders and their relatives (class Arachnida), and the crabs, shrimps, and their relatives (class Crustacea). Arthropods are everywhere, cohabiting with humans: ants, yellow-jackets, spiders, dragonflies. It is only fitting to look at the factors that contribute to the huge army of arthropods before examining one of their members: spiders.

Traits Contributing to the Success of Arthropods

Several adaptations are responsible for the spread of arthropods to nearly every habitat on earth: an exoskeleton, segmentation, the efficient use of oxygen, and a well-developed body plan, including circulatory, sensory, and nervous systems. The exoskeleton is formed via secretion by the epidermis (the outer layer of skin) and is made up mainly of protein and a polysaccharide called chitin. The exoskeleton protects against predators and is responsible for arthropods' increased agility over their wormlike ancestors. The exoskeleton allows precision movement, making possible the flight of the bumblebee and the intricate, delicate manipulations of the spider as it weaves its web. By providing a watertight covering for delicate, moist tissues, the exoskeleton also contributed enormously to the arthropods' invasion of dry land habitats. However, the exoskeleton also creates some unique problems for arthropods. First, it cannot expand as the animal grows, and thus must be shed, or molted, periodically and replaced with a larger one. In the process of molting, arthropods are vulnerable to predators. Second, the weight of the exoskeleton increases exponentially as the animal grows, placing a cap on the size arthropods can reach. No wonder some of the largest arthropods (crabs and lobsters) reside in buoyant watery habitats.

In general, segmentation in arthropods is less distinct than the worms (annelids). Three main segments are evident: the anterior segment forms

The spider's web is made of silk created from fluid proteins produced in glands and excreted through spinners at the rear of the abdomen. (Digital Stock)

the head; the posterior segment holds digestive structures as the abdomen. Between the head and the abdomen is the thorax, consisting of structures used for movement, such as wings and legs.

The efficient gas exchange in arthropods is accomplished by gills in aquatic forms, and by either tracheas or book lungs in terrestrial forms. Arthropods also have a well-developed open circulatory system, by which blood not only travels through vessels but also bathes the internal organs directly.

A well-developed sensory and nerve system is another feature of arthropods. Most have compound eyes and acute chemical and tactile senses. The nervous system consists of a brain with fused ganglia in the head and a series of ganglia running through the body, coordinating movement and other complex behaviors.

Common Characteristics of Spiders

Spiders, scorpions, mites, ticks, and their relatives are members of the class Arachnida. The arachnids comprise about seventy thousand species of terrestrial arthropods. Animals in this class have eight walking legs; most are carnivorous, and many subsist on a liquid diet of blood or predigested prey. Spiders are the most numerous arachnids, accounting for forty-two thousand species belonging to one hundred different families. The number of spiders can be larger than anyone expected. A study in Great Britain counted spiders in a meadow, coming up with 131 spiders per square meter. Within the area of 36,150 square kilometers that composes the Netherlands, there are approximately five trillion spider inhabitants. Put together, these spiders could consume all fifteen million Dutchmen in merely three days.

Many people confuse spiders with insects. Although many similarities exist between spiders and insects, such as the presence of an exoskeleton, the gas exchange system, and the circulation system, three conspicuous traits can serve to distinguish spiders and insects. First, spiders usually have four pairs of legs compared to insects' three. Second, insects have compound eyes whereas spiders have singular eyes with lenses. Third, insects have antennae while spiders do not.

The body of a spider has two distinct parts: the cephalothorax, consisting of the head and breast, and the opisthosoma, or abdomen. The back of a spider is referred to as its dorsal side and the bottom is its ventral side. The eight legs, two jaws, and two feelers (palps) are connected to the cephalothorax. The males have a bulb at the end of their palps, which is used to store and inject semen into the sexual organs of the female. There are usually eight eyes on the cephalothorax, although the number may vary from none to twelve. An extensive nerve system is made up of a brain located in the cephalothorax, and ganglia (the equivalent of nerves in mammals) that run through various sections of the body. The heart is situated at the front upper side of the abdomen. The silk-making spinners are found at the rear of abdomen. These spinners are linked to glands that produce a variety of proteins, which when mixed polymerize to form silk. As the fluid silk is pressed through the spinners, a thread is made. The reproductive organs are located between the book lungs and the spinners. Running though the whole body is the alimentary canal, at the end of which is the excretory system.

Spider Facts

Classification:
Kingdom: Animalia
Subkingdom: Bilateria
Phylum: Arthropoda
Class: Arachnida
Order: Araneae
Suborders: Mygalomorphae (the primitive spiders); Aranaeomorphae (the modern spiders); Mesothelae, with one family of spiders, the Liphistiidae

Geographical location: Every continent except Antarctica

Habitat: Diverse; mostly on land, inside and outside buildings, on or close to the ground, under stones, logs, litter, low or medium foliage, tall shrubs and trees, under bark; some live in freshwater and very few in salt water

Gestational period: Varies; some female spiders can carry sperm for some time after mating until ready to produce an egg sac

Life span: Many live for up to two years

Special anatomy: Eight walking legs, four on each side of the thorax; eight simple eyes, each with a single lens, that are particularly sensitive to movement; an extensive nerve system; silk-making spinners; muscular jaws equipped to inject poison into prey

Various Body Parts and Their Functions

Most spiders are equipped with poison glands to kill prey. The jaws are used to grab and crunch the prey. A pair of syringelike structures, which are hollow and extremely sharp, are found at the end of the jaws. They are used to puncture the body and inject poison into a prey. The venom is produced in special glands and stored in a special bladder, around which is a spiral muscle. This muscle contracts to eject the poison through the syringe into the victim. The poison, made up of proteins, amines and polypeptides, causes paralysis by disrupting the communication between the nervous system and the muscles. The poison and digestive enzymes cause the death of cells and dissolve the contents of the prey. The spider then sucks the prey empty, leaving a shell behind. Many spiders can give a nasty bite comparable to the stinging of a wasp; a spider sting can even be fatal to children and persons with weak constitutions.

Spiders use a trachea, a slit above the spinners that can be opened and closed, for admission of oxygen. Long small tubes run from this slit into the body. Gases are exchanged with the blood by diffusion. Many spiders also have book lungs, which are hollow, leaflike structures through which the blood flows. Many modern spiders have both tracheas and book lungs. With these two systems together providing extra oxygen, the modern spider has an advantage in having quicker and more sustained reaction times than the primitive counterparts with only book lungs. The circulating blood in a spider's body is colorless and called hemolymph. It transports nutrients, hormones, and cells in addition to oxygen. It is also used locally to raise blood pressure during molting and stretching the legs. Spiders have an

The Black Widow: Small but Deadly

The black widow spider, *Latrodectus mactans*, is found in the United States from northern Massachusetts, south to Florida, and west through Louisiana, Texas, Kansas to California, and also throughout Central America. The body length of the female reaches up to 5 inches, but males are smaller. Leg span for the female is 0.75 inches, with that of males slightly longer. The main diet includes beetles, cockroaches, crickets, flies, scorpions, and spiders.

The black widow possesses one of the most powerful venoms. She earns her name from her gruesome habit of devouring the tiny, harmless male spider after mating with him. The black widow's poison is called a neurotoxin—it attacks the nervous system and blocks the transmission of nerve signals to the muscles, causing convulsions, paralysis, and intense pain. Every summer, the male black widow goes searching for a mate. The female simply waits in her web, hanging upside down as usual. When he finds a female, the male approaches very carefully and signals to the female that he is not prey by tapping out a coded message on the web. As soon as the mating is over, however, he must escape or become an easy meal for the female.

open blood circulation system with the heart located in the back of the abdomen. Blood vessels transport the blood to the heart but thereafter the blood flows freely in the open spaces between organs. The heart is an open tube with valves which is hung in a cavity. Elastic muscles around this cavity contract, enlarging the tubes and forcing blood to flow in only one direction. The size of the heart is closely correlated with the size of the trachea system.

A number of nerves extend from the brain to the legs, eyes, and the rest of the body. The brain occupies about 20 to 30 percent of the cephalothorax volume. Spiders have several sensory organs with which to sense and react to their surroundings. They have simple eyes, each with a single lens, which are particularly sensitive to movement. Spiders have neither ears nor sense of taste. However, they are able to detect smell with scent-sensitive hairs located on their legs. With the brain and all sense organs, spiders are sharp hunters.

With its enormous strength, spider silk is an extraordinary material. A thread of silk the thickness of a pencil has enough strength to stop a Boeing 747 flying at full speed. Humans simply do not yet know how to duplicate such a material. Silk threads are produced by several glands located at the spider's abdomen. Every gland produces a thread for a special purpose: glandula ampulleceae for the silk of the walking thread, glandula pyriformes for the attaching threads, glandula acinoformes for the encapsulation of prey, glandula tubiliformes for the thread of cocoons, and glandula coronatae for the adhesive threads. A thread is made up of polymerized protein molecules. The smallest measured thread was only 0.02 micrometer yet a web made up of it is capable of stopping a bee flying at full speed. The thread is also very elastic and can be stretched 30 percent without breaking. Spider webs take a variety of shapes and function to trap prey, produce cocoons, and provide hiding places for the spiders.

Male spiders are often smaller and more colorful than the females. Males can also be recognized by what appears to be a fifth pair of legs. These are actually palps with bulbs for injecting their sperm into a female during mating. During breeding season, males search for females. Once the female is found, the male has to avoid being mistaken for prey by the female. Male spiders of different species use different ways to announce to the female that they are interested in mating. If the signals are right and the female is ready, mating occurs. After the mating, the males of some species must be extremely careful or they will become an easy meal for the female. The females lay their eggs and tend the young. The kingdom of spiders goes on.

—*Yujia Weng*

See also: Arachnids; Arthropods; Circulatory systems of invertebrates; Food chains and food webs; Molting and shedding.

Bibliography

Emerton, J. H. *The Common Spiders of the United States*. Reprint. New York: Dover, 1961. An extensive review of identified spiders, their distribution, and habitat within the United States.

Foelix, Rainer F. *Biology of Spiders*. 2d ed. New York: Oxford University Press, 1996. Thorough and in-depth examinations of anatomy, physiology, and many other aspects of spiders.

Mascord, Ramon. *Australian Spiders*. Rutland, Vt.: C. E. Tuttle, 1970. Provides useful information on spiders found in Australia.

Roberts, Michael J. *Spiders of Britain and Northern Europe*. New York: HarperCollins, 1995. Descriptions of common spiders inhabiting Great Britain and northern Europe.

Simon-Brunet, Bert. *The Silken Web*. Chatswood, New South Wales, Australia: Reed, 1994. Details the construction and functions of diverse forms of spider webs.

SPONGES

Types of animal science: Classification, ecology, reproduction
Fields of study: Anatomy, zoology

Thousands of sponge species occur, most in the oceans. They are sexual hermaphrodites, and also reproduce asexually. Sponges are useful commercially.

Principal Terms

FLAGELLUM: a long cell extension used in locomotion

GEMMULE: an asexual reproductive structure that becomes a new sponge

HERMAPHRODITE: an organism having male and female reproductive organs

OSCULUM: an opening through which a sponge ejects water

OSTIUM: a surface pore through which water enters a sponge

SPICULE: a needlelike structure that is part of a sponge skeleton

SPONGIN: a fibrous skeletal material in soft sponges

Sponges make up the phylum Porifera, the simplest multicellular animals. There are thousands of sponge species. Most inhabit oceans, although freshwater species exist. Each saltwater sponge has a stem that attaches it to a rock or other object on the ocean floor. Sponge remains are found in the oldest, fossil-bearing rocks.

Live sponges can be black, brown, gray, red, purple, or green. They are abundant in earth's oceans, from shallows to huge depths and from the equator to the Arctic. However, sponges are most numerous and varied in tropical to warm temperate habitats. The four sponge groups are the marine Calcarea, with calcium carbonate skeletons; deep sea Hexactinellida (glass sponges), with silica skeletons; marine and freshwater Demospongiae, comprising 95 percent of species, with skeletons made of flexible spongin (as in a bath sponge) and/or silica; and Sclerospongiae, with silica-, spongin- and calcium-containing skeletons.

The Physical Nature of Sponges

Live sponges have outer layers of cells, which provide their color, and inner-layer flagellate cells

Sponges are the simplest type of multicellular organism. (Digital Stock)

Bath Sponge Facts

Classification:
Kingdom: Animalia
Phylum: Porifera
Class: Demospongiae
Family: Spongidae
Genus and species: Spongia adriatica
Geographical location: Tropical and subtropical Adriatic Sea, the Mediterranean, the Gulf of Mexico, and the Caribbean
Habitat: Oceans and seas
Gestational period: Sexually, eggs are fertilized by sperm from a nearby sponge, and resultant larvae become sponges; asexually, pieces of sponge break off, settle down, and grow (regenerate), or buds or gemmules form new bath sponges
Life span: Indefinite and dependent on environment
Special anatomy: As sponges are the most basic form of animal life, there is little anatomy at all

that move water. The simplest sponge is a tube with many pores (ostia) on its surface. Water enters the tube, via ostia, in a current due to flagella attached to inner-layer cells. Flagellate cells absorb oxygen and digest tiny sea creatures. Then water is expelled through an opening, the osculum, atop the tube. Ejection, due to pressure from flagellar movement, moves depleted water away from the sponge, preventing its reuse.

Sponges form groups if a sponge develops young that remain connected to it. As more and more young develop, their body cavities become canal networks. Water then enters via ostia and passes through canals to chambers where flagellate cells absorb food and oxygen. Used water leaves by larger and larger tubes, reentering oceans via an osculum.

Between outer and flagellate cells a sponge has a skeleton made of structures called spicules. When a sponge dies, its flesh decays and the skeleton remains. There are three sponge skeleton types. Calcarea sponge spicules are made of lime. Hexactinellida (glass) sponges have glassy, silica spicules. Some glassy spicules form attractive skeletons, such as Venus' flower basket. In Demospongiae (including freshwater sponges) the skeleton is almost entirely spongin. Spongin skeletons may contain minute spicules of lime, silica, or both. Bath sponge skeletons have no spicules.

Many sponges begin life as fertilized eggs, which divide until becoming free-swimming larvae. Flagella transport larvae until they settle on the ocean bottom and attach to rocks and other objects in order to become adults. Sponge reproduction can also be asexual, via buds or gemmules. Sponges have a great ability to regenerate in order to replace lost body parts or even most of the body. Some sponges, treated so all their cells are separated but left in water, form a new sponge.

Commercial Sponges

Some Demospongiae have soft, elastic, spongin skeletons that absorb large amounts of water. These qualities have long made them useful tools for surgery, military gun-cleaning, and the clean-

Venus' Flower Basket

Venus' flower basket (VFB) sponges inhabit waters up to three thousand feet deep in warm Pacific Ocean regions. They are plentiful off China, Japan, and the Philippines. VFB, a glass (hexactinellid) sponge, has a skeleton of glassy spicules, interconnected into a one-foot-high, three-inch-wide basket or vase, seemingly made from glass threads.

VFB reproduces sexually or asexually. Sexual reproduction by this hermaphrodite uses cells that make sperm or eggs at different times. A VFB may release sperm into the water while a nearby VFB produces eggs. Conversely, it may provide the eggs. Regardless, sperm fertilizes the eggs in a "female" VFB, and larvae become new VFB. In asexual reproduction, VFB pieces break off, settle down, and regenerate a new VFB; sponge buds form on the VFB surface, break off, and become sponges; gemmules may also be released.

ing of automobiles, houses, and bodies. The best such sponges come from the eastern Mediterranean, off the Syrian and Greek coasts. Sponges are also fished for off Tampa Bay, Tarpon Springs, the Florida Keys, and the Bahamas.

In deep waters, suited sponge divers descend into the sea to dredge sponges. In shallow waters, off Florida, glass-bottomed boats from a mother ship are used. A pole ending in a pronged hook loosens sponges sighted and brings them to the surface. On return to the mother ship, sponges are spread on deck until their flesh decays, hung to dry in the rigging, or kept in seaside pens which tides fill and empty, removing sponge tissues and leaving skeletons.

—Sanford S. Singer

See also: Asexual reproduction; Fossils; Marine animals; Marine biology; Reproduction.

Bibliography

Bergquist, Patricia R. *Sponges*. Berkeley: University of California Press, 1978. Describes and classifies sponges, with a solid bibliography.

Esbensen, Barbara Justis. *Sponges Are Skeletons*. New York: HarperCollins, 1993. Explains sponge life and their harvesting and use by humans.

Hartman, Willard D., Jobst W. Wendt, and Felix Weidenmayer. *Living and Fossil Sponges: Notes for a Short Course*. Miami: University of Miami, 1980. Describes many aspects of fossil and contemporary sponges.

Jacobson, Morris K. *Wonders of Sponges*. New York: Dodd, Mead, 1976. Surveys sponge classification, life, natural history, and collection.

Van Soest, R. W. M., B. Picton, and C. Morrow. *Sponges of the North East Atlantic*. New York: Springer Verlag, 2000. An interactive CD-ROM covering the 337 species of sponges found around the northern Atlantic rim. Over 1,500 photographs and drawings, glossary, and literature, distribution, and higher taxa modules.

Wiedenmayer, Felix. *Shallow-Water Sponges of the Western Bahamas*. Basel, Switzerland: Birkhäuser, 1977. This book classifies and describes sponges found in the waters off the Bahamas.

SQUIRRELS

Types of animal science: Classification, ecology, zoology
Fields of study: Ecology, systematics (taxonomy), wildlife ecology, zoology

Squirrels are mammals belonging to the family Sciuridae. Consisting of 272 species, this family represents only about 5 percent of mammal species. Nevertheless, squirrels are very familiar mammals to most people.

Principal Terms

GROUND SQUIRRELS: those members of the family Sciuridae who burrow in the ground

OPPORTUNISTIC OMNIVORE: an animal who includes a variety of plant and animal material in its diet, depending on the availability of different foodstuffs

POLYGYNOUS: mating system in which one male mates with several females during a breeding season

SCIURIDAE: rodent family to which all squirrels belong

TREE SQUIRRELS: those members of the family Sciuridae who nest in trees

Most people are familiar with squirrels, but may not be aware that members of the squirrel family (Sciuridae) can be divided into two basic groups: tree squirrels and ground squirrels. Tree squirrels are arboreal. Not only do they nest in trees, but they also often mate and forage in trees. Ground squirrels are primarily terrestrial. Though some may climb several feet up the trunk of a tree, ground squirrels nest in burrows beneath the soil surface and forage and mate on the ground or in their burrows. Another big difference between ground-dwelling and tree-dwelling squirrels is that ground squirrels hibernate during the winter and tree squirrels do not. Typical ground squirrels include marmots, woodchucks, and chipmunks. Typical tree squirrels include fox and gray squirrels in North America and red squirrels in Europe.

Gray Squirrels

One of the most familiar tree squirrels in North America is the gray squirrel, *Sciurus carolinensis*. Gray squirrels live in deciduous forests and are also abundant in parks and yards in eastern North America. Their common name is somewhat misleading, in that some gray squirrels have a black pelage. There are even a few populations of albino gray squirrels in North America. Nevertheless, most gray squirrels have gray backs with light-colored ventral fur and light to white fur on the backs of the ears. Like other tree squirrels, gray squirrels have large, bushy tails almost equal in length to the squirrel's body. Though males are generally larger than females, there is no pronounced sexual dimorphism. Gray squirrels range in size from 330 to 750 grams.

Gray squirrels are not particularly social. That is, they do not form cooperative groups. Rather, gray squirrels are solitary. They do not defend territories, and home ranges of individual gray squirrels overlap widely. However, they may defend core areas in the autumn to ensure access to food.

Gray squirrels undergo one or two breeding seasons each year, depending on latitude. Squirrels in northern latitudes may only breed once a year, though squirrels in more moderate climates breed up to two times per year. Timing of the breeding season thus also varies with latitude. In the northern United States, gray squirrels begin to

Gray squirrels spend most of the autumn collecting and hoarding nuts to feed on throughout the winter. Unlike ground squirrels, gray squirrels do not hibernate. (Digital Stock)

breed in early to mid February. Gray squirrels are polygynous; one male may mate with several females during a single breeding season. Males compete with each other for access to females; several males may chase a female until one has an opportunity to mate with her. Some males, rather than participate in mating chases, wait behind trees and find opportunities to mate with females during times other than the mating chase. Mating takes place on the ground or in the trees, and thus can be dangerous to both participants, as there is a real risk of falling during mating.

Gestation in gray squirrels lasts approximately forty-four days. Litter sizes range from one to six, with an average of two to three. Gray squirrels are born in a relatively helpless state; they are born hairless and their eyes do not open until about twenty-five days after birth. Gray squirrels nurse for eight to nine weeks, after which time they are weaned. Some litters are at this point abandoned by their mother. Young gray squirrels can fend for themselves at about eighty days of age and begin to build their own leaf nests at about eighteen weeks of age. Sexual maturity is reached at ten months of age. Gray squirrels have an average life span of only eleven to twelve months. However, many individuals live longer than this, even up to ten years in the wild (longer in captivity).

Gray Squirrel Lifestyle

Gray squirrels might best be considered opportunistic omnivores. Commonly known to consume nuts and seeds as well as buds and fruits from hardwood trees, gray squirrels have also been known to consume baby birds, insects, and fungi. Nevertheless, during the autumn and winter, gray squirrels depend almost exclusively on the mast crop from hardwood trees as their food source. Beginning in late summer to early autumn, gray squirrels begin to scatterhoard nuts. Scatterhoarding entails burying

Squirrel Facts

Classification:
Kingdom: Animalia
Phylum: Chordata
Subphylum: Vertebrata
Class: Mammalia
Subclass: Theria
Infraclass: Eutheria
Order: Rodentia
Suborder: Sciurognathi
Family: Sciuridae (squirrels), with forty-nine genera and 267 species

Geographical location: Worldwide, except Australia, Polynesia, and southern South America
Habitat: All habitats except deserts
Gestational period: Forty days
Life span: Ten years in the wild, sixteen years in captivity
Special anatomy: Large bushy tail used for balance and temperature regulation

single nuts in different places around the squirrel's home range and differs from larderhoarding in that gray squirrels do not cache large piles of nuts together in a single location. Memory and an excellent sense of smell allow the squirrels to find buried nuts later in the winter, even when buried under several inches of snow. Gray squirrels can be very selective about which foods to include in their diet. They are known to avoid nuts produced from the red subgenus of oaks, which tend to be high in tannin. When eating acorns from red oaks, squirrels generally consume those parts of the seed that are low in tannin.

Unlike ground squirrels, gray squirrels do not hibernate during the winter. Thus, even in the worst weather, they must leave the safety of their nests to obtain food. During the winter months, gray squirrels will den together in tree cavities, presumably to conserve heat. Females usually den with other females (often in mother-daughter groups) and males usually den with other males. Dens are lined with leaves as insulation. Squirrels also use a variety of anthropogenically produced materials in den construction. Foil-coated fast-food wrappers and laundry lint are not uncommon discoveries in squirrel nests. During warmer months, gray squirrels build leaf nests, called dreys, in the upper branches of hardwood trees.

—Erika L. Barthelmess

See also: Beavers; Gophers; Hibernation; Mammals; Mice and rats; Omnivores; Porcupines; Rodents.

Bibliography

Koprowski, John L. "Natal Philopatry, Communal Nesting, and Kinship in Fox Squirrels and Gray Squirrels." *Journal of Mammalogy* 77 (November, 1996): 1006-1016. Primary research article that examines nesting patterns as a function of relatedness in fox and gray squirrels.

Nowak, Ronald M. *Walker's Mammals of the World.* Vol. 2. 6th ed. Baltimore: The Johns Hopkins University Press, 1999. Exhaustive source of information on mammals in general and squirrels in particular. Summary articles about every species of mammal in the world. Somewhat technical.

Steele, Michael A., et al. "Tannins and Partial Consumption of Acorns: Implications for Dispersal of Oaks by Seed Predators." *The American Midland Naturalist* 130 (October, 1993): 229-238. Primary research article with information about herbivore treatment of acorns containing tannin.

Wilson, Don E., and Sue Ruff, eds. *The Smithsonian Book of North American Mammals.* Washington, D.C.: Smithsonian Institution Press, 1999. Detailed book with descriptions of each species of mammal inhabiting North America. Beautiful pictures included. Highly readable.

STARFISH

Type of animal science: Classification
Fields of study: Anatomy, invertebrate biology, zoology

Starfish are five orders (1,500 species) of marine invertebrate animals making up the echinoderm class Asteroidea. Their main foods include mollusks and coral.

Principal Terms

BIVALVE: a mollusk having two shell halves
CARNIVORE: an animal that eats only other animals
INVERTEBRATE: an animal lacking an internal skeleton

Starfish are carnivorous ocean animals that usually have five starlike arms. They are related to brittle stars, sea urchins, and sea cucumbers, all of which are known as echinoderms. Like these animals, starfish have spiny skins. Starfish make up five orders—1,500 species—of marine invertebrates, the class Asteroidea of the phylum Echinodermata. They have radially arranged arms which hold locomotor tube feet, and which reach diameters from six inches to four feet. They are abundant at all ocean depths and occur in all oceans except near the North and South Poles. Often, a starfish escapes its enemies by breaking off one or more of its arms, which regenerates. In fact, if a starfish is halved, quartered, or cut into smaller pieces, each becomes a new individual.

The starfish's arms are equipped with suction-cup feet, which it uses to pull itself along the ocean floor. (Digital Stock)

The Physical Characteristics of Starfish

Stiff-bodied, starfish move by crawling slowly. Their bodies may have five arms or many more. Starfish skin is leathery and has protective spines, which point upward. The spines are made of lime and develop from the skeleton. The animal also has a large gut, a complex system of body cavities, and a nervous system but no brain.

The central body of a starfish is located where its arms join. Its bottom surface holds a mouth at the point where all the arms meet. A groove also stretches from the mouth to each arm tip. There are rows of holes in the grooves from which tube feet can push out. The feet, with sucker ends, enable crawling. They are supported and moved by an internal hydraulic system inflated with seawater.

Most of the central body is a baglike stomach into which the mouth opens. Starfish also have well-developed senses of touch, smell, and taste, and respond to light. However, they lack the complex behavior patterns of animals having brains.

Starfish Facts

Classification:
Kingdom: Animalia
Phylum: Echinodermata
Class: Asteroidea
Orders: Include Forcipulata, Forcipulatida, and Spinulosida

Geographical location: All ocean depths in all oceans, except near the North and South Poles

Habitat: Ocean bottoms, from shallow water to great depths

Gestational period: No true gestation; sperms and eggs are spawned into oceans, where the eggs are fertilized; larvae drift, settle to the ocean floor, and become adults

Life span: Not known

Special anatomy: Radially symmetrical arms, locomotor tube feet, mouth and stomach cavity that can turn inside out, protective spines

The Life of the Starfish

Starfish spend much of their lives searching for food, mainly clams, mussels, and oysters. To eat bivalve mollusks, a starfish uses its tube feet to open shell halves. It surrounds a mollusk, attaches the tube feet, and uses them to pull in opposite directions. Once the mollusk tires of resisting the starfish's force, the shell opens. Then the starfish pushes its stomach inside out through the mouth and surrounds the mollusk body. Its stomach releases digestive fluid and as the mollusk softens, the starfish eats its flesh and releases the shell. A similar technique is used with other foods, such as coral. Coral polyps (individual corals) are eaten. First, a starfish climbs onto a polyp and presses its stomach out through its mouth. Then, digestive juice softens the polyp's shell and turns the polyp into a soupy liquid, which the starfish eats.

Starfish are themselves eaten by fish and snails. Their most successful predators are giant carnivorous sea snails, such as tritons. A triton rips starfish open and eats their soft tissues. Fish also eat starfish by flipping them over and biting off and eating their soft centers.

Mating occurs in spring or summer, depending on species and habitat. Most starfish have two sexes and mate by secreting sperm and eggs into the ocean (spawning). External fertilization follows and the eggs become larvae. Larvae float in the ocean and sink to its bottom to become adults. Most starfish regenerate body parts. In some starfish, this is also their mode of reproduction: A bud grows and pinches off into a new starfish.

Three Sample Starfish Species

Sunflower starfish (*Pycnopodia helianthodes*), the largest starfish, resemble sunflowers because of their many arms. They grow to diameters up to four feet and may have twenty-five arms. These starfish inhabit the west coast of North America from Alaska to California, from shallows to deep water. They eat bivalves, sponges, coral, worms, crustaceans, other starfish, and small fish. They mate in spring, by spawning.

Crown-of-thorns starfish (*Acanthaster planci*) are star-shaped with many body spikes (thorns). They grow to two-foot diameters and have up to twenty-four arms. A crown-of-thorns starfish also has a large, round midsection, which holds internal organs and many tube feet. Crown-of-thorns starfish inhabit tropical West Pacific and Indian Oceans. Feeding on coral, they live on or near coral reefs, such as Australia's Great Barrier Reef, where they hide during the day and feed at night. A single crown-of-thorns starfish can destroy 1.5 feet of coral reef per week.

European starfish (*Asterias rubens*) are five-armed starfish of the European and African coasts. They grow to 1.5-foot diameters and have tube feet. They feed on bivalves, sponges, corals, worms, crustaceans, other starfish, and small fish.

They find food via chemical signals picked up by tube feet. Their spring mating is by means of spawning.

The main foods of starfish are mussels, oysters, and clams. Starfish are thus serious pests on oyster and clam farms. In addition, they eat coral and can damage reefs. For example, crown-of-thorns starfish sometimes overgrow and damage coral reefs by eating too much coral.

—Sanford S. Singer

See also: Echinoderms; Exoskeletons; Marine animals; Marine biology; Regeneration.

Bibliography

Birkeland, Charles, and John S. Lucas. *"Acanthaster planci": Major Management Problem of Coral Reefs*. Boca Raton, Fla.: CRC Press, 1990. A compilation of information about the crown-of-thorns starfish, with an eye to controlling its predation of coral reefs, emphasizing the starfish's unique morphology, physiology, and life cycle.

Clark, Ailsa McGowan. *Starfishes and Related Echinoderms*. 3d ed. London: Trustees of British Museum (Natural History), 1977. The book contains a good amount of data on starfish and echinoderms.

Hendler, Gordon, John E. Miller, David L. Pawson, and Porter M. Kier. *Sea Stars, Sea Urchins, and Allies: Echinoderms of Florida and the Caribbean*. Washington, D.C.: Smithsonian Institution Press, 1995. Focusing on the echinoderms of the Carribean, a gorgeously photographed guide to starfish of the region and their relatives.

Hurd, Edith Thacher. *Starfish*. New York: HarperCollins, 2000. The book is a simple introduction to starfish appearance, growth, and habits.

STEGOSAURS

Types of animal science: Classification, evolution
Fields of study: Anatomy, evolutionary science, paleontology, systematics (taxonomy)

Stegosaurs were a group of armored, herbivorous dinosaurs characterized by large plates set vertically along the back and spikes on the tail. The plates are thought to have acted as thermoregulatory devices.

Principal Terms

CRETACEOUS: a period of time that lasted from about 146 to 65 million years ago, the end of which was marked by the extinction of the dinosaurs

ECTOTHERMY: a form of metabolism in which internal temperature is regulated by ambient temperature

JURASSIC: a period of geological time that lasted from about 208 to 146 million years ago

ORNITHISCHIA: one of the two main dinosaur groups, characterized by a pelvis in which the pubis is swung backward

PUBIS: one of the three bones that make up the pelvis (the others are the ischium and ilium)

Stegosaurs are a group of quite large (up to eight meters long), quadrupedal, ornithischian dinosaurs. That is, they have a pelvis in which the pubis points backward, and thus they are allied with all other herbivorous dinosaurs except the gigantic sauropods. Their closest relatives are the other armored dinosaurs, the ankylosaurs, with which they are grouped as Thyreophora ("shield bearers"), characterized by rows of plates along the back and sides of the body. The earliest stegosaurs are represented by fragmentary remains from the Middle Jurassic of England, but by the Late Jurassic they are well known from complete skeletons from Africa, Asia, Europe, and North America. The Late Jurassic was the most successful time for stegosaurs and they are particularly well known from articulated skeletons of *Stegosaurus* of this age from North America. By the Early Cretaceous their distribution had contracted to Europe, Africa, and China, and by the Late Cretaceous they are absent from all continents except India. India was separate from all the other continents from the Middle Jurassic onward as it drifted away from Africa and toward Asia, and this isolation may have protected the stegosaurs as they became extinct everywhere else.

Anatomy and Lifestyle

The best known stegosaur is *Stegosaurus* ("roofed reptile"), from the Late Jurassic of North America,

Stegosaur Facts

Classification:
Kingdom: Animalia
Subkingdom: Bilateria
Phylum: Chordata
Subphylum: Vertebrata
Class: Reptilia
Order: Ornithischia
Suborder: Thyreophora
Geographical location: North America, Europe, China, India, and Africa
Habitat: Terrestrial habitats
Gestational period: Unknown
Life span: Unknown
Special anatomy: Large, quadrupedal dinosaurs in which the head was small, the back was covered by a double row of large vertical plates, and the tail bore large spikes

Current theory suggests that the plates arranged along the backs of stegosaurs served some kind of thermoregulatory function; blood could be drawn up into the plates to cool, or to be warmed by exposure to the sun. (©John Sibbick)

and thus it is commonly used to typify the entire group. *Stegosaurus* had a small head with an unusually small brain, even for a dinosaur, and a narrow, horn-covered beak at the tip of the snout. The sides of the jaws were lined with leaf-shaped serrated teeth but these were not arranged in batteries and there is no evidence of the sophisticated grinding apparatus that was developed in some other herbivorous dinosaurs. However, *Stegosaurus* was a large animal and must have needed large quantities of plant food, so it is likely that it was a low-level browser that chopped up vegetation and then quickly passed it back to a large stomach, where it would have fermented, perhaps with the help of gastroliths (stomach stones) to further break it down.

The most distinctive feature of *Stegosaurus* is the row of vertical plates along the back. There has been some disagreement as to their relationship to each other, as they were set in soft tissue and not attached to the skeleton; however, it is generally agreed that they formed two parallel rows in which the plates were staggered. The plates are not optimally positioned for defense and it has been suggested that they were devices for thermoregulation, acting as radiators to gain or lose heat and help maintain a constant internal body temperature in animals that may have had an ectothermic metabolism (an internal temperature regulated by the ambient temperature, as in most modern reptiles). Experiments have shown that the plates were ideally shaped and positioned to do this. The surface of the plates is covered by fine grooves, indicating the presence of numerous blood vessels, and the plates themselves are hollow, implying that they were richly supplied with blood. Thus the animal could have flushed blood over the plates to cool itself if its internal temperature had risen too much, or to warm the blood by exposure to the sun if its internal temperature was dropping.

The proportions of the legs in *Stegosaurus* show that it was not a fast runner, and thus as a large and relatively slow-moving herbivore it would have needed some means of defense against predators. This would undoubtedly have been the spike-bearing tail. Two pairs of spines projected laterally from the tip of the tail and could have inflicted severe injuries on an attacking carnivorous dinosaur as the tail was lashed from side to side while the animal backed toward the attacker.

—*David K. Elliott*

See also: *Allosaurus*; *Apatosaurus*; *Archaeopteryx*; Dinosaurs; Evolution: Animal life; Extinction; Fossils; Hadrosaurs; Ichthyosaurs; Paleoecology; Paleontology; Prehistoric animals; Pterosaurs; Sauropods; *Triceratops*; *Tyrannosaurus*; Velociraptors.

Bibliography

Alexander, R. McNeill. *Dynamics of Dinosaurs and Other Extinct Giants*. New York: Columbia University Press, 1989. Informative coverage of how large animals such as stegosaurs operated including discussion of tail mechanics and thermoregulation.

Bakker, Robert T. *The Dinosaur Heresies*. New York: William Morrow, 1986. An idiosyncratic look at the dinosaurs with extensive discussion of defense and thermoregulation in stegosaurs.

Benton, Michael J. *Vertebrate Palaeontology*. 2d ed. London: Chapman and Hall, 1997. General vertebrate paleontology text that devotes one chapter to the stegosaurs and their relatives.

Currie, Philip J., and Kevin Padian. *Encyclopedia of Dinosaurs*. San Diego, Calif.: Academic Press, 1997. Excellent coverage of all aspects of dinosaur biology.

Norman, David. *The Illustrated Encyclopedia of Dinosaurs*. New York: Crescent Books, 1985. Although old, this book has wonderful illustrations and an excellent text with extensive coverage of stegosaurs.

STORKS

Type of animal science: Classification
Fields of study: Anatomy, conservation biology, physiology, population biology, wildlife ecology

Storks are wetland birds that are physically striking in appearance and noted for their spectacular flight. They are found throughout the world, but their numbers have decreased dramatically due to habitat destruction by humans.

Principal Terms

FLEDGLING PERIOD: period after hatching, during which a nestling grows flight feathers and learns to fly
INDICATOR SPECIES: a species monitored by biologists as a means of ascertaining the health of the ecosystem in which it lives
NOMADIC: moving about from place to place according to the state of the habitat and food supply
SCAVENGER: an animal that feeds on the dead carcasses of other animals
THERMALS: rising currents of warm air

Storks are large wading birds that live near swamps, marshes, lakes, and rivers. There are nineteen species in the world. North America is home to two species, the wood stork of the southeastern United States and the jabiru of Mexico. Many species live in Africa, Asia, and Europe, and a few species also can be found in South America and Australia. The white stork is most commonly known. It summers in Europe, frequently nesting on rooftops, and is the source of the legends of storks delivering infants.

Physical Characteristics of Storks

Storks stand from 2.5 to 5 feet tall and weigh between five and thirteen pounds. The smallest species is the Abdim's stork. The saddlebill and the marabou stork are the largest. Storks have powerful wings and a wide wingspan. The marabou's wingspan is the greatest of all land birds, reaching nearly ten feet. Plumage can be white, gray, or black. Species with dark plumage often have iridescent shades of purple, green, or blue. Males and females look alike, but the male is larger.

Storks have long, slender legs that may be red, white, gray, or black. They have three long toes that are webbed at the base, and a shorter back toe. The strong beak is long and pointed and is straight or slightly curved.

Stork Behavior

Storks are carnivores and hunt for food in wetlands, along bodies of water, and in grassy plains. They feed on insects, worms, fish, frogs, reptiles, and small birds and mammals. The adjutant storks are scavengers, primarily seeking carrion. Many storks are tactile foragers, walking slowly through shallow water, groping with an open bill that snaps shut reflexively when it comes into contact with prey. Others are visual foragers, snatching insects or stirring the water to disturb prey, then seizing it with a thrust of the beak.

Storks do not vocalize. Instead, they communicate with bill clattering and snapping, rasping, and hissing. Nestlings make a high-pitched braying which decreases with age.

Storks fly with their necks stretched out. The legs dangle during takeoff, and then extend out behind in flight. Storks glide on thermals, soaring to high altitudes. They sometimes perform aerial acrobatics such as diving and flipping.

Some stork species migrate seasonally, others are nomadic, and some are nonmigratory. The availability of prey largely determines the migration and movement of stork colonies. Prey availability is in turn closely tied to climate and rainfall patterns. Storks will adapt the timing of migration and breeding and the choice of colony nesting site according to these climatic conditions, postponing or skipping a breeding cycle if food availability is low.

Excepting one or two species, storks live in colonies, building their nests of sticks, reeds, and vines in trees, on rooftops, or on the ground. Some mate for life; others may pair for just one season. Courtship begins with a color change of the birds' bill, face, legs, and skin. The male performs various aggressive, ritualized displays to attract the female, who then performs appeasement displays. Both members of the pair participate in nest building, incubation, and caring for nestlings. Clutch size is two to five eggs, and incubation averages thirty days. Parents feed the young by regurgitating food into the nest. The fledgling period varies from six weeks to four months.

Stork Facts

Classification:
Kingdom: Animalia
Phylum: Chordata
Subphylum: Vertebrata
Class: Aves
Order: Ciconiiformes (wading birds)
Suborder: Ciconiae (storks, ibises, spoonbills)
Family: Ciconiidae (storks)
Genus and species: Ten genera, one subgenus, and twenty-one species, including *Mycteria americana* (American wood stork), *M. cinerea* (milky stork), *M. ibis* (yellow-billed stork), *M. leucocephala* (painted stork); *Anastomus oscitans* (Asian open-bill stork); *A. lamelligerus* (African open-bill stork); *Ciconia nigra* (black stork), *C. abdimii* (Abdim's stork), *C. episcopus* (woolly-necked stork), *C stormi* (Storm's stork), *C. maguari* (Maguari stork), *C. ciconia* (white stork), *C. boyciana* (oriental white stork); *Ephippiorrhyncus asiaticus* (black-necked stork), *E. senegalensis* (saddle-bill stork); *Jabiru mycteria* (jabiru stork); *Leptoptilos javanicus* (lesser adjutant stork), *L. dubius* (greater adjutant stork), *L. crumeniferus* (marabou stork)
Geographical location: Every continent except Antarctica; most species are found in the Eastern Hemisphere
Habitat: Wetlands; some species require distinct wet and dry seasons
Gestational period: Incubation averages thirty days
Life span: Seven to thirty years, varying by species; longer in captivity
Special anatomy: Long, slender legs; long neck; long, pointed bill

Marabou storks, like vultures, have naked heads and necks, instead of feathers that would be spattered with gore from the flesh they eat. (Robert Maust/Photo Agora)

Conservation

Since 1950, the overall population of storks has declined by half, due to the destruction of wetland habitats from deforestation, drainage, farming and pesticide use. In 2000, the Storm's stork, greater adjutant, and oriental wood stork were listed as endangered by the International Union for Conservation of Nature and Natural Resources. The lesser adjutant, milky stork, black-necked stork, painted stork, and wood stork are considered vulnerable or near threatened. In the United States, the wood stork serves as an indicator species for the health of the Florida Everglades ecosystem. Solutions to the declining world stork populations include habitat improvement and protection, captive breeding and reintroduction of species, and artificial establishment of nesting and colony sites.

—Barbara C. Beattie

See also: Beaks and bills; Birds; Domestication; Feathers; Flight; Migration; Molting and shedding; Nesting; Respiration in birds; Wildlife management; Wings.

Bibliography

Garcia, Eulalia. *Storks: Majestic Migrators*. Milwaukee: Gareth Stevens, 1996. A discussion of the physical characteristics and behavior of storks for the juvenile audience.

Hancock, James. *Birds of the Wetlands*. San Diego, Calif.: Academic Press, 1999. Organized geographically and beautifully illustrated, this book describes the status of the world's wetland habitats and their resident bird species.

Hancock, James, James A. Kushlan, and M. Philip Kahl. *Storks, Ibises, and Spoonbills of the World*. San Diego, Calif.: Academic Press, 1992. This authoritative illustrated reference includes detailed descriptions of each species, including identification, distribution, behavior, breeding, and conservation.

Klinkenberg, Jeff. "Coming Back on Its Own Terms." *National Wildlife* 36, no. 3 (April/May, 1998): 52-59. This article discusses the environmental difficulties faced by the wood stork in the Florida Everglades, its adaptation, and future.

SWAMPS AND MARSHES

Types of animal science: Ecology, geography
Fields of study: Conservation biology, ecology, environmental science, limnology, wildlife ecology

Swamps and marshes are shallow wetlands that are characterized by emergent vegetation and seasonal water level fluctuations. Swamps and marshes are among the most productive ecosystems on earth and often provide prime habitat for birds. Because swamps and marshes are often perceived as worthless, they have been extensively drained and their loss is of great concern to conservationists.

Principal Terms

CONSUMER FOOD CHAIN: a simplified description of the grazing and predator/prey relationships within an ecosystem

EMERGENT VEGETATION: aquatic vegetation that grows tall enough to be visible above the water

INVERTEBRATE: a simple animal lacking a backbone

PRIMARY CONSUMER: an organism that gets its nourishment from eating primary producers, which are mostly green plants and algae

PRIMARY PRODUCTIVITY: production of biomass mainly by green plants

Swamps and marshes differ from ponds and lakes in two primary ways: they are shallower, with seasonally fluctuating water levels, and they have little or no open water, being dominated by emergent vegetation. The predominant vegetation in marshes comprises grasses, sedges, and rushes. Marshes tend to have shallower water than swamps, and are more apt to dry out completely during the drier part of the year. The predominant vegetation in swamps comprises trees and shrubs. Not only is the water generally deeper in swamps, it is often more permanent, although water levels still tend to fluctuate seasonally.

Although swamps and marshes are usually thought of as freshwater wetlands, saline swamps and marshes also exist, either in desertlike environments where evaporation rates are high, or in coastal areas. Coastal swamps and marshes tend to fluctuate in depth and salinity as the tides change. Coastal swamps are often referred to as mangrove swamps or mangals. Coastal marshes are often called estuaries. Most of the following information will focus on freshwater swamps and marshes.

Insects and Other Invertebrates

The smallest of the invertebrates are primary consumers, who are at the base of the consumer food chain in wetlands. When algae is abundant in the water, primary consumers also proliferate. One of the most common of these is the water flea (*Daphnia*), which is just barely visible to the naked eye. When viewed under the microscope, the reason for its name becomes apparent, as it looks remarkably like a flea. Many species of water flea exist, varying greatly in size, head shape, swimming appendages, and other traits. Water fleas and other small invertebrates are preyed upon by larger invertebrates, tadpoles, and small fish.

The bulk of the invertebrates are insects, in either their adult or juvenile forms. Many flying insects, such as mosquitoes and dragonflies, spend their early life in the water. Dragonfly nymphs are especially vicious predators, preying upon almost anything small enough for them to grab in their strong jaws, including small fish. Adult dragonflies and their cousins, the damselflies, are commonly seen flying around marshes and swamps.

Other insects, such as water striders, backswimmers, water boatmen, and diving beetles, are also abundant in the still waters, making their entire lives in or on the water. Even a few spiders (which are arachnids, not insects) have adapted to the aquatic way of life, able to stay underwater for extended periods by trapping bubbles of air next to their bodies.

Another common invertebrate group is the shellfish. Freshwater clams live buried in the sediment and filter food out of the water, while mussels can form dense assemblages on rocks and other debris in the water. A few clams and mussels from tropical and semitropical parts of the world have inadvertently been introduced to some temperate wetlands with devastating effects. Snails can also be found in many marshes and swamps, where they feed on algae growing on rocks and on the submerged stems and leaves of plants.

Fish, Reptiles, and Amphibians

The occurrence of fish is most often associated with water depth. In shallow, seasonal marshes or swamps, fish are often absent. In deeper marshes and swamps, they can be abundant and provide food for other animals, especially birds. These fish can range from the unique, bottom-dwelling catfish to more active fish such as perch and bass. Many smaller, less noticeable species also occur, some of which are near extinction due to loss of unique habitat.

The most common reptile in marshes and swamps is the turtle. Most turtles are predators, although some feed primarily on plant material. The snapping turtle has a reputation for eating almost anything, plant or animal. Although turtles can be observed swimming in the water, they are most often seen sunning themselves on warm rocks just in reach of the water, where they can quickly escape from potential predators. Among the more dangerous reptiles of swamps and marshes are the alligators, crocodiles, and caimans. These are exclusively predatory, and although generally not very aggressive toward humans, they do attack on occasion. Snakes are also predators. Garter snakes grab their prey us-

Swamps and marshes provide habitat for a wide variety of wetlands species, such as this sandhill crane. (AP/Wide World Photos)

ing their mouth and gradually swallow it. The cottonmouth is a poisonous viper that injects venom into its prey to incapacitate it before swallowing. Although snakes are more often thought of as terrestrial, they are expert swimmers.

Amphibians include salamanders and frogs, both of which spend their early life entirely in the water, where their eggs are laid. Frog eggs develop into tadpoles with tails and no legs; as they develop, the tail disappears and legs start to grow. Salamander larvae have legs from the beginning and they never lose their tail. When young, frogs and salamanders feed mostly on algae and small invertebrates in the water. As adults, frogs typically eat insects, and salamanders eat a variety of invertebrates, including worms and grubs.

Birds and Mammals

Near the top of the food chain, a large variety of birds either live in swamps and marshes or come there to hunt. Common residents include ducks, geese, herons, and egrets. In these shallower waters, "puddling" ducks predominate, characterized by feeding behavior where they reach for vegetation on the bottom with only their rumps protruding from the surface. Herons and egrets frequently wade slowly and deliberately, searching for frogs or fish to quickly grab and eat. Many songbirds take advantage of the habitat; some, such as the marsh wren, even make their nests in the bullrushes near the margin of the water. Predatory birds such as hawks, eagles, falcons, owls, and osprey come to prey on other birds, snakes, fish, or smaller mammals.

Various mammals also take advantage of the aquatic bounty. Mink and otters freely swim in search of fish. Others, like raccoons, tend to hunt for food on the margins. Rodents of various kinds eat greens and seeds, abundant because of the water supply, and foxes and coyotes hunt the rodents. Other mammals, such as deer, come for the water, and sometimes eat the tender vegetation as well. Beaver may even be responsible for the development of a new swamp or marsh by building a dam across a creek.

Human Destruction of Swamps and Marshes

In spite of the great richness of life present in swamps and marshes, human society, in general, views these ecosystems as unsightly and useless. Consequently, many wetlands have been drained to make way for farmland, roads or other developments. Because so much has been lost, environmental laws in the United States now prohibit further wetland destruction, unless new wetlands are formed to replace those that are lost. These laws are a recognition of the ecological importance of swamps and marshes.

The primary productivity of swamps and marshes is only surpassed by tropical rain forests and algal beds and reefs. Shallow water and ample light allow rich plant and algae growth, which supports the rich variety of organisms found here. Swamps and marshes are especially important to migrating waterfowl, who need them for their abundant food supplies. As wetlands have been lost, waterfowl numbers have been reduced. Their high productivity also makes them important in absorbing excess carbon dioxide from the atmosphere, thus reducing global warming.

—*Bryan Ness*

See also: Ecosystems; Food chains and food webs; Forests, coniferous; Forests, deciduous; Habitats and biomes; Lakes and rivers; Marine biology; Rain forests; Tidepools and beaches.

Bibliography

Bransilver, Connie, Larry W. Richardson, Jane Goodall, and Stuart D. Strahl. *Florida's Unsung Wilderness: The Swamps*. Englewood, Colo.: Westcliffe, 2000. A combination of excellent photographs with accompanying text. Focuses primarily on the environmental concerns of our dwindling wetlands.

Scheffer, M. *Ecology of Shallow Lakes*. Boca Raton, Fla.: Chapman and Hall, 1998. A theoretically based book to help ecologists understand the dynamics of shallow lake ecosystems. Mathematically based, but concepts are usually discussed with a minimum of math and lots of diagrams.

Streever, Bill. *Bringing Back the Wetlands*. Darlinghurst, Australia: Sainty & Associates, 1999. A very readable book on wetlands and the dangers they face today. Includes firsthand accounts of destruction, preservation and restoration of wetlands.

Tiner, Ralph W. *In Search of Swampland: A Wetland Sourcebook and Field Guide*. Piscataway, N.J.: Rutgers University Press, 1998. A guide to the identification of wetlands. Well illustrated and designed for readers with minimal background. Focuses primarily on the northeastern United States.

Vileisis, Ann. *Discovering the Unknown Landscape: A History of America's Wetlands*. Washington, D.C.: Island Press, 1999. A history of wetlands and human interactions with them in the United States. Focuses on the importance of wetlands and the growing recognition of this in more recent times.

Wetzel, Robert G. *Limnology: Lake and River Ecosystems*. 3d ed. San Diego, Calif.: Academic Press, 2001. A classic and comprehensive text; no better general reference is currently available for freshwater ecology. Includes extensive information on swamps and marshes.

SWANS

Types of animal science: Anatomy, classification, reproduction
Fields of study: Anatomy, ornithology, zoology

Swans are the largest, most beautiful waterfowl. Some species were hunted almost to extinction for their feathers, before legislation protected them.

> **Principal Terms**
>
> COB: a male swan
> CYGNET: a newly hatched swan
> MANDIBLE: one of the two halves of a swan's bill
> PEN: a female swan

Swans, the largest members of the waterfowl family of birds, have snow-white feathers and graceful necks. There are eight swan species, which inhabit rivers, lakes, and ponds on every continent except Antarctica. They eat both on land and in water; their diet consists of land plants and grasses, aquatic plants, fish, and invertebrates. Swans must dig much of their aquatic food from bottom mud, after dipping their necks far into the water to reach it.

Many American swans nest around the Arctic Ocean and Hudson Bay. In winter, they migrate in a V-formation to the Carolinas at airspeeds up to sixty miles per hour, filling the air with sounds ranging from deep bass notes to those like clarinets. Swans that live in more temperate climates, such as the Great Plains of the United States, do not migrate. Eastern hemisphere swans include European whistling swans and Bewick's swans. A scarlet-billed black swan inhabits Australia. The initial ugliness of cygnets, which only lasts for a

Swans are highly monogamous birds, forming pair-bonds that last their entire lives. (Corbis)

year, was the inspiration for Hans Christian Andersen's fairy tale "The Ugly Duckling."

Physical Characteristics of Swans

Swans reach a maximum length (from beak to tail tip) of six feet, and weigh up to thirty-five pounds. Their plumage is usually white, white plus black, or rarely all black. Whooping (whooper), Bewick's, whistling, mute, and trumpeter swans are white. Black-necked swans have black areas on white bodies. Wingspans range from six to ten feet.

Swan necks are long and slender, usually held in a graceful S-curve. Trumpeters, whoopers, and whistlers, however, hold their necks straight upright. Swans have two short legs with huge, scale-covered, webbed feet. The tips of their strong bills are broad and flat, for optimum use in tearing underwater plants, a major part of their diet. Swans often feed by poking their heads underwater. They are not graceful on land, but fly, swim, and dive very well. All swans, including "mute" swans, hiss, bark, whistle, whoop, and cluck, especially when migrating.

The Lives of Swans

Swans form flocks to breed, nest, feed, and migrate. Depending on species, a flock has between a few dozen and thousands of individuals. Swans also form family units, comprising a set of parents with their offspring, within flocks. All swans choose mates for life. On mating, swans build nests from plants, twigs, and down, which they use over and over. Then the pens lay five to ten pale-colored eggs and incubate them until they hatch, after eighteen to thirty-nine days.

Cygnets, covered with gray to brown down, leave their nests a few days after hatching, following their parents to learn to swim and find food. Although the pen is usually responsible for the eggs, once the cygnets are born, the cob may feed and protect them. They fly for the first time at age four months. Most Northern Hemisphere swans breed yearly. Southern Hemisphere swans breed every two years. A swan is mature at age three or four and can live for twenty to thirty years in the wild or forty to fifty years in captivity.

Swan Facts

Classification:
Kingdom: Animalia
Subkingdom: Bilateria
Phylum: Chordata
Class: Aves
Order: Anseriformes
Family: Anatidae (ducks, geese, and swans)
Tribe: Anserini
Genus and species: Three genera and twelve species, including *Cygnus columbianus* (whistling swan), *C. cygnus* (whooping swan), *C. bewickii* (Bewick's swan), *C. melancoryphus* (black-necked swan), *C. buccinator* (trumpeter swan), *C. olor* (mute swan); *Chenopis atrata* (black swan); *Coscoroba coscoroba* (Coscoroba swan)

Geographical location: Every continent except Antarctica

Habitat: Rivers, lakes, ponds, swamps, and moist forest areas

Gestational period: Incubation of twenty-five to forty-five days

Life span: Twenty to thirty years in the wild; forty to fifty years in captivity

Special anatomy: Long necks in S-curves or straight upright; short legs with scaled, webbed feet; broad, flat, strong bills

Swan Conservation

South Americans, Europeans, Asians, and North Americans hunted trumpeters and other swans almost to extinction by eating swan meat, and using their feathers for pens, hat decorations, and women's scarves. By the 1930's, many species were in trouble; for example, under one hundred trumpeter swans were alive at that time. Legislation and protective actions have enlarged swan populations by allowing them to increase by natural reproduction.

—Sanford S. Singer

See also: Beaks and bills; Birds; Feathers; Flight; Migration; Molting and shedding; Nesting; Respiration in birds; Wildlife management; Wings.

Bibliography

Bellrose, F. C. *Ducks, Geese, and Swans of North America*. Rev. ed. Harrisburg, Pa.: Stackpole, 1981. A comprehensive research guide to the waterfowl of North America. Includes species identification, migration ranges and patterns, adult and juvenile sizes, breeding territories and behavior, population trends, and food habits for all species.

Fegely, Thomas D. *Wonders of Geese and Swans*. New York: Dodd, Mead, 1976. This book examines the swan's physical appearance, habits, and behavior.

Pfeffer, Wendy. *Mute Swans*. Parsipany, N.J.: Silver Press, 1996. Describes the characteristics and life cycle of mute swans.

Price, Alice L. *Swans of the World: In Nature, History, Myth, and Art*. Tulsa, Okla.: Council Oaks Books, 1994. This book explores the swan's natural history, as well as humans' perceptions of them in myth, history, and art.

Scott, Dafila. *Swans*. Stillwater, Minn.: Voyageur, 1995. Descriptions of all eight orders of swans, with distribution, population, and individual features. Includes maps and numerous photographs.

SYMBIOSIS

Type of animal science: Ecology
Fields of study: Evolutionary science, immunology, invertebrate biology, zoology

All animals live in close association, or symbiosis, with other species. Most symbioses are based on nutritional interrelationships involving competition or cooperation. Some animals cannot survive without their symbiotic partners, while others are harmed or killed by them.

Principal Terms

COMMENSALISM: symbiotic associations based chiefly on some form of food sharing, which may also involve shelter, protection, or cleaning

HOST: by convention, the larger of two species involved in a symbiotic association

INTERMEDIATE HOST: an animal species in which nonsexual developmental stages of some commensals and parasites occur

MUTUALISM: a type of commensalism in which both symbiotes benefit from the association in terms of food, shelter, or protection

PARASITE: a symbiote that must live in intimate contact with its host to survive; a parasite may be pathogenic or beneficial to the host

PARASITE MIX: all the individuals and species of symbiotes living in a host concurrently

RESERVOIR HOST: a host species other than the one of primary interest in a given research study

SYMBIOSIS: all forms of evolved, nonaccidental, nontrivial, interspecies associations, excluding predator-prey relationships

SYMBIOTE: a species involved in any form of symbiotic association with another species

Understanding the ways in which different species of animals interact in nature is one of the fundamental goals of biology. Predator-prey relationships, competition between species for limited resources, and symbiosis are the major forms of species interactions, and these have profoundly influenced the diversity and ecology of all forms of life. Significant advances have been made in understanding how organisms interact, but in studies of symbiosis (which literally means "living together") one finds the most complex, interesting, and important examples of both cooperation and exploitation known in the living world.

Symbiosis involves many types of dependent or interdependent associations between species. In contrast to predator-prey interactions, however, symbioses are seldom rapidly fatal to either of the associating species (symbiotes) and are often of long duration. With the exception of grazing animals that do not often entirely consume or destroy their plant "prey," most predators quickly kill and consume their prey. While a predator may share its prey with other individuals of the same species (clearly an example of "living together"), such intraspecific behavior is not considered to be a type of symbiosis. Fleas, some ticks, mites, mosquitoes, and other bloodsucking flies are viewed as micropredators rather than parasites.

All organisms are involved in some form of competition. The abundance and availability of environmental resources are finite, and competition for resources occurs both between members of the same species and between individuals and populations of different species. When the number of individuals in a population increases, the intensity of competition for limited food, water, shelter, space, and other resources necessary for survival and reproduction also increases. Thus, competition plays a major role in populations of free-living animals (those not inhabiting the body of other organisms) and in populations living on or in other animals. For example, both tapeworms and whales must compete for resources, and both have evolved habitat-specific adaptations to accomplish this goal. Whales compete with whales, fishes, and other predators for food; tapeworms compete with tapeworms and other symbiotes (such as roundworms) for food and space; and tapeworms and whales compete with each other for food in the whale's gut.

"Symbiosis" is a term used to describe nonaccidental, nonpredatory associations between species. When used by itself, the term "symbiosis" does not provide information on how or why species live together, or the biological consequences of their interactions. Recognizably different forms of symbioses all have one or more characteristics in common. All involve "living together"; most involve food sharing; many involve shelter; and some involve damage to one or both symbiotes.

Hosts and Symbiotes

Host species may be thought of as landlords. Hosts provide their symbiotes (also called symbionts) with transportation, shelter, protection, space, some form of nutrition, or some combination of these. Host species are generally larger and structurally more complex than their symbiotes, and different parts of a host's body (skin, gills, and gut, for example) may provide habitats for several different kinds of symbiotes at the same time. The three primary categories of symbiosis most commonly referred to in popular and scientific works are commensalism, mutualism, and parasitism.

Symbiotes that share a common food source are known as commensals (literally, "messmates"). In the usual definition of commensalism, one species (usually referred to as the commensal, although both species are commensals) is said to benefit from the relationship, while the other (usually referred to as the host) neither benefits nor is harmed by the other. Adult tapeworms which live in the intestinal tracts of vertebrate hosts provide a classic example of commensals. Adult tapeworms share the host's food, usually with little or no effect on otherwise healthy hosts. As in all species, however, too large a tapeworm population may result in excessive competition, lower fitness, or disease in both the host and the tapeworms. For example, the broad fish tapeworm of man, *Diphyllobothrium latum*, may cause a vitamin B12 deficiency and anemia in humans when the worm burden is high. In addition to tapeworms, many human symbiotes called "parasites" are, in fact, commensals.

External commensals (those living on the skin, fur, scales, or feathers of their hosts) are called epizoites. A good example of an epizoite is the fish louse (a distant relative of the copepod), which feeds on mucus of the skin and scales of fishes. Another type of commensalism is called phoresis (phoresy), which involves passive transportation of the commensal (phoront) by its host. Examples of phoreses include barnacles carried by whales and sea turtles, and remoras (sharksuckers), which, in the absence of sharks, may temporarily attach themselves to human swimmers. In inquilinism, the transported commensal (inquiline) shares, or more accurately, steals, food from the host, or may even eat parts of the host. Perhaps the best-known inquilines are the glass- or pearlfishes, which take refuge in the cloacae of sea cucumbers and often eat part of the host's respiratory system. A unique type of commensalism, known as symphilism, is found in certain ants and some other insects (hosts) which "farm" aphids (symphiles) and induce them to secrete a sugary substance which the ants eat.

Mutualism

The most diverse type of commensalism is mutualism. In some works, particularly those dealing with animal behavior, mutualism is used as a synonym of symbiosis; hence, the reader must use caution in order to determine an author's usage of these terms. As used here, mutualism is a special case of commensalism, a category of symbiosis. The relationship between mutuals may be obligatory on the part of one or both species, but it is always reciprocally beneficial, as the following examples illustrate. Some species of hermit crabs place sea anemones on their shells or claws (sea anemones are carnivores which possess stinging cells in their tentacles). Hermit crabs without anemones on their shells or claws may be more vulnerable to predators than those with an anemone partner. Hermit crabs, which shred their food in processing it, lose some of the scraps to the water, which the anemones intercept, and eat. Thus, the crab provides food to the anemone, which in turn protects its provider. Such relationships, which are species-specific, are probably the result of a long period of coevolution.

A different type of mutualism, but one having the same outcome as the crab-anemone example, is found in associations between certain clown fishes and sea anemones. Clown fishes appear to be fearless and vigorously attack intruders of any size (including scuba divers) that venture too close to "their" anemone. When threatened or attacked by predators, these small fishes dive into an anemone's stinging tentacles, where they find relative safety. Anemones apparently share in food captured by clown fishes, which have been observed to drop food on their host anemone's tentacles.

Cleaning symbiosis is another unique type of mutualism found in the marine environment. In this type of association, marine fishes and shrimp of several species "advertise" their presence by bright and distinctive color patterns or by conspicuous movements. Locations where this behavior occurs are called "cleaning stations." Instead of being consumed by predatory fishes, these carnivores approach the cleaner fish or shrimp, stop swimming, and sometimes assume unusual postures. Barracudas, groupers, and other predators often open their mouths and gill covers to permit the cleaners easy entrance and access to the teeth

Remoras attach themselves to larger fish for transport. (Digital Stock)

and gills. Cleaners feed on epizoites, ectoparasites, and necrotic tissue that they find on host fishes, to the benefit of both species. Some studies have shown that removal of cleaning symbiotes from a coral reef results in a significant decrease in the health of resident fishes.

Parasitism

Parasitism is a category of symbiosis involving species associations that are very intimate and in which competitive interactions for resources may be both acute and costly. The extreme intimacy (rather than damage) between host and parasite is the chief difference between parasitism and other forms of symbiosis. Parasites often, but not always, live within the cells and tissues of their hosts, using them as a source of food. Some types of commensals also consume host tissue, but in such cases (pearl fishes and sea cucumbers, for example) significant damage to the host rarely occurs. Commensalism is associated with nutritional theft.

Some, but not all, parasites harm their hosts, by tissue destruction (consumption or mechanical damage) or toxic metabolic by-products (ammonia, for example). Commonly, however, damage to the host is primarily the result of the host's own immune response to the presence of the parasite in its body, cells, or tissues. In extreme cases, parasites may directly or indirectly cause the host's death. When the host dies, its parasites usually die as well. It follows that the vast majority of host-parasite relationships are sublethal. A number of parasites are actually beneficial or crucial to the survival of their hosts. The modern, and biologically reasonable, definition of parasitism as an intimate type of symbiosis, rather than an exclusively pathogenic association between species, promotes an ecological-evolutionary understanding of interspecies associations. Most nonmedical ecologists and symbiotologists agree that two distinct forms of intimate associations, or parasitisms (with many intermediate types), occur in nature. The most familiar are those involving decreased fitness in humans and in their domestic animals and crops.

Among animal parasites, malarial parasites, hookworms, trypanosomes, and schistosomes (blood flukes) cause death and disease in millions of people each year. The degree to which these parasites are pathogenic, however, is partly the result of preexisting conditions of ill health, malnutrition, other diseases, unsanitary living conditions, overcrowding, or lack of education and prevention. Parasites which frequently kill or prevent reproduction of their hosts do not survive in an evolutionary sense, because both the parasites and their hosts perish. Both members of intimate symbiotic relationships constantly adapt to their environments, and to each other. Over time, evolutionary selection pressures result in coadaptation (lessening of pathogenicity) or destruction or change in form of the symbiosis.

Nonpathogenic or beneficial host-parasite associations are among the most highly evolved of reciprocal interactions between species. The extreme degree of intimacy of the symbiotes (not lack of pathogenicity) distinguishes this type of parasitism from mutualism. Parasitic dinoflagellates (relatives of the algae that cause "red tides") are found in the tissues of all reef-building corals. These photosynthetic organisms use carbon dioxide and other waste products produced by corals. In turn, the dinoflagellates (*Symbiodinium microadriaticum*) provide their hosts with oxygen and nutrients that the corals cannot obtain or produce by themselves. Without parasitic dinoflagellates, reef-building corals starve to death. Similar host-parasite relationships occur in termites, which, without cellulose-digesting parasitic protozoans in their gut, would starve to death.

Studying Symbiosis

Early studies of symbiosis focused primarily on discovery and description of commensals and parasites found in humans and their domestic food animals. Malarial parasites and some of the important trematodes (flatworms known as flukes) and nematodes (roundworms) of humans were described by the ancient Greeks, and references to the guinea worm (*Dracunculus medinen-*

sis) are found in the Bible, where it is called "the fiery serpent." Some of the dietary conventions or laws observed in modern cultures have the side effect of preventing harmful symbioses, although it is still debated whether these proscriptions actually have their basis in early observations. It is widely known, for example, that pork products, if eaten at all, should never be consumed without thorough cooking. Swine are intermediate hosts for two very pathogenic human parasites, the trichina worm (the nematode *Trichinella spiralis*) and the bladderworm (the infective stage of the tapeworm *Taenia solium*). Much research in parasitology involves the description of symbiotes, particularly those of potential medical, veterinary, or agricultural importance.

The life cycles of many commensals and parasites are extremely complex and often involve two or more intermediate hosts living in different environments, as well as free-living developmental stages. Except for symbiotes of medical importance, relatively few complete life cycles have been worked out. Knowledge of life cycles remains as one of the most important areas of research in parasitology and is usually the phase of research following the description of new species.

Scientists have long recognized that "chemical warfare" (antibiotics, antihelminthics, insecticides) against microbial and animal parasites, and their insect and other vectors, provides only short-term solutions to the control or eradication of symbiotes of medical importance. Research attempts are being made to find ways of interrupting life cycles, sometimes with the use of parasites of other parasites. This research requires sophisticated ecological and biochemical knowledge of both the host-parasite relationship and the parasite mix. Studies of the parasite mix are ecological (parasite-parasite and host-parasite competition), immunological (host defense mechanisms and parasite avoidance strategies), and ethological (host and symbiote behavioral interactions) in nature. Investigators involved in this kind of research must be well trained in many of the biological disciplines, including epidemiology (the distribution and demographics of disease).

Immunology is the most promising modern research area in parasitology. Not only have specific diagnostic tests for the presence of cryptic (hidden or hard to find) parasites been developed, but also vaccines may be discovered that can protect people from such destructive protozoan diseases as malaria. Malaria has killed more humans than any other disease in history, and it currently causes the death of more than one million people, and lowers the quality of life for millions of others, each year.

The Interrelationship of Species

All species are involved in complex interrelationships with other species that live in or on their bodies, or with which they intimately interact behaviorally or ecologically. Such interactions may play a minor role in the life and well-being of one or both of the associates, or they may be necessary for the mutual survival of both. In relatively few symbiotic relationships, one or both species may suffer damage or death. Pathogenic associations are relatively rare, because disease or death of one symbiote generally results in corresponding disease or death of the other. Such relationships, which cannot persist over evolutionarily long periods of time, may nevertheless cause catastrophic loss of life in nonadapted host populations.

Domestic animals cannot live in some parts of the world, such as the central portion of Africa, because they have little or no resistance to parasites of wild species, which are the normal hosts and are not harmed. Native species have coadapted with the parasites. This situation presents a moral dilemma to humans. In the face of human needs for space and other resources, should native animals be displaced or killed? Or should human populations proactively slow their reproductive rates? History shows that humanity has often chosen to take the former course.

The common view that animals which live in other animals are degenerate creatures that take advantage of more deserving forms of life is understandable but inaccurate. Symbiotes are highly specialized animals that do not live cost-free, or always to the detriment of their hosts. Symbiotic relationships between species have vastly increased the diversity, complexity, and beauty of the living world.

—Sneed B. Collard

See also: Coevolution; Competition; Ecological niches; Ecology; Ecosystems; Evolution: Historical perspective; Habitats and biomes; Immune system; Invertebrates; Plant and animal interactions; Population growth; Predation.

Bibliography

Boothroyd, John C., and Richard Komuniecki, eds. *Molecular Approaches to Parasitology*. New York: Wiley-Liss, 1995. Describes the uses of molecular tools to study parasites, focusing on recent advances.

Caullery, Maurice. *Parasitism and Symbiosis*. London: Sidgwick and Jackson, 1952. One of the classic books dealing with all the types of symbiosis in animals and plants. This nonmedical text includes a wealth of information. Examples of the ways in which species interact, and the consequences and evolution of these interactions, are emphasized from an ecological point of view. Well illustrated, with an extensive bibliography. Highly recommended for advanced high school and college students with some background in invertebrate zoology.

Limbaugh, Conrad. "Cleaning Symbiosis." *Scientific American* 205 (August, 1961): 42-49. The classic early description and analysis of cleaning associations of animals living primarily in kelp beds and coral reefs. Beautifully written and illustrated. Limbaugh perished in a diving accident in France, but this pioneering work set the stage for most further research in the field.

Margulis, Lynn. "Symbiosis and Evolution." *Scientific American* 225 (August, 1971): 48-57. Describes the origin and evolution of early organisms that, as highly adapted parasites, have become incorporated into, and are now essential components of, modern cells. Shows the importance and consequences of symbiosis at the cellular level. Well illustrated; short bibliography. Written for an informed general audience.

Margulis, Lynn, and Dorion Sagan. *Slanted Truths: Essays On Gaia, Symbiosis, and Evolution*. New York: Copernicus, 1997. Discusses the symbiosis of all living species in the context of the Gaia hypothesis.

Noble, Elmer, Glenn Noble, Gerhard Schad, and Austin MacGinnes. *Parasitology: The Biology of Animal Parasites*. 6th ed. Philadelphia: Lea & Febiger, 1989. An entry-level college text that is exceptional in giving equal emphasis to classical and modern aspects of parasitology. Includes discussion of terminology, history of parasitology, life cycles, ecology, physiology, and immunology. Profusely illustrated and cross-referenced. Useful glossary and bibliography. Recommended for readers interested in any aspect of human and animal symbiosis.

Toft, Catherine Ann, Andre Aeschlimann, and Liana Bolis, eds. *Parasite-Host Associations: Coexistence or Conflict?* New York: Oxford University Press, 1991. Discusses the costs and benefits of parasitism from an interdisciplinary perspective.

Whitefield, Philip. *The Biology of Parasitism: An Introduction to the Study of Associating Organisms*. Baltimore: University Park Press, 1979. Written specifically to illustrate the wide variety of symbiotic associations that exist between organisms. Provides an excellent, integrated discussion of terms used in the field and different ways of classifying symbioses. Although written at the college level, the book will be extremely useful to all readers interested in symbiotic phenomena and the nature of interspecies relationships. Well illustrated; excellent bibliography.

Zann, Leon P. *Living Together in the Sea*. Neptune City, N.J.: T. F. H., 1980. This fascinating and highly recommended book covers all major, nonmolecular aspects of symbioses in marine animals. The book's outstanding feature is hundreds of superb color photographs illustrating examples of different symbiotic relationships. The brief narrative portion of the book is written in a style and degree of complexity suitable for high school and college students and will be understandable to general readers who are unfamiliar with marine organisms or habitats. Comprehensive bibliography.

Zinsser, Hans. *Rats, Lice, and History*. Reprint. New York: Bantam Books, 2000. One of the best popular books ever written on the subject of infectious diseases and their role in shaping human cultures and history. The book includes discussions of both microbial and animal parasites. Very highly recommended for the general reader and postgraduate biology students.

SYSTEMATICS

Type of animal science: Classification
Fields of study: Evolutionary science, systematics (taxonomy)

Systematics is the subdivision of biology that deals with the identification, naming, and classification of organisms and with understanding the evolutionary relationships among them. Systematics attempts to incorporate everything that is known about organisms in its effort to produce classification systems that may reveal the evolutionary relationships among various groups. The system of classification that has resulted from this discipline divides all organisms into five kingdoms: Monera, Protista, Fungi, Plantae, and Animalia.

Principal Terms

BINOMIAL NOMENCLATURE: the two-word system used for naming every individual species; the wolf, for example, is *Canis lupus*

EUKARYOTE: an organism whose cells have internal structures, such as a nucleus and a mitochondrion, separated from one another and from the rest of the cytoplasm by membranes

KINGDOM: the broadest category of organisms; the system currently used recognizes five kingdoms—Monera, Protista, Fungi, Animalia, and Plantae

NOMENCLATURE: the part of systematics that deals with establishing a valid name for a species, according to specific guidelines

PHYLUM (pl. PHYLA): the second-broadest category of classification; each kingdom is divided into phyla

PROKARYOTE: an organism that has no internal membranes in its cells; the only membrane in a prokaryote is the cell membrane

TAXON (pl. TAXA): the basic unit of taxonomy; any of the categories of classification to which an organism may be assigned

Systematics is concerned with the identification, naming, and classification of living organisms and with the relationships that exist among them. In its broadest sense, systematics gathers and summarizes everything that is known about organisms: their morphological, physiological, psychological, ecological, evolutionary, developmental, molecular, and genetic characteristics. The goal of the systematist is to develop a natural classification that will reveal evolutionary relationships among the various groups. A natural classification strives to show true evolutionary relationships. In this kind of system, organisms placed in a given category should share evolutionary origins. Although it is nearly impossible to be absolutely certain of the evolutionary ancestry of a group of organisms, one of the aims of classification is to approximate the natural relationships as closely as possible.

The earliest systems of classification, which go at least as far back as the ancient Greeks, grouped organisms into categories based on artificial standards. A system based on artificial criteria might, for example, use the length of the tail as the main standard to separate animals into different categories. Thus, under such a system, short-tailed animals such as elephants, pigs, and chihuahua dogs would be grouped together; horses, cats, and skunks would be in a separate group.

Binomial Nomenclature

The system of classification used by scientists today considers the species the basic unit of classifi-

Louis Agassiz

Born: May 28, 1807; Motier, Switzerland

Died: December 14, 1873; Cambridge, Massachusetts

Fields of study: Marine biology, paleontology, systematics (taxonomy)

Contribution: Agassiz's classification and description of fossil and living fish established his reputation as a major scientist. Later he vigorously opposed theories of evolution.

Jean Louis Rodolphe Agassiz studied at the universities of Zurich, Heidelberg, and Munich. His volume describing and classifying a collection of Brazilian fish earned him a Ph.D. from the University of Erlangen in 1829; the next year he received a doctor of medicine degree from Munich. In 1832, he went to Paris, where his work impressed the noted scientists Alexander von Humboldt and Georges Cuvier. Humboldt arranged a professorship for Agassiz at Neuchâtel, Switzerland. Cuvier gave Agassiz the material on fossil fish that he had collected.

Louis Agassiz made his name classifying fossil and living fish, but he is more widely known for his rejection of Darwin's theories of evolution. (Library of Congress)

Agassiz spent the next ten years studying fossil fish collections in museums and private collections across Europe, publishing his findings in a six-volume *Recherches sur les poissons fossiles* (1833-1843; research on fossil fish). Using the comparative methods taught by Cuvier, Agassiz carefully analyzed and classified more than seventeen thousand ancient fish. As a classifier, Agassiz meticulously distinguished between closely related forms, a procedure tending toward the multiplication of species. Convinced that an all-powerful deity had planned the entire range of creation, he rejected the idea that a hereditary connection extended from ancient to modern fish, insisting that each species represented a separate and special idea of God. While at Neuchâtel, he also published a series of papers describing how Ice Age glaciers had transformed the topography of Europe.

After Harvard University offered him a professorship of zoology and geology in 1847, Agassiz settled permanently in the United States. In 1859, he published his *Essay on Classification*, explaining his philosophy of nature, his belief in special creation, and his views on zoological classification. By that year, he had raised enough money to build a Museum of Comparative Zoology at Harvard. Agassiz planned to have the museum show the relationship of species, while insisting on the distinct and separate creation of each.

When Charles Darwin's *On the Origin of Species* appeared in 1859, Agassiz attacked it vigorously, calling the work untrue in its facts and unscientific in its methods. His rejection of Darwin attracted contemporary religious leaders uneasy with ideas that seemed to contradict those of the Bible. Younger scientists were less impressed by Agassiz's position, arguing that his concept of glacial periods supported Darwin's mechanism of change, and reading the geological order in which Agassiz showed species succeeding each other as evidence of an evolutionary progression. Agassiz's work on classification used paleontology, embryology, and geographical associations to demonstrate the true relationship of organisms. Agassiz believed this relationship had originated within the mind of God, but many of Agassiz's successors found it verification of an evolutionary sequence of change.

—*Milton Berman*

cation. To most individuals, the concept of a species is intuitively obvious. Dogs, cats, horses, cows, and pigs each constitute a separate, distinct group of organisms. Dogs breed with dogs to produce puppies, which in turn grow up and are able to have puppies of their own. These two concepts, as simple and obvious as they seem, form the cornerstone of the classification system in use today. This species concept was first introduced by the English naturalist John Ray in the late seventeenth century. Today, a species is usually defined as a population or group of populations of organisms that share similar characteristics and that, ideally, are capable of breeding to produce fertile offspring.

The earliest systematists were the ancient Greeks. Although many classification systems have been proposed over the centuries, the system used today to classify all living organisms is based on one developed in the eighteenth century by a Swedish scientist named Carolus Linnaeus. The beginning of modern classification is considered to date from the publication, in 1758, of the tenth edition of Linnaeus's book *Systema Naturae*. Linnaeus simplified all the previously existing systems of classifying living things and developed a two-word system for naming each organism. This two-word system, called binomial nomenclature, consists of two names ("binomial" simply means "two names") that are assigned to one, and only one, species. These two names are somewhat similar to a first and last name. For example, for the human species, the name is *Homo sapiens*, which means "wise man." *Homo*, the first word in the name, is similar to a last name, such as Jones. This is the name of the genus to which a species belongs. A genus may include a group of related species, as a family includes a group of related individuals. The second word, *sapiens*, is equivalent to a first name, such as Mary. This is the species name. The same two-word name is used for a given species in all languages, thus giving the system consistency and universality.

The names for a species are often in Latin or Greek. When words from a different language are used, Latin or Greek endings are added for consistency. Specific sets of rules must be followed in naming a species. For example, for naming animals, the rules used are outlined in the *International Code for Zoological Nomenclature*, known among zoologists as "the code." The code is published by the International Commission of Zoological Nomenclature. The Commission, in turn, answers to the International Union of Biological Sciences. Similar sets of rules have been outlined for scientists naming plants, bacteria, and viruses. The value of these specific sets of rules for naming organisms is that inconsistency is minimized.

Carolus Linnaeus gave his name to the Linnaean system of binomial nomenclature still used to identify species today. (Library of Congress)

Other examples of binomial species names are, for the grizzly bear, *Ursus horribilis* (meaning "horrible bear"), for the common earthworm, *Lumbricus terrestris* ("worm of the earth"), and for the housefly, *Musca domestica* ("fly of the house"). Sometimes species are named to honor a discoverer or to commemorate a famous personage, but

Carolus Linnaeus

Born: May 23, 1707; Råshult, Småland, Sweden
Died: January 10, 1778; Uppsala, Sweden
Fields of study: Entomology, ornithology, systematics (taxonomy), zoology
Contribution: Linnaeus, the most influential taxonomist of the eighteenth century, introduced the binomial system of nomenclature for the scientific naming of plant and animal species.

Carolus Linnaeus became interested in botany and the natural sciences as a youth. He attended the University of Lund and the University of Uppsala, where he received his degree in medicine. In 1730, Linnaeus was appointed lecturer in botany. Two years later, the Uppsala Academy of Sciences gave Linnaeus the opportunity to conduct an extensive field trip in wild and virtually unexplored regions of Lapland. In 1736, Linnaeus visited England and Paris, where he met many distinguished botanists, including Sir Hans Sloane, Johann Jakob Dillenius, and the three Jussieu brothers. Linnaeus completed the *Hortus Cliffortianus* while in Holland and then returned to Sweden. In 1738, he established a very successful medical practice in Stockholm. In 1741, he was appointed to the chair of medicine at Uppsala. One year later he accepted the chair of botany.

Linnaeus published the results of his exploration of Lapland in Amsterdam as the *Flora Lapponica* (1737). An English translation was published by Sir James Edward Smith as *Lachesis Lapponica* (1811). These publications brought Linnaeus to the attention of the scientific community, but his worldwide reputation was established by the *Systema Naturae, or The Three Kingdoms of Nature Systematically Proposed in Classes, Orders, Genera, and Species* (1735) and the *Genera Plantarum* (1737). The *Species Plantarum*, which provided a complete description of the specific names, was not published until 1753.

The first edition of the *Systema Naturae* contained the basic principles of the Linnaean system, although Linnaeus revised the text many times to accommodate new information and ideas. Plants and animals were grouped into species, genus, order, class, and kingdom. The international scientific community generally accepted the first edition of the *Species Plantarum* (1753) and the fifth edition of the *Genera Plantarum* (1754) as the starting point for naming flowering plants and ferns. Although the Linnaean classification system is based mainly on flower parts and is not natural, in the sense called for by Aristotle, it was useful and efficient in a period in which thousands of new plant species were being discovered. Using the binomial system, botanists could quickly place a new plant into a named category. The binary nomenclature standardized by Linnaeus became the universally accepted method of naming plants and animals. Indeed, the simplicity and success of the Linnaean system made it difficult for scientists to replace it when more natural systems were subsequently proposed.

Linnaeus thought that his work on the reproductive organs of plants was his major contribution to botany. His classification system organized plants into classes according to the number and character of the stamens; classes were divided into orders by the number and character of the pistils. Nevertheless, Linnaeus believed that the process of naming and classifying plants and animals was the most important aspect of biology and the essential foundation of science.

In 1761, Linnaeus was granted a Swedish patent of nobility, which entitled him to be called Carl von Linné. When Linnaeus died in 1778, he was buried with the honors proper to royalty, but his widow later sold his collections and books to Sir James Edward Smith. The Linnean Society of London acquired these materials in 1829.

—*Lois N. Magner*

occasionally the name may be based on some less serious criteria. For example, a scientist who identified a group of related beetles named them with words created by scrambling the letters in his mistress's name.

The System of Categories

The classification system in use today groups species into progressively broader categories, beginning with the species, the narrowest major category. Each of these categories is called a taxon;

there are seven major taxa. Organisms within each taxon are believed to share certain evolutionary ties. Related species are grouped within a genus (plural, genera), related genera are grouped in a family, families in orders, and orders in classes. One or more classes may then be grouped in a phylum (plural, phyla), and finally, related phyla are placed in a kingdom. Each of these seven major taxa may be further subdivided. Thus, when the human species is fully classified, it looks something like this:

Kingdom: Animalia (animals)
Subkingdom: Metazoa (animals composed of many cells)
Phylum: Chordata (contain a notochord, a dorsal, hollow nerve cord, and gills at some stage of the life cycle)
Subphylum: Vertebrata (jaws and paired limbs)
Superclass: Tetrapoda (two pairs of legs, lungs, a bony skeleton)
Class: Mammalia (hair, warm-blooded, nurse their young)
Subclass: Theria (young develop with a placenta or in a pouch)
Infraclass: Eutheria (placenta only)
Order: Primates (each limb has five fingers or toes with nails)
Suborder: Anthropoidea (upright, larger skull)
Superfamily: Hominoidea (no tail, no cheek pouches)
Family: Hominidae (large brain, flat face, small canine teeth)
Genus: Homo (even teeth, grinding molars)
Species: sapiens (toolmakers, thinkers, organizers, speakers)

The Whittaker System

Many different systems of classification have been developed since Linnaeus published *Systema Naturae*. As more information about species has been discovered and as more species have been identified, biologists have continually found it necessary to revise the classification systems. The most widely accepted scheme currently used was proposed by Robert H. Whittaker, of Cornell University, in 1969. The Whittaker system of classification divides all organisms into five kingdoms: Monera, Protista, Fungi, Plantae, and Animalia. Three criteria are used to determine where to place various types of organisms. Basically, to assign any group to the appropriate kingdom, the following three questions are asked, in this order: Is the organism a prokaryote or a eukaryote? Is the organism unicellular or multicellular? What is the organism's mode of nutrition? Although there are other systems of classification still in use, the simplicity and logic of the Whittaker system have contributed to its broad acceptance by biologists.

By applying these criteria, the five kingdoms can be derived easily. In response to the first criterion ("Is it a prokaryote or a eukaryote?"), all organisms that are considered prokaryotic are placed in the kingdom Monera. The Monera includes all bacteria, including the blue-green algae, or cyanobacteria. All monerans are prokaryotes—that is, the cells have no membrane-bound structures such as a nucleus, mitochondria, or chloroplasts.

All remaining organisms—the eukaryotes—are then classified according to the two remaining criteria. All eukaryote organisms that are unicellular (single-celled) are placed in the kingdom Protista. Protists are a very diverse group of organisms, united by their common unicellular organization. Examples of protists include amoeba, diatoms, *Paramecium*, *Toxoplasma*, and *Plasmodium*. Many of these, such as diatoms, are generally harmless and sometimes useful organisms. Others, such as *Toxoplasma* and *Plasmodium*, are termed pathogenic—that is, they can cause disease.

All remaining organisms are both eukaryotic and multicellular. These eukaryotes are now divided according to the third criterion, the organism's mode of nutrition. There are three basic modes of nutrition that an organism can use: absorptive, photosynthetic, and ingestive. In absorptive nutrition, the organism absorbs small molecules from the surrounding medium directly into the cell. These organisms have no digestive systems, and they need to live in an environment that will provide them with all the small organic molecules they need to survive. In a photosynthetic mode of nutrition, the organism converts the energy from the sun and carbon dioxide from the surroundings into usable nutrients. Finally, in ingestive nutrition, the organism engulfs or swallows food materials, then digests this food in specialized organs of the body.

Organisms whose mode of nutrition is absorptive are classified in the kingdom Fungi. The image of the kingdom Fungi generally elicits a negative response in humans—the first things that come to mind are often bread mold, spoiled potatoes, ringworm, and athlete's foot. Yet many members of the Fungi are beneficial. For example, penicillin was originally produced only from the fungus *Penicillium*. Many other antibiotics were originally discovered and commercially prepared from fungi. Other examples of useful fungi are the common mushroom used in cooking and those used in the preparation of blue cheeses.

The two remaining modes of nutrition are used to separate the remaining organisms. All eukaryotes using the photosynthetic mode of nutrition are grouped in the kingdom of Plantae (the plant kingdom). This kingdom includes the algae, mosses, ferns, liverworts, shrubs, and trees. Finally, eukaryotes that use ingestion as a way of obtaining nutrients are placed in the kingdom Animalia, or the animal kingdom. The animal kingdom includes the familiar animals such as dogs, pigs, cows, and elephants. Also included are some less obvious members, such as sponges, corals, jellyfish, and barnacles.

One might wonder how "natural" the five-kingdom system of classification is. Separating prokaryotes from eukaryotes is a fundamental, as well as evolutionary ancient, division. Although the other divisions are less clear, the consensus is that the Protista arose from the Monera and that the Protista in turn gave rise to each of the other kingdoms independently.

The Process of Classification

In practice, systematics may be subdivided into steps, or phases, through which the systematist takes a given species in the process of assigning it to its appropriate slot in a classification. These phases are taxonomy, nomenclature, and classification. Taxonomy (which means "the naming of taxa") is the phase of systematics that deals with the identification, differentiation, and naming of species. The process of identification and differentiation of a species involves making certain that the species is indeed different from all others and not merely a variation of a known species. This is often not an easy chore. Although some organisms can be differentiated quite easily, others are much more difficult. There are certainly differences between a husky and a miniature poodle, for example, yet they are members of the same species. Yet the timber wolf, which bears many physical similarities to the husky, is placed in a different, although related, species.

To accomplish the most accurate differentiation possible, the taxonomist draws on knowledge from many different disciplines in biology. Information from such different fields as molecular genetics and ecology may be used in establishing the differences between two groups.

Once a species is identified as new, it must be given a binomial name. The naming of species, or of any other taxa, is known as nomenclature. This is the legalistic phase of systematics. Once an organism is determined by a taxonomist to belong to a new species, a valid name for this new species must be established. In doing this, the careful scientist engages in extensive research to establish that the name assigned to a new species meets all the rules outlined for that particular group. When that research is completed, a description of the species is published in a scientific journal, such as the *Proceedings of the Biological Society of Washington*, or in a more specialized journal for that particular group (for example, *Crustaceana* for new species of crustaceans). Publishing information is an important part of the job of the systematist. One of the major goals of the systematists is to make cer-

John Ray

Born: November 29, 1627; Black Notley, Essex, England

Died: January 17, 1705; Black Notley, Essex, England

Fields of study: Environmental science, systematics (taxonomy)

Contribution: John Ray was the first naturalist to develop a taxonomic system applicable to both plants and animals that used species as its fundamental unit.

John Ray's father was the village blacksmith and his mother was an herbalist and healer. Her influence probably stimulated Ray's early interest in the medicinal virtues of plants. Ray (who also spelled his surname "Wray" until 1670) attended the grammar school in nearby Braintree. In 1644, he matriculated at St. Catherine's Hall, at the University of Cambridge, with the aid of a fund that had been established for needy scholars. He transferred to Trinity College in 1646.

Ray earned his bachelor's degree in 1648 and was elected to a fellowship at Trinity in 1649. Ray took holy orders in 1660, but he lost his fellowship in 1662 because, as a Puritan, he refused to subscribe to the Act of Uniformity, which required all clergy to take an oath of canonical obedience. Ray became totally dependent on the generosity of friends, who allowed him to spend the rest of his life studying natural history and writing numerous books. Ray was elected to the Royal Society in 1667.

In 1660, Ray published a catalog of the plants that grew in the vicinity of Cambridge. This led to studies of other areas in Britain. During an expedition to Wales and Cornwall in 1662, Ray entered into a partnership with his friend and pupil Francis Willughby (1635-1672). The two naturalists shared the goal of conducting a comprehensive study of the natural history of all living things. Ray agreed to catalog the plant kingdom while Willughby investigated the animal kingdom.

From 1663 to 1666, Ray and Willughby enhanced their knowledge of flora and fauna while touring the European continent. When they returned to England, they began composing a series of books describing the entire animal and plant kingdoms. Ray published a *Catalogus Plantarum Angliae* (a catalog of English plants) in 1670. Unfortunately Willughby died only two years later. Ray published *F. Willughbeii . . . Ornithologia* (1676; the ornithology of F. Willughby) and *F. Willughbeii . . . de Historia Piscium* (1685; history of the fishes), listing Willughby as the author, even though Ray had done much of the work. Ray's *Methodus Plantarum Nova* (1682; a new method of plant classification) established the taxonomic importance of the distinction between monocotyledons and dicotyledons.

Ray's most important contribution to botany was his three-volume *Historia Generalis Plantarum* (1686-1704; a general history of plants). In this work Ray established the species as the ultimate unit of taxonomy. Ray suggested that a breeding test could be used to determine whether plant specimens were members of the same species. Urging naturalists to aspire to a more sophisticated and natural system, he attempted to incorporate all structural characteristics, including internal anatomy, into his taxonomic system. Although Ray's system of taxonomy was not truly natural, it did approach that goal more closely than a strictly artificial system like that of Carolus Linnaeus.

Another example of Ray's cataloging tendencies was his *Collection of English Proverbs Digested into a Convenient Method for the Speedy Finding Any One upon Occasion* (1670), an early example of folklore research which was probably intended to assist clergymen in the writing of their sermons. This book was still forming the basis of English proverb collections into the late nineteenth century. Ray's work in the fields of natural history and folklore illustrates the permeable boundaries between disciplines of his age.

In the 1690's, Ray published several books on religion. This was not really a departure from his scientific work because Ray always believed that his taxonomic studies reflected the order and purpose inherent in God's design. *The Wisdom of God Manifested in the Works of the Creation* (1691) was Ray's most popular and influential book. Indeed, the period between the publication of *The Wisdom of God* and *On the Origin of Species* (1859) saw the flowering of an approach to science known as natural theology, largely based on Ray's work.

—*Lois N. Magner*

tain that the newly organized knowledge is available to all biologists.

The classification phase of systematics involves the assigning of organisms to specific taxa. Classification provides a filing system that allows the logical grouping of species. Like the process of identification and differentiation of a species, classification requires an extensive knowledge of biology. The systematist involved in classification takes into account virtually everything known about a given species in order to arrive at the most logical, natural classification. Because of the breadth of knowledge required in their field, some biologists involved in systematics feel that theirs is the ultimate synthesizing branch of biology.

Humans are almost instinctive sorters and classifiers, tending to sort everything into groupings, classes, or types. These are mostly groupings of convenience: Information is easier to store, retrieve, and exchange in this way. Thus, people may be classified as short or tall, heavy or thin, blonde, redhead, or brunette. Biologists classify living things for similar reasons—that is, to facilitate the exchange of information. The classification of organisms also satisfies the curiosity of biologists to know how many different kinds of organisms are found on earth. Of the estimated 10 million species living on earth, only about 1.5 million have been identified. Finally, and most important, classifying organisms by their similarities gives scientists insights into their relationships to one another, as well as their relationship to extinct groups of organisms.

—Alina C. Lopo

See also: Amphibians; Arthropods; Birds; Chordates, lower; Evolution: Historical perspective; Fish; Insects; Invertebrates; Mammals; Reptiles; Vertebrates.

Bibliography

Arms, Karen, and Pamela S. Camp. *Biology*. 3d ed. Philadelphia: Saunders College Publishing, 1987. A good introductory-level textbook. Includes a chapter on the classification of organisms, and ten additional chapters on the major groups of living organisms. Self-quizzes and questions for discussion.

Claridge, Michael F., H. A. Dawah, and M. R. Wilson, eds. *Species: The Units of Biodiversity*. New York: Chapman and Hall, 1997. A collection of symposium papers discussing the practical aspects of the concept of species in the daily work of biologists.

Futuyma, Douglas. *Evolutionary Biology*. 3d ed. Sunderland, Mass.: Sinauer Associates, 1998. A fine evolutionary biology text that discusses speciation in detail, with particular emphasis on evolutionary mechanisms and some discussion of population genetics. Written for upper-level students, but background information is provided. References at the ends of the chapters.

Kaul, T. N. *Introduction to Mushroom Science*. Enfield, N.H.: Science Publishers, 1997. Covers fungi systematics, among other topics.

Margulis, Lynn, and K. V. Schwartz. *Five Kingdoms: An Illustrated Guide to the Phyla of Life on Earth*. 3d ed. New York: W. H. Freeman, 1999. A heavily illustrated outline of the five-kingdom system. Excellent for all levels. Author Margulis was influential in gaining acceptance for Whittaker's five-kingdom classification system.

Purves, William K., and Gordon H. Orians. *Life: The Science of Biology*. 6th ed. Sunderland, Mass.: Sinauer Associates, 2001. A useful introductory college-level text. Covers systems of classification and taxonomy and each of the five major kingdoms. Although college level, this text will also be useful for the high school student.

Wolfe, Stephen L. *Biology: The Foundations*. 2d ed. Belmont, Calif.: Wadsworth, 1983. A clearly written, concise college textbook that will be useful to high school students as well. The chapter on classification and the diversity of life discusses the origins and hierarchy of classification in straightforward terms. A useful, brief introduction to each of the five kingdoms. The chapter also includes brief descriptions of the five-kingdom system of classification. Numerous useful illustrations; end-of-chapter questions.

TAILS

Types of animal science: Anatomy, evolution, physiology
Fields of study: Anatomy, conservation biology, developmental biology, entomology, evolutionary science, herpetology, invertebrate biology

The tail is a prolongation of the animal body that serves the animal as a means of defense, locomotion, stabilization, or ornament.

Principal Terms

ARBOREAL: living in trees
BURROWING INSECTIVORE: an insect-eating animal that usually lives in nests formed by digging holes or tunnels in the ground
INVERTEBRATE: animals without backbones, such as insects, frogs, and snakes
PREHENSILE TAILS: tails that are adapted for seizing and holding
VERTEBRATE: animals with backbones, such as mammals
VISCERA: any internal body organ, such as intestines or entrails

The tail is the prolongation of the backbone, beyond the trunk of the body, of any animal, insect, or fish. The tail of the vertebrate is composed of flesh and bone but does not contain any viscera. For many aquatic animals (such as fish and amphibians) as well as animals that use water as part of their living environment (such as crocodiles, otters, and whales), the tail is fundamental to their locomotive ability. Squirrels and other arboreal animals use the tail to keep balance and as a rudder when they jump from branch to branch, while others, such as the spider monkey and the chameleon, use the tail as an extra limb to increase their mobility through the branches of the rain forest. The tail may also be used as a means of defense for porcupines, as a warning signal in rattlesnakes, as a hunting weapon in alligators and scorpions, an ornamental sexual attractant in peacocks, or even a communication tool for dogs. Most birds do not carry a tail; instead, the prolongation has been fused into the short pygostyle bone, which serves as the holder of tail feathers and assists the birds in flying.

The shape, morphology, and structure of the tail vary according to the nature and behavior of the specific animal. Burrowing insectivores, like all other burrowing animals, usually have no tail. In contrast, climbing and running species have very large tails. Slow-moving animals, such as hedgehogs, have short tails, while chameleons have coiled tails.

Tails in Invertebrates

A number of lizards have thick tails that are covered by large, spiny, hard scales. The tail is often used as a defensive measure against predators such as snakes, especially when the head and body of the lizard are wedged between rocks. Moreover, lizards are capable of shedding their tails, which wriggle in a way that may confuse their predator, thus giving them enough time to escape. Each vertebra of such a tail has a fracture line along which it splits when the tail muscles contract. A unique lizard is the Solomon Islands skink, whose prehensile tail muscles are bound both to the vertebrae and to a fibrous sheath of collagen located under the skin, thus creating a much stronger and more flexible tail. Stimulation of the nerve contraction in the severed position keeps the tail moving for several seconds after severing occurs. Normally the tail splits off in one place, but in a few lizard species, such as the glass snakes

Many salamanders will wiggle their tails to distract predators; when the predator grabs at the tail, the salamander drops the appendage altogether and makes its escape. (Rob and Ann Simpson/Photo Agora)

(*Ophisaurus*), tails may be broken in more than one place. The stump usually heals very quickly and a new tail regenerates, although it is not as long and not as elaborate as the original. Bioengineers in the late 1990's conducted research trying to isolate the gene that is responsible for the regeneration of the severed tails.

Amphibians such as tadpoles and salamanders also lose and, in many cases, regenerate their tail, which has a spinal cord. Tadpole tails have a stiff rod for support, called the notochord, while the salamander's tail has a backbone with vertebrae. No tail is regenerated in the salamander if the spinal cord is severed, unlike the tadpoles, where the tail is reformed regardless of the fate of the severed spinal cord.

Tails in invertebrates are used in characteristic, unique ways among the different animals. Iguanas use their tails like large oars to swim in water, while tucking its legs close to its body. When threatened, the armadillo lizard puts its tail in its mouth, rolls over, and assumes the shape of a tight ball. Day geckos use their tails as a means of support while jumping from tree to tree to avoid their main predator, the falcon. The skink has a very short tail, whose shape is very close to that of its head. This confuses its enemies, since they do not know the direction the skink is going to take while escaping.

Tails in Vertebrates

The squirrel owes its name to its tail. The name is derived from the Greek word *skia* (meaning "shadow") and *oura* ("tail"), indicating that the tail is large enough to shade the rest of the animal body. Unlike the bat, which is the only mammal that truly flies, the flying squirrel is the only vertebrate that glides. Using its strong and sturdy back feet to jump from the top of a tree, it flattens its tail and spreads the loose folds of the skin so that it can glide in air. Just before landing, the gray squirrel lowers its tail first, then quickly lifts it and lands on its hind feet. When in danger, the red squirrel attempts to scare its predator by flicking its tail while using a series of noises, such as whistling, chattering, and chirping, and stomping its hind feet.

Other nonmammals, such as snakes, crocodiles, and turtles, may lose their tails to predators or to accidents, although not voluntarily. In fact, some snakes, such as the African python *Calabaria* and the oriental venomous *Maticora*, wave their thick, colored tails toward their enemy while retreating slowly. Both male and female diamondback rattlesnakes have the ability to rattle their tail ninety times per second. During the motion, the sonic muscles pump calcium out of the myoplasm fifty times faster than the locomotor muscles do. As a result, the filaments in the sonic muscles release each other and get ready for the next contraction much more quickly than the locomotor muscles.

Parrots (Psittaciformes) are the most popular birds that possess colorful and widely variable tails. In some species the tail is short, square, or rounded; in others it is long and pointed, but no parrots have forked tails. Birds that fly long distances tend to have longer tails, sometimes longer than the total length of their body. Climbing parrots usually possess rounded wings and blunt tails.

Long, elaborate tails are considered by evolutionists as unusual ornaments to win mates and use in elaborate courtship rituals. Wildlife scientists believe that the more attention-getting displays also give an indication of which bird will make a good parent. In agreement with Charles Darwin's theory of sexual selection, fe-

Tail Evolution

Various discoveries in Africa and Pakistan in the 1980's and 1990's have provided evidence that supports Charles Darwin's evolutionary idea that land animals may have the ability to evolve into aquatic animals, such as whales. The findings suggest that whales were actually land-dwelling mammals, once upon a time. The very well-preserved fossil whale, *Ambulocetus natans*, had a long mammalian tail, in addition to feet and legs. This is in strong disagreement with the creationist idea that discounts any possible slow transformation of a land mammal to an aquatic species. This observation also supports the principle that some anatomical features may either evolve or become atrophic as evolution progresses.

Recent evidence indicates that Ediacarans may have been the ancestors of modern life forms. These species emerged about six hundred million years ago and were the first large, complex life species. Although basic organs such as eyes, mouths, heads, and tails did not appear to be part of the original anatomy, paleontologists suggest that Ediacarans survived well into the Cambrian period, when evolution started taking place, and all the above organs slowly emerged and developed with time. The prehistoric ancestors of sturgeons, the largest and longest lived fish, appear to have been similar to sharks in appearance, with submarine-like bodies, strongly upturned heterocercal tails, and pronounced flattened snouts. Fossils of *Anomalocaris*, a large aquatic Cambrian predator, found in China and the Burgess Shale of North America, have provided concrete evidence that they must have had fanlike tails, which were used effectively while swimming.

Although fundamental, tails in several species, such as in stingrays, are not necessary for the animal to survive. Even the human being is believed to have had a tail, which with time has proven unnecessary. In the case of elephants, the trunk is much more useful than the tail. Adjustment to the environment has been established by experimental results conducted by Alan Harvey, who studied the development of baby crabs that were made to grow in and out of a shell. The tails of the latter grew to be symmetrical, while those in the shell became more asymmetrical, because their left sides had more room to grow than the right sides.

male animals of some species develop a preference for armaments that now have purely ornamental function, while others show preference for a certain trait which males eventually have to adopt if they are to mate. Male swallows that have long tails have a much higher degree of paternity and produce more biological offspring as compared to similar birds that are short tailed, indicating a distinct positive correlation between male tail length and paternity in this species.

The function of the heterocercal tail in shark locomotion has been given two explanations. The first one suggests that as a result of the lift created by beating the tail, the net force acting on the tail is directed dorsally and anteriorly. In the so-called Thomson's model, the tail generates a net force directed through the shark's center of gravity.

Sea animals use their tails in peculiar ways. In the depths of the species-rich Amazon River, electric fish and catfish predominate. Among the unusual incidents observed, electric fish appear to eat the tails of other fish. Eels plant their tails in burrows they dug in the sand underwater and let their bodies wave in the current, while waiting for food, such as drifting tiny crustaceans, fish eggs, and plankton, to reach them.

—*Soraya Ghayourmanesh*

See also: Anatomy; Beaks and bills; Bone and cartilage; Brain; Circulatory systems of invertebrates; Circulatory systems of vertebrates; Claws, nails, and hooves; Digestive tract; Ears; Endoskeletons; Exoskeletons; Eyes; Fins and flippers; Immune system; Kidneys and other excretory structures; Lungs, gills, and tracheas; Muscles in invertebrates; Muscles in vertebrates; Nervous systems of vertebrates; Noses; Physiology; Reproductive system of female mammals; Reproductive system of male mammals; Respiratory system; Sense organs; Skin; Teeth, fangs, and tusks; Tentacles; Wings.

Bibliography

Adler, Tina. "Record-Breaking Muscle Reveals Its Secret." *Science News* 150 (July 27, 1996): 53. This short article provides the biochemical explanation of why diamondback rattlesnakes' tails can rattle up to ninety times per second. The role of calcium production appears to be significant.

Coates, Michael J., and Martin J. Cohn. "Fins, Limbs, and Tails: Outgrowths and Axial Patterning in Vertebrate Evolution." *BioEssays* 20, no. 5 (May, 1998): 371. This general article discusses the development and evolution of organs that are essential for the animal locomotive capability.

Gould, Stephen Jay. "Hooking Leviathan by Its Past: Two Tales of Tails Confirm the Theory of the Whale's Return to the Sea." *Natural History* 103 (May, 1994): 8. This article correlates the discovery of prehistoric whale fossils with Charles Darwin's theory that land animals may have the ability to evolve with time into aquatic animals, such as whales.

Major, Peter F. "Tails of Whales and Fins, Too." *Sea Frontiers* 33 (March/April, 1987): 90-96. This is a review article that discusses the nature and function of tails in various sea animals, including the whales.

Stegermann, Eileen. "Sturgeon: The King of Freshwater Fishes." *The Conservationist* 49 (August, 1994): 18-23. This article discusses the evolution of the sturgeon from a prehistoric ancestor that was similar to the shark in appearance and the strongly upturned tails that were used for locomotive purposes when swimming.

Stewart, Doug. "The Importance of Being Flashy." *International Wildlife* 25 (September/October, 1995): 30-37. This article discusses the effect of males' big antlers or flashy tails in mating with females. A comparison of this observation is correlated to Charles Darwin's theories of sexual selection.

Wright, Karen. "When Life Was Odd." *Discover* 18 (March, 1997): 52-57. This article discusses the role of the Ediacarans, the most ancient, primitive, tailless, large, and complex animal structure, as the ancestors of modern-life species, and their slow evolution into the Cambrian period.

TASMANIAN DEVILS

Types of animal science: Anatomy, classification
Fields of study: Anatomy, developmental biology, ecology, evolutionary science, physiology, zoology

The Tasmanian devil is the world's largest carnivorous marsupial. It is now found only on the island of Tasmania, where it is reasonably common throughout the island and is strictly protected. Its noisy and screechy sounds make it seem much more ferocious than its size might indicate.

Principal Terms

ADAPTIVE RADIATION: the process by which many species evolve from a single ancestral species in adapting to new habitats

DINGO: the wild dog brought to Australia by the aborigines

MARSUPIAL: primitive mammals in which the young are born alive but in a very immature state

MARSUPIUM: the pouch in which the immature young remain attached to nipples until they complete their development

PLACENTAL MAMMAL: a mammal that possesses a placenta, an organ used to nourish the developing embryo

The Tasmanian devil belongs to a primitive group of mammals, the marsupials, which are found today primarily in Australia, Tasmania, New Guinea, southern Mexico, Central America, and South America. The name "marsupials" refers to the animal's marsupium, the pouchlike structure to which the immature young move after they are born, becoming attached for several months as they complete their development. The well-known pouch of kangaroos is the classic example of a marsupium, although it should be noted that a few marsupials do not have pouches and in some, including the Tasmanian devil, the pouch opens downward. During their long geological separation from the rest of the world, marsupials in Australia and South America underwent adaptive radiation, which produced an enormous diversity of forms as species became adapted to various habitats. As a result, many marsupials resemble placental mammals although they are not closely related. Thus, there are marsupials that resemble flying squirrels, moles, woodchucks, cats, and dogs. The Tasmanian devil, although a marsupial, has many similarities in structure and behavior to a dog.

Physical Characteristics

The devil belongs to a family of marsupials known as the Dasyuridae, which are found in Tasmania, Australia, New Guinea, and some nearby islands. The group includes the Tasmanian wolf or tiger, the numbat or banded anteater, and shrewlike, catlike, and ratlike forms. The Tasmanian devil is distinctive even within this distinctive group of marsupials. It received its name, "The Devil," from the early European settlers to Tasmania, who were awed by its fierce behavior and loud vocalizations. The devil actually is a stocky but short-limbed animal, doglike in shape, and only weighs between 4.5 and 12 kilograms, standing only about 30 centimeters tall. Its dark, mostly black color also contributes to its "evil" reputation. The animal does have a voracious appetite. It has a large head and very powerful jaws with long canine teeth. The devil is the largest carnivorous marsupial, extremely strong for its size; it can feed on animals larger than itself, including small kangaroos. It usually rests during the day and seeks food at night. Its food is varied and may include amphibians, lizards, rodents, insects, and poultry. The devil may prey on live animals or may feed on carrion. Its nightly movements may cover distances up to ten miles or more. The legs are short and the animal usually moves along slowly, although it is capable of running quickly for short distances.

The Tasmanian devil is a doglike marsupial. Its name derives not from its appearance, but from its fierce nature and loud cries. (Courtesy of Talune Wildlife Park)

Early Life

As is true of all marsupials, the young are born alive after a relatively short gestation period—about three weeks. Breeding usually takes place in early spring. The newborns travel from the birth canal to the pouch, which is backward-opening in this species, and attach themselves to a nipple, where they remain secure for about four months.

For the next several months, the young will venture outside of the pouch at times and, finally, become weaned and independent by the end of the year. Breeding usually commences between the second and third year. Reproduction is somewhat limited, as the mother has only four nipples and usually more than four young are born. It is not unusual for only two young to survive to weaning.

The Tasmanian devil has had a persecuted history, especially with the advent of European settlers. Its fierce appearance, loud screeches, and occasional predation on domestic livestock and poultry caused it to be hunted, trapped, and poisoned to the point of near extinction. Its value in cleaning up carcasses was not appreciated. The Tasmanian devil finally became protected by law in 1941. Going from outlaw to icon, the Tasmanian devil was selected as the symbol of the Tasmanian National Parks and Wildlife Service.

—*Donald J. Nash*

See also: Fauna: Australia; Kangaroos; Marsupials; Opossums; Reproduction; Reproductive systems of female mammals.

Tasmanian Devil Facts

Classification:
Kingdom: Animalia
Subkingdom: Bilateria
Phylum: Chordata
Subphylum: Vertebrata
Class: Mammalia
Subclass: Theria
Order: Marsupiala
Family: Dasyuridae
Genus and species: Sarcophilus harrisii

Geographical location: Tasmania, but formerly was also found in Australia

Habitat: From coastal areas to the mountains, in shrubs and forests

Gestational period: Twenty-one days, then remains in marsupium for five to six months

Life span: Between six and eight years

Special anatomy: A powerfully built, solid animal, with a large head, short tail, and marsupium

Bibliography

Darling, Kathy. *Tasmanian Devil*. New York: Lothrop, Lee, and Shepard, 1992. A picture book with numerous photographs, introducing the devil and its behavior, habitat, and other characteristics.

Kalman, Bobbie. *What Is a Marsupial?* New York: Crabtree Publishing, 2000. An informative account of the physical characteristics and behavior of the main types of marsupials, and the differences between them and placental mammals in reproduction, gestation, and birth.

Odgers, Sally, and Darrel Odgers. *Tasmania: A Guide*. 2d ed. New York: Simon & Schuster, 2000. A colorful guide to the geography and wildlife of Tasmania, including tales of Tasmanian devils and kangaroos.

Paddle, Robert. *The Last Tasmanian Tiger: The History and Extinction of the Thylacine*. New York: Cambridge University Press, 2001. An interesting account of a close relative of the Tasmanian devil, outlining problems faced by native marsupials on the island.

TEETH, FANGS, AND TUSKS

Type of animal science: Anatomy
Fields of study: Anatomy, conservation biology, developmental biology, entomology, herpetology, invertebrate biology

Teeth, fangs, and tusks are hard, resistant structures found in the jaws of vertebrates. Teeth are used primarily for catching and chewing food, while fangs and tusks are specialized for defense.

Principal Terms

ELAPIDS: a snake classification which includes cobras and rattlesnakes that have short, fixed front fangs

HETERODONT: having two or more types of teeth, such as molars and incisors

HOMODONT: having teeth all of the same type

INCISOR TEETH: teeth that are located in the front of the mouth and whose function is to tear, hold, and cut the prey

TOXIN: any substance, such as the venom in snakes or spiders, that is toxic to an animal

VIPERIDAE: poisonous terrestrial or semiaquatic snakes

VIPERS: a snake classification that includes copperheads and rattlesnakes, which have long, movable front fangs

Teeth, fangs, and tusks are found in vertebrates, and are used for obtaining and masticating food and for defense. Fangs and tusks are elongated canine or incisor teeth, which serve to deliver venom (fangs) or as defensive weapons (tusks). Tusks, which are made of ivory, have been valued for making jewelry and decorative objects, which has led to the endangerment of the elephants and walruses that carry them.

Teeth

Teeth are hard, resistant structures that are found on the jaws, as well as in or around the mouth and pharynx of vertebrates. Teeth were formed through the evolution of bony structures found in primitive fish. A tooth consists of a crown and one or more roots. The crown is the visible, functional part, while the root is attached to the tooth-bearing bone. However, many living organisms, such as birds, turtles, whales, and many insects, do not have teeth.

Although the teeth of many vertebrates have been adapted for special uses, their function is to catch and masticate food and defend against predators or enemies, as well as other specific purposes. Rodents and rabbits, for instance, have curved incisors that are deeply embedded in the jaws and grow longer with age. Several types of apes have enlarged canines for defense, while the sawfish, which is the only animal with teeth completely outside its mouth, uses its teeth to attack its prey. All perissodactyls, a group of herbivorous animal species characterized by an odd number of toes on the hind foot, including horses, rhinoceroses, and tapirs, have evolved specialized forms of teeth that are adapted for grinding. Generally, lizards are insectivores and have sharp tricuspid teeth that are adapted for grabbing and holding. Mollusk and crustacean feeders, such as the caiman lizard, have blunt, rounded teeth along the jaw margin or on the palate.

Tooth-bearing animals may be heterodont or homodont. Most mammals are heterodont and carry two or more types of teeth, such as the incisors and the molars. The purpose of the incisors is to tear and bite into the food, while the molars crush and grind the food. Cats do not have flat-crowned crushing teeth; instead only the stabbing and anchoring canine teeth cut up the food, which is then swallowed. In the case of elephants, the upper second incisors have developed into ivory tusks, which are the longest and heaviest teeth in any living animal. On the other hand, fish and most reptiles are homodonts—their teeth are all of about the same size, and their purpose is to catch prey. This is the main reason why their teeth are regularly replaced during their life span.

Snakes have teeth that curve back toward the throat. Thus, as soon as the prey is caught, it is pushed into the throat. In poisonous snakes, teeth called fangs have a canal through which the poison (venom), normally stored in glands that are in the roof of the mouth, may be ejected.

Fangs

Fangs are long, pointed teeth used by many animals as a means of self-defense, for securing their prey, or other reasons. Many snakes use their hollow fangs as hypodermic needles to puncture their victim and, in the case of a venomous snake, to inject toxic venom. Because of the snake's muscular elasticity, the fang tips at penetration average 112 percent further apart than their bases at rest. The wound resulting from penetration of the flesh is called a snakebite. A nonvenomous snakebite is usually similar to a puncture wound which, when untreated, may become infected and in extreme cases may cause gangrene. That of a venomous snake is much more serious and there is a potentially lethal effect, which depends on several factors such as the size of the victim, the bite location, the quantity of venom that has been injected, the speed of venom absorption into the victim's blood circulation, and the speed with which first aid and the antidote are given.

Several venomous snakes, such as the ringhals and the black necked cobra (types of African cobras), have the ability to spit. A fine stream of venom is forced out of each fang which, instead of having to go through a straight canal that ends in a long opening, is forced through a different canal that turns sharply forward to a small round opening on the front surface. Contraction of the muscle that surrounds the poison gland leads to the spitting of the venom, which is harmful to human eyes unless washed quickly. Front-fanged snakes include pit vipers such as rattlesnakes, fixed-fanged snakes such as brownsnakes and cobras, seasnakes, and true vipers (Viperidae). All have hollow, tubelike fangs created by the extension of the dentine across the anterior seam. Evolutionary herpetologists have postulated that this anterior seam may have been open several tens of thousands of years ago.

Only poisonous snakes possess fangs and venom glands, which are considered to be an evolutionary result of salivary glands found in primitive fish. Poisonous snakes bite the victim with their fangs and proceed to inject the venom in the wound in their effort to kill their prey. The venom contains toxins, which are chemical compounds that have the potential of attacking the blood and the nervous system with lethal consequences. The proteinaceous enzymes that are found in the venom are also used to digest the eaten animal. These snakes are generally subdivided into vipers and elapids. Vipers, which include copperheads and rattlesnakes, have long, movable front fangs, while elapids, such as cobras and rattlesnakes, have short front fangs that are fixed. Occasionally some venomous snakes have a fang on the upper jaw in the rear of the mouth, which makes them less harmful to large animals since the venom injection occurs at a much slower pace. The devouring or swallowing of the prey usually takes place only after the venom has taken full effect and the animal is dead. Generally, small snakes embed their fangs in the prey for a longer duration than larger snakes.

The analysis and determination of the composition of the hard dental tissues of the Viperidae has been conducted using classical microscopy, scanning electron microscopy, transmission electron microscopy, X-ray diffraction, and infrared spectroscopy. The results have shown a thin, calcified outer layer, composed of very small needlelike crystals that are randomly distributed. The calcified outer layer contains pores and collagen fibers that are incompletely mineralized, especially in the wall of the poison canal. Chemical analysis of the dentine has indicated a poorly mineralized apatite with a high level of carbonate content.

All six species that belong to the marine fish classified under the genus *Chauliodus* (order Salminoformes) are called viperfish. They are all small in size, the largest being the Pacific viperfish (*C. macouni*), which is no more than one foot long. They are characterized by their long fangs, which protrude from the upper and lower jaws and are used to securely grab their prey. The tigerfish of the Zambezi river, with its two rows of long and very sharp teeth, has been seen as more fearsome than the true piranha or the mako shark.

Tusks

Evolutionary theories suggest that the tusks of both the walrus and the wild boar are enlarged canine teeth, while in the case of the pig, the lower incisor has been modified with time into an organ that is used for digging purposes. In male Indian elephants and African elephants of both sexes, the tusk is the upper incisor, which continues to grow throughout their lifetime. The female Indian elephant has either no tusks or very small ones. Male Ceylonese elephants, found in Sri Lanka, generally have no tusks, while Sumatran elephants, found in Indonesia, bear the longest tusks. Elephant tusks from Africa are typically six feet long, conical at the end, and weigh approximately fifty pounds each. Indian elephants have slightly smaller tusks. Each tusk from the largest pair, recorded and exhibited at the British Museum, is twelve feet long, has a barrel circumference of eighteen inches and weighs close to 150 pounds.

Studies by Raman Sukumar, an ecologist at the Indian Institute of Science in Bangalore, and Milind Watve, a microbiologist, indicate that male Asian elephants with longer tusks are prone to

Elephant tusks evolved from the upper incisors, whereas walrus and boar tusks are enlarged canine teeth. A genetic link has been shown between the length of a male elephant's tusks and his resistance to parasites, which increases his genetic fitness. (Digital Stock)

host many fewer parasites. This is in agreement with the theory of the evolutionary biologist William D. Hamilton, of the University of Oxford, who proposed in 1982 that males that carry genes resistant to parasites have the ability to be healthier and therefore live longer. At the same time, these species develop secondary sexual characteristics which enable females to select males that will produce better offspring.

Anatomically, the tusk is composed of several layers, the innermost of which grows the latest. One third of the tusk is embedded in the bone sockets of the elephant's skull. At the beginning, the head end of the tusk has a hollow cavity that becomes almost fully solid with aging. Only a narrow nerve channel runs through the center of the tusk to its end.

Ivory is a type of dentin that is the major component of the elephant tusk and is desirable worldwide for its beauty and durability. There are generally two types of ivory, soft and hard. The ivory isolated from the tusks from East African elephants is soft, while that found in the West African elephants is hard. Hard ivory is usually darker in color and is straighter than the soft type, which has a fibrous internal texture, is less brittle, and is a more opaque white. The demand for ivory has led to a large number of elephant slaughtering incidents and a dangerous decline of the African elephant population, beginning in the late nineteenth century and continuing up to the twenty-first. The discovery by archaeologists in the early 1980's of tusk material in a Greek ship that was sunk around 1400 B.C.E. revealed that ivory was a trade commodity even during the Bronze Age.

Elephants belong to the order Proboscidea, whose early ancestors were not larger than the average pig. It is believed that during the process of evolution, the lower jaw elongated beyond the upper and eventually turned into tusks. As a result, the nose and upper lips developed into an elongated cover to the projecting lower jaw. During the Eocene Epoch (between fifty-four and forty million years ago) the upper tusks were lost, and a downward-hooked, tusk-tipped mandible developed. The mandible and its tusks became more shovel-like during the Miocene era (twenty-six to seven million years ago). It is believed that the tusks assumed a cylindrical shape not very much later. It appears that tusks were also part of the anatomy of the woolly mammoth, as seen in the specimen that died about twenty thousand years ago and was discovered in a Siberian excavation in the 1990's.

—*Soraya Ghayourmanesh*

See also: Anatomy; Beaks and bills; Bone and cartilage; Brain; Circulatory systems of invertebrates; Circulatory systems of vertebrates; Claws, nails, and hooves; Digestive tract; Ears; Endoskeletons; Exoskeletons; Eyes; Fins and flippers; Immune system; Kidneys and other excretory structures; Lungs, gills, and tracheas; Muscles in invertebrates; Muscles in vertebrates; Nervous systems of vertebrates; Noses; Physiology; Reproductive system of female mammals; Reproductive system of male mammals; Respiratory system; Sense organs; Skin; Tails; Tentacles; Wings.

Bibliography

Bagla, Pallava. "Longer Tusks Are Healthy Signs." *Science* 276 (June 27, 1997): 1972. A discussion of the study by Raman Sukumar and Milind Watve shows that the longer the tusks an Indian elephant has, the less the parasites acquired and the healthier the animal.

Bower, Bruce. "Bronze Age Trade Surfaces from Wreck." *Science News* 126 (December 8, 1984): 359. This article discusses the discovery of the wreck of a Bronze Age Greek cargo ship which sank in the Mediterranean around 1400 B.C.E., which contained the earliest historical evidence that ivory was valuable to human cultures.

Chapple, Steve. "Fish with Fangs." *Sports Afield* 216 (June/July, 1996): 144. This article relates the experience of the author with the tigerfish of the Zambezi River and emphasizes the devastating ability of the fishes' very sharp teeth.

Coppola, M., and D. E. Hogan. "When a Snake Bites." *Journal of the American Osteopathic Association* 94, no. 6 (June, 1994): 766. This article deals with the mechanism and effect of the snakebite and the after effects, with a special coverage of the supportive treatment of the patient and description of the physiologic effects.

Hayes, W. K. "Ontogeny of Striking, Prehandling, and Envenomation Behavior of Prairie Rattlesnakes (*Crotalus v. viridis*)." *Toxicon* 29, no. 7 (1992): 867-875. This article compares the rattlesnake's pattern of striking and poisoning of deer mice with particular emphasis on the quantity of venom used versus the length of the snake. The experimental procedure for the determination of the amount of venom used is described.

TENTACLES

Types of animal science: Anatomy, classification
Fields of study: Anatomy, invertebrate biology, marine biology

A tentacle is a slender, leglike or armlike protrusion from the body of an animal. It is used for protection, as an organ of touch, or to capture food.

Principal Terms

BUD: protuberance used in asexual reproduction
BUDDING: bud development into a complete organism
CARRION: dead animals
FUNNEL: an opening in a cephalopod mantle, providing oxygen and propulsion
GAMETE: sperm or an egg
NEMATOCYST: poison sting cell
RADULA: tonguelike, toothed organ that grinds food and drills holes in shells of prey

Tentacles are slender, leglike or armlike protrusions from the body of a living organism. They are used for protection, as organs of touch, or to capture food. They are often seen in coelenterates, an animal phylum that includes jellyfish, anemones, and coral polyps. The name *coelenterate* comes from the Latin for "hollow intestine."

Coelenterates are hollow tissue sacs with two layers of cells in their walls. The cells carry out digestion, excretion, and reproduction. There is one body opening, a mouth. Food enters it, is digested, and used by the cells. Wastes also leave through the mouth, surrounded by long, slender sense organs called tentacles, which also grab food and pull it into the mouth.

Box Jellyfish: Poison Tentacles

Box jellyfish (*Chironex fleckeri*), native to Australia's north coast and Southeast Asian coastlines, are the most poisonous creatures on earth. They get their name from their box-shaped, translucent bodies, which are approximate 1.25 feet wide. Sixty tentacles, up to 15 feet long and 0.25 inches in diameter, hang from the box.

The box holds an eye and a mouth. Prey—plankton, small fish, and shrimp—are caught in tentacles and pulled into the mouth. Tentacles hold huge numbers of sting cells called nematocysts. When something comes in contact with the tentacles, the nematocysts inject a poison which paralyzes and kills their prey, and can cause extreme pain and even death to large animals or humans. Severe stings will kill within four to six minutes after a human is stung.

Worm and Mollusk Tentacles

Some worms have tentacles. For example, sandworms (family Nereidae) inhabit shallow ocean waters worldwide. Most live in sand or mud burrows. They have colorful bodies ranging from one inch to three feet long and are in the same phylum as earthworms. A sandworm body has several hundred segments. Each segment has two muscular parapodia, both attached to bundles of bristly chaetae. The first segment of a sandworm holds two light-sensitive tentacles, four eyes, and two sensory palps. The second segment has eight more tentacles called cirri. The palps and cirri help sandworms find crustaceans, small fish carrion, and other prey.

Many cephalopods have tentacles. For example, the nautilus (family Nautilidae) is found in South Pacific and Indian oceans. All nautilids have soft bodies with spiral, brown and white shells.

Octopuses have suckers on their tentacles which are used to catch, hold, and explore objects around them. (Digital Stock)

Their mouths have beaks and radulas (toothed tongues) to eat carrion and crustaceans on ocean bottoms. Around the nautilid mouth are almost one hundred short tentacles that are used to sense objects, feed, and move. At one side of the arms, an opening called a funnel allows water to enter and carry oxygen to gills. A nautilid can also spray a jet of water from the funnel, to swim.

Octopus and Squid Tentacles

Octopuses (family Octopodidae) have eight slender, flexible tentacles. On the underside of each arm is a row of suckers. Sensors in suckers serve in defense, detect prey, capture it, and identify its texture, shape, and taste. Arms join at their bases into a bulb-shaped head/body.

An octopus squeezes water out of the mantle cavity and moves its tentacles to swim. It captures mollusk and crab prey by wrapping tentacles around them and using its suckers to tear them to pieces. In the center of the arm juncture is its mouth, which has a radula and a beak to continue shredding prey.

The giant squid (*Architeuthis harveyi*) inhabits the Atlantic Ocean depths. Its torpedo-shaped body is up to sixteen feet in diameter and twenty-five feet long from head to tail. This cephalopod tapers toward its posterior and has at the tail end two fins for swimming and steering. Two eyes provide excellent vision. In front of the eyes are eight round, elongated tentacles with suckers and hooks on their undersides. Two longer tentacles lack suckers and hooks, except at the tip. The tentacles are thirty-six feet long, and from tail to tentacle tip giant squid often exceed sixty feet in length. They are earth's largest invertebrates. Squid hide near the ocean bottom and ambush or pursue prey. The two long tentacles are shot forward to seize their victims, which are passed along to the mouth. A beak in the mouth crushes and tears up prey.

Squid and octopus are eaten by humans worldwide. Nautilids, sandworms, and related organisms help maintain the balance of nature, eating carrion and helping to keep the ocean clean. Other tentacled organisms are eaten by fish that humans use as food. On the other hand, squid and octopuses eat food fish and crabs, competing with humans, and jellyfish can kill via the nematocysts in their tentacles.

—*Sanford S. Singer*

See also: Anatomy; Circulatory systems of invertebrates; Digestive tract; Eyes; Fins and flippers; Invertebrates; Jellyfish; Marine animals; Marine biology; Mollusks; Muscles in invertebrates; Octopuses and squid; Physiology; Poisonous animals; Sense organs; Skin; Worms, segmented.

Bibliography

Hunt, James C. *Octopus and Squid*. Monterey, Calif.: Monterey Bay Aquarium Foundation, 1997. A survey of world cephalopods, written by a marine biologist for a popular audience.

León, Vicki. *A Tangle of Octopuses: Plus Cuttlefish, Nautiluses, and a Giant Squid or Two.* Parsippany, N.J.: Silver Burdett Press, 1999. Describes the physical characteristics, behavior, diet, life cycle, and intelligence of these cephalopods.

Taylor, Leighton R., and Norbert Wu. *Jellyfish*. Minneapolis: Lerner, 1998. Describes the life cycles, habitats, and physical structure of jellyfishes.

Yonge, Charles M., and T. E. Thompson. *Living Marine Molluscs*. London: Collins, 1976. A classical work in mollusks, including good illustrations and a bibliography.

TERMITES

Types of animal science: Behavior, classification
Fields of study: Anatomy, entomology, invertebrate biology, physiology

Termites are cellulose-eating social insects that live in colonies of hundreds to millions of individual members produced by a single king and queen. Termites are highly beneficial to the ecosystem because they break down dead plant material, but some species can become serious pests when they colonize manmade wooden structures.

Principal Terms

ALATES: recently molted winged adult termites
CARTON: cardboardlike material composed of wood fragments, saliva, and fecal matter, used for constructing termite nests
CELLULOSE: fibrous polysaccharide that chiefly constitutes the cell walls of plants
PHEROMONE: chemical substance produced by an animal that usually elicits certain behavioral responses in other animals of the same species
PROTOZOAN: mobile, one-celled animal
REPRODUCTIVES: sexually mature male and female
SYMBIOTIC: having a mutually beneficial relationship

Termite fossils date from about 130 million years ago, but they probably evolved much earlier from a primitive, wood-eating, roachlike ancestor. There are about 1,900 termite species divided among six families. Five of the families are considered primitive or lower termites because, like their primitive wood-eating roach relatives, they harbor symbiotic protozoa in the hindgut that digest cellulose. Without these protozoans, the termites would starve to death. The higher termite family, Termitidae, is the largest family, containing about 75 percent of all termite species. Higher termites may be able to digest cellulose themselves, or bacteria in the gut may secrete enzymes to aid in digestion.

Caste System and Nests

Termites are notable for their highly organized societies. Because most termites are effectively deaf and blind, they communicate through touch, smell, and taste. Most species are divided into castes of reproductives, workers, and soldiers. Normally, there is only one reproducing pair, the primary reproductives, or queen and king. Secondary reproductives are present, too, in case the queen or king dies.

The sterile workers and soldiers are of both sexes. Workers care for the eggs and nymphs, provide food for the nymphs, soldiers, and reproductives, and construct, repair, and maintain the nest. Soldiers have evolved modified heads and jaws for defending the nest. The heads are large and hard with powerful, scissorlike mandibles or long, tubular snouts that squirt sticky chemicals. Some soldiers have both formidable jaws and chemical weapons.

Termites are vulnerable to desiccation, changes in temperature, and hungry ants, birds, aardvarks, and other predators. They maintain a moist, temperature-controlled, safe environment by constructing nests. Dry-wood termites never touch the soil but nest in the wood they feed upon, gnawing out tunnels and chambers inside living tree trunks and branches, rotting logs, or furniture and wooden buildings. Subterranean termites must maintain contact with the soil for food sources such as grass and humus or for moisture.

Termite Facts

Classification:
Kingdom: Animalia
Subkingdom: Bilateria
Phylum: Arthropoda
Subphylum: Uniramia
Class: Insecta
Subclass: Pterygota
Order: Isoptera (termites)

Geographical location: Every continent except Antarctica

Habitat: Majority of species inhabit tropical rain forests but many live in temperate and subtropical zones, deserts, or mountains up to an altitude of 2,500 meters (8,200 feet)

Gestational period: Varies but averages approximately two weeks

Life span: The primary queen and king may live to seventy years, although ten or twenty years is more common; workers and soldiers live from two to five years

Special anatomy: Isoptera means "equal wings"—adult reproductives have two oval-shaped pairs of overlapping wings of nearly equal length; small to medium-sized, very soft, usually light-colored bodies; thorax fused with the abdomen; workers have specialized mouthparts for chewing wood, while soldiers' mouthparts have been modified for fighting; except for reproductives, termites usually lack eyes

Many species are master builders, constructing elaborate nests of carton in trees with covered runways leading to the ground or mounds in the soil complete with ventilation shafts, towering chimneys up to 9 meters (29.5 feet) high, and even fungus gardens to supplement their cellulose diet. Although the architecture varies widely, most nests provide an inner chamber for the egg-laying queen and her king, and areas for brood chambers and food storage.

Life Cycle

Unlike insects such as butterflies, termites undergo incomplete or gradual metamorphosis. From the time they hatch, immature termites looks like pale, wingless, miniature versions of adults. To allow for growth, these nymphs molt periodically, shedding and then eating their outer skin. All nymphs start out the same. The correct balance among the castes appears to be maintained by bodily secretions

Termites, like ants, create large mounds for their colonies to live in. (Digital Stock)

containing hormones that are transferred by licking, but the mechanism is not yet understood.

New colonies are usually formed when alates swarm during certain seasons. Workers prepare by digging tunnels to the surface, with exit holes and sometimes launching platforms. Once the alates leave, soldiers prevent them from returning. They are weak fliers and usually descend within a few hundred meters of the original nest. Wings are shed after landing. Females attract males by raising the abdomen and emitting a pheromone. Before mating, pairs locate a likely site for a nest and seal the entrance with fecal matter. The first batch of eggs is usually small. The king and queen take care of the eggs until there are enough older nymphs to take over. After a few years, the queen's ovaries and abdomen increase in size, and her egg laying accelerates. She may grow to 11 centimeters (4.3 inches) long and produce up to 36,000 eggs per day.

—Sue Tarjan

See also: Arthropods; Bees; Communication; Digestion; Digestive tract; Home building; Insect societies; Insects; Metamorphosis; Nesting; Wasps and hornets.

Bibliography

Behnke, Frances L. *A Natural History of Termites*. New York: Charles Scribner's Sons, 1977. Written for young adults, this is a fascinating and comprehensive study.

Choe, Jae C., and Bernard J. Crespi, eds. *The Evolution of Social Behavior in Insects and Arachnids*. New York: Cambridge University Press, 1997. Recent research into social evolution and behavior of insects (including termites) and arachnids. For the scholarly.

Harris, W. Victor. *Termites: Their Recognition and Control*. London: Longmans, 1961. First report in English detailing the severity of termite damage to manmade structures worldwide. Impressive photographs.

Krishna, Kumar, and Frances M. Weesner, eds. *Biology of Termites*. 2 vols. New York: Academic Press, 1969-1970. Definitive general reference of termite biology.

Telford, Carol, and Rod Theodorou. *Through a Termite City*. Des Plaines, Ill.: Heineman Interactive Library, 1998. Although designed for a juvenile audience, this delightful and informative photographic depiction of life in a termite mound appeals to all ages.

TERRITORIALITY AND AGGRESSION

Type of animal science: Ethology
Fields of study: Ethology, invertebrate biology, zoology

Aggressive behavior and territoriality are common features of animals. Territories may differ in function across species, but general trends occur. Territoriality is best viewed as a means by which individuals maximize their own reproductive success rather than as a mechanism of population regulation.

Principal Terms

ADAPTIVE FUNCTION: the reason that a characteristic evolved by means of natural selection

CONSPECIFICS: members of the same species

DOMINANCE HIERARCHY: a social system, usually determined by aggressive interactions, in which individuals can be ranked in terms of their access to resources or mates

HOME RANGE: an area that an animal frequently uses but does not defend

POPULATION REGULATION: long-term stability of population size at a level that prohibits overexploitation of resources

RESOURCE-HOLDING POTENTIAL: the ability of an individual to control a needed resource relative to other members of the same species

STRATEGY: a behavioral action that exists because natural selection favored it in the past (rather than because an individual has consciously decided to do it)

TERRITORIALITY: the active defense of an area that is required for survival and/or reproduction

Any field or forest inhabited by animals contains countless invisible lines that demarcate territories of individuals of many different species. Humans are oblivious to these boundaries yet have quick perception of human property lines; other animals are equally oblivious to human demarcations. Most organisms, in fact, appear to attend only to the territorial claims made by members of their own species. If separate maps of individual territories could be obtained for each species in the same habitat and superimposed on one another, the resulting hodgepodge of boundaries would show little consensus on the value of particular areas. Yet basic similarities exist in why and how different species are territorial.

Causes of Territoriality

The existence of aggression and territorial behavior in nature hardly comes as a surprise. Even casual observations at a backyard bird feeder reveal that species that are commonly perceived as friendly can be highly aggressive. The observation of birds at feeders can lead to interesting questions concerning territorial behavior. For example, bird feeders usually contain much more food than any one bird could eat: Why, then, are aggressive interactions so common? Moreover, individuals attack conspecifics more often than birds of other species, even when all are eating the same type of seeds.

Aggressive defense of superabundant resources is not expected to occur in nature; however, bird feeders are not a natural phenomenon. Perhaps the aggressive encounters that can be observed are merely artifacts of birds trying to forage in a crowded, novel situation, or perhaps bird feeders intensify aggressive interactions that occur less frequently and less conspicuously in nature. While the degree to which aggression observed at feeders mirrors reality is open to question, the observation of a greater intensity of interactions between conspecifics definitely reflects a natural phenomenon. Members of the same species are usually more serious competitors than are members of different species because they exploit exactly the same resources; members of different species might only share a few types of resources. Despite the ecological novelty of artificial feeders, noting which individuals win and lose in such an encounter can provide valuable information on the resource-holding potential of individuals that differ in various physical attributes such as body size, bill size, or even sex. For organisms that live in dense or remote habitats, this type of information can often be obtained only by observations at artificial feeding stations.

Territorial defense can be accomplished by visual and vocal displays, chemical signals, or physical encounters. The sequence of behaviors that an individual uses is usually predictable. The first line of defense may involve vocal advertisement of territory ownership. One function of bird song is to inform potential rivals that certain areas in the habitat are taken. If song threats do not deter competitors, visual displays may be employed. If visual displays are also ineffective, then residents may chase intruders and, if necessary, attack them. This sequence of behaviors is common in territorial interactions because vocal and visual displays are energetically cheaper than fighting and involve less risk of injury to the territory owner.

It may be less obvious why fighting is a necessary component in territorial interactions for both territory owners and intruders. Without the threat of bodily injury, there is no cost to intruders that steal the resources of another individual. This would severely hamper an owner's ability to control an area. On the other hand, if intruders never physically challenge territory owners, then it would pay for all territory owners to exaggerate their ability to defend a resource. Thus, physical aggression may be essential. Animals do not frequently kill their opponents, however, so there must be something that limits violence. Various species of animals possess formidable weapons, such as large canine teeth or antlers, that are quite capable of inflicting mortal wounds. Furthermore, a dead opponent will never challenge again. Yet fights to the death are rare in nature. When they do occur, some novel circumstance is usually involved, such as a barrier that prohibits escape of the losing individual. Restraint in normal use of weapons, however, probably does not indicate compassion among combatants. Fights to the death may simply be too costly, because they would increase the chance that a victor would suffer some injury from a loser's last desperate attempts to survive.

Functions of Territory

Territories can serve various functions, depending on the species. For some, the area defended is only a site where males display for mates; for others, it is a place where parents build a nest and raise their offspring; for others, it may be an all-purpose area where an owner can have exclusive access to food, nesting sites, shelter from the elements, and refuge from predators. These different territorial functions affect the area's size and the length of time an area is defended. Territories used as display sites may be only a few meters across, even for large mammal species. Territorial nest sites may be smaller still, such as the densely packed nest sites guarded by parents of many colonial seabirds. All-purpose territories are typically large relative to the body size of the organism. For example, some passerine birds defend areas that may be several hundred meters across. Although all three types of territories may be as ephemeral as the breeding season, it is not uncommon for all-purpose territories to be defended year around.

The abundance and spatial distribution of needed resources determine the economic feasibility of territoriality. On one extreme, if all required resources are present in excess throughout the habitat, territory holders should not have a reproductive advantage over nonterritory holders. At the other extreme, if critical resources are so rare that enormous areas would have to be defended, territory holders might again have no reproductive advantage over nonterritory holders. If needed resources, however, are neither superabundant nor extremely rare and are somewhat clumped in the habitat, territoriality might pay off. That is, territorial individuals might produce more offspring than nonterritorial individuals.

Studies of territoriality raise more questions than biologists can answer. Researchers investigate how large an area an individual defends and whether both sexes are equally territorial. They seek to determine whether the territories of different individuals vary according to quality. The density of conspecifics may influence territoriality; on the other hand, territoriality itself may serve to regulate population size, although evidence suggests that this is an incidental effect.

All-purpose territories vary considerably in size, depending on the resource requirements of the individuals involved and the pattern of temporal variation in resource abundance. In some organisms, individuals only defend enough area to supply their "minimum daily requirements." In others, individuals defend a somewhat larger area—one that could still support them even when resource levels drop. In others, individuals defend territories that vary in size depending on current resource levels. For example, pied wagtails (European songbirds) defend linear territories along riverbanks that are about six hundred meters long during the winter. The emerging aquatic insects they consume are a renewable resource, but renewal rates vary considerably during the season. Rather than adjusting territory size to match the current levels of prey abundance in the habitat, wagtails maintain constant territory boundaries. This inflexibility persists even though territories that extend for only three hundred meters could adequately support an individual for about one-third of the season. In contrast, the territory size of an Australian honey eater varies widely during the winter. Nectar productivity of the flowers visited by honey eaters varies considerably during the season. By adjusting territory size to match changing resource levels, individual birds obtain a relatively constant amount of energy each day (about eighteen kilocalories).

Territorial Roles

In some species, only males are territorial. In other species, both sexes defend territories, but males defend larger territories than females do. In some mammals in which both sexes are territorial, males are aggressive only to other males, and females are only aggressive to other females. In these species, male territories are sufficiently large to encompass the territories of several females. Presumably, these males have increased sexual access to the females within their territories. Perhaps the most curious example of sex-specific territorial behavior is observed in a number of coral reef fish, in which all individuals in the population are initially female and not territorial. As the individuals grow older and larger, some develop into males. Once male, they engage in territorial behavior.

Within a species, significant variation in territory quality exists among individuals. Studies on numerous species have demonstrated a relationship between territory quality and an individual's resource-holding potential. For example, larger individuals tend to control prime locations more often than smaller individuals. In addition, possession of higher-quality territories often results in increased reproductive success. For some species, this occurs because individuals with better territories obtain mates sooner or obtain more mates than individuals with poorer territories. In other species, possession of superior territories increases the survival chances of the owner.

As the density of conspecifics increases, the ability of individuals to control territories decreases. In some species, the territorial system may break down completely, with all individuals scrambling for their share of needed resources in a chaotic fashion. In other species, the territorial system is replaced by a dominance hierarchy. All competitors may remain in the area, but their access to resources is determined by their rank in the hierarchy. For example, elephant seal males can successfully defend areas containing from eighty to a hundred females from other males. Very dense clusters of females, however (two hundred or more), attract too many males for one male to monopolize. When this happens, one male—usually the largest male—dominates the rest and maintains disproportionate access to females.

Territoriality undeniably has an adaptive function: to increase the survival and reproductive success of individuals. Territoriality can also have several possible incidental effects, one of which was once considered to be an adaptive function: serving as a means of population regulation. The reasoning behind this hypothesis is simple. The number of territories in a habitat would limit the number of reproducing individuals in a population and would thereby prevent overpopulation that could cause a population crash. Support for this hypothesis would include demonstration that a significant number of nonbreeding adults exist in a population. Indeed, for several species, experimental removal of territory owners has revealed that "surplus" individuals quickly fill the artificially created vacancies. In most of the species studied, however, these surplus individuals are primarily males. Population growth can be curbed only by limiting the number of breeding females, not the number of breeding males. Furthermore, the population regulation argument assumes that some individuals abstain from reproduction for the good of the population. If such a population did exist, a mutant individual that never abstained from reproducing would quickly spread, and its descendants would predominate in future generations.

Territoriality in the Field

Territoriality is typically investigated in the field using an observational approach. Initial information collected includes assessing the amount of area used by each individual, how much of that area is defended from conspecifics, and exactly what is being defended. It is relatively easy to discern the spatial utilization of animals. For many species, all that is required is capturing each individual, marking it for field identification, and

While many animals engage in aggressive behavior and may wound each other severely, fights to the death are rare in nature. (Digital Stock)

watching its movements. For species that range long distances, such as hawks or large mammals, and species that are nocturnal, radio telemetry is frequently used. This methodology requires putting radio transmitters on the individuals to be followed and using hand-held antennas, or antennas attached to cars or airplanes, to monitor movements. For fossorial species (animals that are adapted for digging), animal movements are often determined by repeated trapping. This method involves placing numerous baited live traps above the ground in a predetermined grid.

Knowing the spatial utilization of an animal does not document territoriality. Many types of animals repeatedly use the same regions in the habitat but do not defend these areas from conspecifics. Such "home ranges" may or may not contain areas that are defended (that is, territories). Territorial defense can be readily documented for some animals by simply observing individual interactions. These data often need to be supplemented by experiments. Behavioral interactions might only occur in part of the organism's living space because neighbors do not surround it. For these individuals, researchers play tape-recorded territorial vocalizations or place taxidermy mounts of conspecifics in different locations and note the response of the territory holder. For other species, such as fossorial rodents, direct estimates of territory size cannot be obtained because aggressive interactions cannot be observed; as a result, territory boundaries must be inferred from trapping information. Regions in which only the same individual is repeatedly trapped are likely to be areas that the individual defends. This is an indirect method, however, and can be likened to watching the shadow of an organism and guessing what it is doing.

It is often difficult to determine exactly what an animal is defending in an all-purpose territory where organisms use many different types of resources. Which resource, that is, constitutes the "reason" for territorial defense? On the other hand, several resources may contribute in some complex way. For many species these things simply are not known. This uncertainty also complicates estimates of territory quality. For example, red-winged blackbirds in North America have been particularly well studied for several decades by different investigators in various parts of the species range. Males defend areas in marshes (or sometimes fields), and some males obtain significantly more mates than others. Biologists think that males defend resources that are crucial for female reproduction. Some males may be more successful at mating than others because of variation in territory quality. Yet the large number of studies done on this species has not yielded a consensus on what the important resources are, whether food, nest sites, or something else.

Theoretical investigations of territorial behavior often employ optimality theory and game theory approaches. Optimality theory considers the benefits and costs of territorial defense for an individual. Benefits and costs might be measured simply as the number of calories gained and lost, respectively. Alternatively, benefits might be measured as the number of young produced during any one season; costs might be measured as the reduction in number of future young attributable to current energy expenditures and risks of injury. For territorial behavior to evolve by means of natural selection, the benefits of territorial behavior to the individual must exceed its costs.

Game theory analyses compare the relative success of individuals using alternative behaviors (or "strategies"). For example, two opposing strategies might be "defend resources from intruders" and "steal resources as they are encountered." In the simplest case, if some individuals only defend and other individuals only steal resources, the question would be which type of individual would leave the most offspring. Yet defenders interact with other defenders as well as with thieves, and the converse holds for thieves. By considering the results of interactions within and between these two types of individuals, a game theory analysis can predict the conditions under which one strategy would "win" or "lose" and how the success of each type of individual would vary as the frequency of the other increases in the population. A complete understanding of territoriality involves not only empirical approaches in the field but also the development of testable theoretical models. Considerable advances have been made recently merging these two methodologies. Future investigations will no doubt include experimental control over resource levels that will allow definitive tests of predictions of alternative theoretical models.

Territoriality and Aggression

The importance of investigating any biological phenomenon might be measured by its contribution to understanding nature in general and humankind in particular. By these criteria, aggression and territoriality may be among the most important topics that could ever be studied. Among animals in general, some species are highly aggressive in defending their living space, and others ignore or tolerate conspecifics in a nearly utopian manner. Some animals are territorial during only part of the annual cycle, and some only in specific areas that they inhabit; others remain aggressive at any time and in any place. Thus, a main goal for researchers is to unravel the ecological and evolutionary conditions that favor aggressive behavior and territoriality.

Aggression and territorial behavior appear to have evolved in various organisms because, in the past, aggressive and territorial individuals outreproduced nonaggressive and nonterritorial ones. An implicit assumption of behavioral biologists is that animals other than humans do not interact aggressively because of conscious reasoning, nor are they consciously aware of the long-term consequences of aggressive acts. Should these consequences be detrimental, natural selection will eliminate the individuals involved, even if this means total extinction of the species. Humans are different. They are consciously aware of their actions and of the consequences of such actions. They need only use conscious reasoning and biological knowledge of aggressive behavior to create conditions that can reduce conflict between individuals and groups.

—Richard D. Howard

See also: Adaptations and their mechanisms; Communication; Competition; Ecology; Emotions; Ethology; Instincts; Mammalian social systems; Mating; Pheromones; Population fluctuations; Reproduction.

Bibliography

Alcock, John. *Animal Behavior.* 7th ed. Sunderland, Mass.: Sinauer Associates, 2001. A highly readable source of information on all aspects of animal behavior. The chapter on the ecology of finding a place to live covers many aspects of animal spacing, including territoriality.

Allen, Colin, and Marc Bekoff. *Species of Mind: The Philosophy and Biology of Cognitive Ethology.* Cambridge, Mass.: MIT Press, 1997. Looks at the comparative, evolutionary, and ecological aspects of nonhuman cognition, including aspects of territoriality and aggression.

Davies, Nicholas B., and John R. Krebs. *An Introduction to Behavioral Ecology.* 4th ed. Boston, Mass.: Blackwell Scientific Publications, 1997. The chapter on competing for resources covers many aspects of territorial behavior, and chapter 7 on fighting and assessment gives an introduction to game theory analysis.

Dennen, J. van der, and V. S. E. Falger, eds. *Sociobiology and Conflict: Evolutionary Perspectives on Competition, Cooperation, Violence, and Warfare.* New York: Chapman and Hall, 1990. This collection of essays on conflict, enmity, and warfare chiefly focuses on humans, but includes a wide-ranging survey article on intergroup competition between species.

Howard, Eliot. *Territory in Bird Life*. New York: Atheneum, 1962. Provides one of the earliest views on the function of territorial behavior. This account is interesting both in its own right and in comparison with later ideas.

Ratcliffe, Derek A. *The Peregrine Falcon*. 2d ed. San Diego, Calif.: Academic Press, 1993. An in-depth study of the peregrine falcon, a bird notable for both its territoriality and its aggressive nature.

Wilson, Edward O. *Sociobiology*. Cambridge, Mass.: The Belknap Press of Harvard University Press, 1975. Offers information on territorial behavior for nearly every type of animal. Chapter 12, "Social Spacing, Including Territory," covers the general theoretical aspects of territorial behavior; later chapters treat various animal taxa individually.

THERMOREGULATION

Type of animal science: Physiology
Fields of study: Biophysics, physiology

Temperature regulation in animals is a process that may utilize either environmental or physiological sources of heat to maintain conditions conducive to life. By learning about temperature regulation, scientists have gained insights into interactions between animals and their environments or between the functional components of their bodies.

Principal Terms

CONVECTION: a transfer of heat from one substance to another with which it is in contact

COUNTERCURRENT MECHANISM: a heat exchange system in which heat is passed from fluid moving in

ECTOTHERM: an animal that regulates its body temperature using external (environmental) sources of heat or means of cooling

ENDOTHERM: an animal that regulates its body temperature using internal (physiological) sources of heat or means of cooling

HELIOTHERM: an animal that uses heat from the sun to regulate its body temperature

HOMEOSTASIS: the maintenance by an animal of a constant internal environment

HOMEOTHERM: an animal that strives to maintain a constant body temperature independent of that of its environment

OPTIMUM TEMPERATURE: the narrow temperature range within which the metabolic activity of an animal is most efficient

POIKILOTHERM: an animal that does not regulate its body temperature, which will be the same as that of its environment

THERMOGENESIS: the generation of heat in endotherms by shivering or increased oxidation of fats

Body-temperature regulation by animals is essential for life. The maintenance of life relies on the sum of all chemical reactions or metabolic activity in an organism. These reactions are facilitated by catalysts, substances not directly involved in a reaction as either a product or reagent but essential for accelerating the process or allowing the reaction to proceed under conditions compatible with life. For example, a reaction that, in a test tube, might require exceedingly high temperatures will proceed, if catalyzed, at normal body temperatures. Biological catalysts are complex proteins called enzymes. These are fragile molecules and are quite temperature-sensitive. If exposed to excessively high or low temperatures, they will be denatured and lose their functional properties.

Homeostasis is the maintenance of a constant internal environment, one suitable for proper enzymatic activity. Homeostatic mechanisms involve three components: a sensor (or receptor) that reacts to changes in environmental conditions, a coordinator (or integrator) that responds to information from the sensor, and one or more effectors (activated by the coordinator), which elicit appropriate, regulatory responses.

Temperature sensors are scattered throughout the bodies of most animals, but those specifically associated with temperature regulation in vertebrates (animals with backbones) are found in the hypothalamic region of the brain. Coordinators are found within the brain (or its equivalent in simpler animals), again in the hypothalamus of more advanced types. Effectors may be any structure capable of affecting temperature.

Animals generally function at temperatures between 4 and 40 degrees Celsius. Peak metabolic efficiencies, however, exist over a much narrower range, called the optimum temperature. This temperature varies by the animal and its habitat. Optimum temperatures often approach lethal limits, the highest temperature an animal can tolerate. This necessitates precise control of temperature in order to avoid exceeding those limits. Within lower temperature ranges, some animals can alter metabolic requirements in order to adapt to changing temperatures without sacrificing efficiency. This process, which involves complex biochemical and cellular adjustments, is called "temperature compensation." Animals that utilize metabolic mechanisms to maintain constant, relatively high body temperatures are often referred to as being "warm-blooded." Others, whose body temperatures are not regulated or are regulated primarily by behavioral means, are called "cold-blooded." That these terms are imprecise and irrelevant becomes obvious when one considers that the temperature of a desert-dwelling "cold-blooded" lizard or insect may often exceed that of any bird or mammal. On the other hand, the core temperature of some hibernating mammals may be reduced to being anything but "warm."

Most invertebrates (animals lacking backbones) as well as many fishes, amphibians, and some reptiles, do not regulate body temperatures; they are called poikilotherms. They monitor environmental conditions, attempt to seek out areas where temperatures are suitable, and avoid those where they are not. Their temperatures are essentially identical to environmental temperatures. If excessively high temperatures are unavoidable for more than short periods, death may occur. Low temperatures are seldom fatal (unless below freezing) but will result in diminution of metabolic functions, causing the animal to become torpid, or inactive. Since these animals are vulnerable, they will seek shelter, which is why insects, for example, are rarely encountered during colder months.

Ectotherms

Animals that regulate body temperatures fall into two categories. Those that utilize environmental sources of heat are called ectotherms (animals that "heat" their bodies using external sources). Those that utilize physiological temperature control mechanisms are called endotherms (animals that "heat" their bodies using internal sources). Since endotherms (birds and mammals) strive to keep temperatures constant, they may also be called homeotherms (animals that maintain constant temperatures). All regulators must invest considerable energy in the process. To minimize that expenditure, they utilize microhabitats in which regulatory mechanisms are not necessary. Ectotherms use behavior, enhanced by physical or physiological mechanisms, to take advantage of environmental conditions. A principal source of heat for most ectotherms is sunlight; temperature regulators that rely on the sun are called heliotherms (animals that "heat" their bodies using the sun). Lizards from temperate zones (areas with moderate and/or seasonal climates) are the most efficient ectotherms and may serve as models to illustrate the process. Tropical species, which live in constant, warm environments, tend to be poor regulators.

Sunlight and heat may be assimilated directly by basking lizards or indirectly by convection from sun-heated surfaces. Basking occurs when an animal exposes itself to sunlight by seeking unshaded perches. Position and posture are critical. Lizards will orient themselves in order to expose the greatest amount of surface to the sun. This involves a position in which the animal is broadside to the sun. Surface area is further enhanced by flattening the body dorsoventrally (top to bottom). Similarly, animals may absorb heat from the substrate. Lizards flatten themselves against a warm surface to maximize the area through which heat is assimilated. Area is critical in elevating temperatures, either by basking or convection, but does not increase proportionately with volume as animals increase in size. Thus, large ectotherms require disproportionately more energy and time to raise their temperatures than animals with similar proportions but smaller dimensions. This explains why the first animals to emerge in the spring or early morning tend to be small. Also, since dark colors absorb more radiation (heat and light), cold animals will stimulate pigment cells and are invariably much darker than those at optimum temperatures. That these mechanisms work effectively is illustrated by observations of active lizards at near-freezing temperatures at high elevations in the Andes of South America. When these lizards are captured, body temperatures of 31 degrees Celsius are recorded. In another study, lizards active at –4 degrees Celsius have been found to have body temperatures above 10 degrees Celsius. Some investigators have observed lizards, buried in sand during the night, emerging slowly, exposing only their heads. Since many lizards have large blood sinuses in their heads, it has been suggested that they can raise their body temperatures while minimizing exposure to predators. It is unlikely that this is effective, as heat gained would be rapidly lost to the substrate by convection. Only if the ground were warmer than air and only until body temperature reached that of the ground would this mechanism be operative.

In ectotherms, cooling is a much more difficult proposition. Without access to a source of "cold," ectotherms can do little more than minimize heat absorption. Coloration is lighter to increase reflection, orientation is toward the sun, posture involves lateral (side-to-side) compression, and animals will "tiptoe," lifting themselves away from warm substrates. If these are inadequate, animals must seek shelter. Many desert-dwelling lizards exhibit activity cycles that peak twice each day (morning and evening) to avoid cold nights and hot midday periods.

Endotherms

Endotherms use physiological effectors to raise or lower temperatures. If cold, they will generate heat (thermogenesis) by rapid muscular contractions (shivering) or increased oxidation of fats. Simultaneously, devices minimizing heat loss will be implemented. These include lowered ventilation (breathing) rates; since inhaled air is warmed during passage through the respiratory tract, heat is lost with each expiration. Also, superficial blood vessels narrow (vasoconstriction), reducing flow of warm blood to the skin, from which heat is lost by convection. Attempts to insulate skin are illustrated by "goose bumps." Though ineffective in sparsely haired humans, this reaction to cold is quite effective in mammals with thick body hair or fur. Muscles attached to hair follicles contract and draw hairs into an upright position, and the ends droop, trapping dead air between matted ends and skin. A fine undercoat in many species enhances the process. Dead air is an excellent barrier to heat flow. A similar device affecting feathers exists in birds.

When hot, endotherms keep muscular activity to a minimum, increase ventilation rates (panting), and expand superficial blood vessels (vasodilation). Rates of heat dissipation in some mammals are enhanced by sweating. Sweating and panting rely on evaporative cooling, the same principle involved in using radiators to prevent hot automobile engines from overheating. Endotherms adapted to hot climates produce concentrated urine and dry feces to conserve water, since much is lost in cooling.

Many of these mechanisms are surface-area related. Consequently, endotherms in hot climates, especially large species with relatively poor surface-to-volume ratios, often possess structures, such as elephant's ears, to increase area through which heat may be dissipated. On the other hand, endotherms occupying cold habitats are designed to minimize exposed surfaces. For example, arctic hares have short ears and limbs compared to the otherwise similar jackrabbits of warmer climes. In addition, cold-adapted endotherms may decrease rates of heat loss from poorly insulated appendages by means of countercurrent mechanisms. Heat from blood in arteries flowing into a limb is passed to venous blood returning to the body. This minimizes the amount of heat carried into a limb, whose surface-to-volume ratio is very high. It also functions to warm the returning blood, which prevents cooling of the body core. The appendages themselves are very cold; portions may even be at below-freezing temperatures. Actual freezing is prevented by special fats in the extremities.

Studying Thermoregulation

Specific methods vary according to the subject, approach, and discipline in question. Anatomy (study of structure), using both micro- and macroscopic methods, often centers on surface-related phenomena. For example, studies investigating the vascularization (blood supply) of whale flukes, whose physiology is difficult to study, have indicated that these are quite capable of dissipating heat and have led to the knowledge that these animals, even in cold water, because of their large size and poor surface-to-volume ratios, have potential problems with overheating. The role of blubber was reevaluated in this light and is now recognized as being one of fat storage with little to do with insulation. Furthermore, with new technologies in electron microscopy, anatomists have been able to describe, often for the first time, the complex structural components of organs (and even cells) that are active in thermoregulation.

Physiological studies of function are of two major types. One involves measurements of activity under different thermal regimes; for example, patterns of locomotion or digestion (involving specifically neural and muscular or neural, muscular, and glandular entities, respectively) may be observed at different temperatures. Often, these include observations of performance on treadmills or of rates at which food items are processed in controlled laboratory settings. On a different scale, metabolic activity itself might be linked to temperature by measuring rates of oxygen consumption in special metabolic chambers or utilization rates of products necessary for particular chemical reactions. These types of investigations have led to the determination of optimum and lethal temperatures in many species.

A second type of physiological study deals with actual thermoregulation. The ability to monitor body temperatures continuously, even in small animals, by means of radiotelemetry has made possible whole series of experiments in which animals' thermal responses to induced or natural conditions can be evaluated. Investigations of this type have provided insights into, for example, adaptive hypothermia (significantly reduced body temperatures) in small endotherms such as bats and hummingbirds. These species drastically reduce their core temperatures when inactive in order to conserve energy otherwise rapidly lost as heat through their relatively large surface areas.

Since laboratory work often fails to simulate natural conditions adequately, observations of animals in nature have been instituted. These seek to evaluate thermoregulation in the contexts of ethology (the study of behavior) and ecology (the study of organisms' relationships with their environments). These types of studies frequently entail prolonged observations until patterns of behavior or habitat use emerge and can be quantified and evaluated. The use of rapid-reading thermometers or implanted radiothermisters facilitates understanding of the often-subtle modifications in thermoregulatory behavior or microhabitat use characteristic of many animals. Relating recorded temperatures to changes in posture, position, orientation, activity level, and ambient temperatures of substrate and air has, for example, led to an appreciation of how efficiently some ectotherms regulate temperature and the complexity of the mechanisms involved.

Applications of Thermoregulation Research

Long restricted by concepts of "warm-blooded" versus "cold-blooded" animals, investigators did not begin in-depth explorations of thermoregulation until the twentieth century. Most early efforts grew out of medical studies dealing with dynamics of human temperature regulation, especially in the context of pathological states associated with fever or trauma-induced hypothermia. Monitoring these conditions led to an appreciation of how complex temperature regulation is and how many of the body's systems are involved. These studies, in turn, led to investigations of similar mechanisms in animals. Initially, most dealt with laboratory animals, but pioneering investigations into thermoregulation by animals in natural habitats soon opened whole new vistas. These studies were subsequently extended to "cold-blooded" species, which in turn led to an appreciation of how effective behavioral temperature regulation could be. In the 1970's, suggestions that at least some dinosaurs may have been homeotherms stimulated further interest in this field of study.

Most heat exchange with the environment occurs through skin or respiratory systems; muscular systems generate heat as a by-product of contraction; digestive and urinary systems regulate elimination of wastes, which influences retention or loss of heat-bearing water; cardiovascular systems transport heat; and nervous and endocrine systems regulate the entire complex. In addition, all cells require a proper thermal environment and may affect heat production by altering rates of oxidative metabolism. Therefore, a more complete understanding of thermoregulation has enhanced scientists' awareness of both normal and pathological functions in most body systems. Specific medical applications of these studies include induced hypothermia during surgery-related trauma and treatment of accident-related hypothermia using mechanisms first observed under natural conditions in animals.

Studies of temperature-regulating mechanisms, both behavioral and physiological, have also provided insights into relationships between animals and their environments. Thermoregulatory needs have been used to explain behavioral and ecological phenomena for which causative agents were previously unknown. From a practical perspective, this knowledge is useful in developing management tools to sustain disrupted or endangered ecosystems. Appropriate techniques must be developed with a thorough knowledge of the dynamics in any given system, and this must be based on biological criteria rather than human perceptions. For example, reforested areas have often been managed as crops, with all the attendant problems of monocultures (areas cultivated for plants of only one species). Among these is the lack of biodiversity (variety of life-forms). When efforts began to take into consideration microhabitat requirements, often related to temperature regulation, varieties of plants—many with little or no commercial value in themselves—were planted. This resulted in managed areas becoming capable of supporting many different species.

Finally, a more complete knowledge of structures related to thermoregulation has been applied by paleontologists (scientists who study fossils) to the study of dinosaurs. Long thought to be "sluggish," lizardlike ectotherms, dinosaurs are now thought by many investigators to have been more like mammals and birds in their physiological capabilities. This image is more in tune with their domination of the earth for some hundred million years.

—Robert Powell

See also: Cold-blooded animals; Ethology; Habitats and biomes; Kidneys and other excretory structures; Metabolic rates; Osmoregulation; Warm-blooded animals; Water balance in vertebrates.

Bibliography

Avery, Roger A. *Lizards: A Study in Thermoregulation*. Baltimore: University Park Press, 1979. This book effectively summarizes principles of thermoregulation by using lizard models to illustrate adaptations to various habitats. Somewhat technical, but appropriate for advanced high school students. Nicely illustrated; bibliography.

Bakker, Robert T. *The Dinosaur Heresies: New Theories Unlocking the Mystery of the Dinosaurs and Their Extinction*. New York: Kensington, 1986. A marvelously entertaining book treating dinosaurs as homeotherms. Line drawings shed a whole new light on dinosaurs as active, dynamic animals. The book provides insights into homeothermy and the impact it can have on all aspects of animals' lifestyles. Literature-cited section. Written for a popular audience.

Dukes, H. H. *Dukes' Physiology of Domestic Animals*. 11th ed. Ithaca, N.Y.: Comstock, 1993. A comprehensive textbook for veterinary students, covering all aspects of domestic animal physiology including thermoregulation.

Gans, Carl, and F. Harvey Pough, eds. *Physiology C: Physiological Ecology*. Vol. 12 in *Biology of the Reptilia*. New York: Academic Press, 1982. Technical, but provides very complete coverage of thermoregulation in reptiles, with six articles by different authors covering various aspects of the topic. Extensive literature-cited sections at the end of each article.

Hickman, Cleveland P., Larry S. Roberts, and Frances M. Hickman. *Integrated Principles of Zoology*. 11th ed. Boston: McGraw Hill, 2001. A textbook written with exceptional clarity, one of the best in a field of many good general zoology books. The chapter on homeostasis, osmotic regulation, excretion, and temperature regulation covers general principles of thermoregulation, discusses concepts of ectothermy and endothermy, and describes mechanisms and adaptations in both ectotherms and endotherms. Diagrams illustrate examples. Selected references are given at the end of the chapter. Glossary. College level, but suitable for advanced high school students.

Johnston, Ian A., and Albert F. Bennett, eds. *Animals and Temperature: Phenotypic and Evolutionary Adaptation*. New York: Cambridge University Press, 1996. A comprehensive review of temperature adaptation in animals.

Schmidt-Nielsen, Knut. *Desert Animals: Physiological Problems of Heat and Water*. New York: Dover, 1979. This engaging book provides considerable insights into thermoregulation by animals living in possibly the most inhospitable climate on earth. The use of case studies describing mechanisms by individual species is quite useful. Coverage is fairly technical but suitable for the interested nonscientist. Complete literature-cited section.

_______. *How Animals Work*. Cambridge, England: Cambridge University Press, 1972. This well-written text discusses in some detail various mechanisms of temperature regulation in vertebrates. Adequately illustrated. The strength of this work is its integration of temperature regulation into the total context of animal physiology. Complete literature-cited section.

TIDEPOOLS AND BEACHES

Type of animal science: Geography
Fields of study: Marine biology, oceanography, zoology

Beaches are muddy, sandy, or pebbly shores of oceans or lakes. They protect these bodies of water and are esthetically pleasing recreation sites. Tidepools are pools of sea water left behind when tides ebb. Their presence is due to beach and coast topography.

Principal Terms

BACKSHORE: the horizontal beach part farthest from the ocean

BARRIER ISLANDS: islands seaward of estuaries, parallel to the shore and separated from it by a lagoon

DELTA: an accumulation of sediment at a river mouth

ESTUARY: a stream-fed bay

FORESHORE: the sloping beach part closest to the ocean, which is continually altered by waves, tides, and longshore currents

INTERTIDAL FLAT: a beach part located in shallow water just seaward of the foreshore, made of mud and/or sand, holding burrowing animals

LONGSHORE CURRENT: a current parallel to a seashore, from shoreline through breakers, created by waves approaching the coast at a small angle to the shore

TIDE: the regular rise and fall of sea level due to gravitational fields of the moon and sun

TIDEPOOL: a pool of water originally or permanently composed of water left behind at ebb tide

The term "beach" indicates the muddy, sandy, or pebbly shore of a body of water such as an ocean or lake. Most beaches are located around earth's seacoasts. They are composed of mud, sand, and pebbles, sediments that accumulate in arrangements and forms dependent on coastal actions, sediment sources, and rates of sediment deposition. There are three main beach types: sediment strips edging rocky coasts; borders of sediment accumulations along rivers; and sediment regions parallel to coasts and associated with barrier islands.

At most times a beach has a seaward, sloping foreshore lacking vegetation and a landward, horizontal backshore that can hold some hardy plants. Near the land's end, waves rise and break along the foreshore. Sediment is moved both along this shore and perpendicular to it. During storms, waves erode beaches, moving sediment back into oceans. This temporarily leaves only foreshore. In the ensuing calm, waves move sediment landward again and rebuild the beaches.

Due to very rapid changes in wave sizes, the appearance and sediment types of a foreshore often change significantly from day to day. Backshores, normally unaffected by waves, are altered most by winds. There are also variations in beach form along and perpendicular to the beach shoreline. Most common are undulating foreshores that vary from beach to beach and time to time. Beach sands in areas having mild climates are mostly quartz and feldspar. In tropical regions, beaches are often made of calcium-containing remains of marine organisms.

Natural Forces and Beaches

Beaches develop along coastlines due to natural processes including waves, longshore currents, tides, climate, and gravity. Waves, in their contin-

Currents That Affect Beaches

Longshore currents are those parallel to a sea shore and extending from the shoreline through the breakers. They occur because waves approach coasts at a small angle and bend in the shallows. The speed of a longshore current is related to wave size and angle of approach to a shore. During quiet weather, longshore currents move at under half a mile per hour, but during storms they can move ten to twelve times that speed. The combined actions of waves and longshore currents transport a lot of sediment along shallows bordering a shore. Longshore currents move in either direction along a beach, depending on the direction of wave approach, itself a result of wind direction. Thus, waves suspend sediment and longshore currents transport it along beaches.

Tides, the regular rise and fall of sea level due to the gravitational fields of the moon and the sun, cause daily changes in ocean levels of one to fifty feet. Tidal currents transport large amounts of sediment and erode rock, and tidal rise and fall distributes wave energy across shores by changing water depth. In estuaries, tides create the speeds needed to move sand. On open coasts, that is, on beaches, tides do not move the water fast enough for sediment transfer. However, the rise and fall of tides along open coasts indirectly affect sediment movement, because their landward movement or retreat causes shorelines to move. This changes the region where waves and longshore currents operate. Beach slope is also crucial, with gently sloped beaches having the largest shoreline changes during tide cycles.

ual beachward motion, vary in size at different coastal areas and different times. They interact with the ocean bottom as they move into shallows, suspending sediment and moving it landward. Large waves suspend sediment from deeper water and move larger sediment particles than small waves. For example, small waves only move sand, while large waves can even move boulders. During storms, large waves return beach sediment to deep water. Waves also erode coastal bedrock by abrasion from their suspended pebbles and larger rock fragments.

The important effects of climate on beach development begin with rainfall, which creates runoff streams, transports sediment to seacoasts, and causes differences in the volume and types of sediment sent to coasts in different world areas. Temperature is also quite important because it causes weathering of coastal sediment and rock. This is most extensive in cold areas, where water freezes in rock cracks, causing their fragmentation. Wind action is also important because of its relation to wave size. Coasts with fairly constant, strong winds have high-energy waves. Onshore winds around the earth's coasts produce sand dunes wherever sediment is available and can accumulate. Gravity acts indirectly in wind and wave production, and directly in the down-slope movement of sediment.

Most beaches cannot support large plants, though some hardy plants are found on their backshores. Near beach waterlines, organic matter such as decaying seaweed can be found. However, a bit further seaward, mud and sand flats (intertidal flats) hold burrowing animals including worms, clams, mussels, and burrowing shrimp. The worms ingest sediment and eat the organic matter it holds. Other burrowing organisms, such as clams, use tubes to reach into the water above them to filter out food when they are covered by the tide. Crustaceans and starfish also use intertidal flats to seek prey.

Intertidal flats are important feeding grounds for wading birds such as sandpipers, terns, and plovers. In temperate climates these birds remain all year. Hundreds of thousands of them also make seasonal migrations to specific beaches located between their summer and winter habitats. During such migrations the birds rely on intertidal flats for food along the way. For example, hundreds of thousands of sandpipers stop, each July and August, on the intertidal flats of Canada's Bay of Fundy, each eating tens of thousands of burrowing shrimps before they migrate to South America.

Tidepools

Tidepools are pools of various types and sizes originally or permanently composed of water left behind by the ocean when the tides go out. Their presence is due to the topography of beaches and coasts. Some tidepools occupy beach crevices or fissures with seaward ends open to the sea and landward ends abutting cliffs, caves, or boulders. Others occur in rocky basins with high rims on the seaward side. The rims hold back water when the tide ebbs. Seaweed lines a tidepool's walls or bottoms. Sponges, hydroids, anemones, sea slugs, insects, mussels, fishes, jellyfish, and starfish live in many of the pools, either temporarily or permanently.

Fish and jellyfish may become temporary tidepool residents when swept into them by tides. Almost as suddenly these visitors, who enter the pools to seek prey, leave with one of the next few tidal cycles. Permanent tidepool inhabitants include seaweeds. The seaweed species depend on the position on the beach of a given tidepool, the acidity or alkalinity of the pool, and its salt content. Tidepools high on a beach contain only plants. This is because they are almost entirely isolated from the sea and so their temperatures are too high to allow animal survival. Even plants may have some trouble living in such pools.

Lower on the beach, tidepools provide far more stable conditions because they are connected to the ocean, filling and only partly emptying during tidal cycles, so both plants and animals live in them. Yet they, too, are affected by the duration of the ocean's presence or absence, and the inhabitants of a pool in the middle of the foreshore are very different from those of low-foreshore pools, which are separated from the sea only very briefly.

Other Types of Beach Sites

Beaches form along depositional coasts, that is, wherever sediment accumulates due to longshore currents, waves, and tides. Locations other than long, open coastal areas are associated with river deltas, estuaries, lagoons, and barrier islands. Deltas are accumulations of sediment at river mouths. They vary in size and shape. However, all require that more sediment is deposited at a river mouth than is carried into the ocean, and that shallow sites are available for sediment accumulation. Delta size is proportional to the size of the river involved and delta shape is dictated by the strengths of the river, tide, and waves. Deltas can be river-dominated (such as the Mississippi River Delta) when waves and tides do not affect water discharge or accretions of sediment. Such deltas are irregularly shaped. A wave-dominated delta experiences sediment erosion that simplifies and smoothes its edges. Tide-dominated deltas are funnel shaped.

Waves may erode coastal bedrock and eventually destroy beach habitats altogether. (Corbis)

Irregularly shaped depositional coasts are rich in stream-fed bays, or estuaries. They receive a lot of sediment due to coastal runoff. Seaward of estuaries are sandy barrier islands that parallel the shoreline, formed by waves and longshore currents. The islands are separated from the mainland by lagoons, which are long, narrow bodies of water. Barrier islands contain well-developed beaches, dunes, and tidal flats on their landward sides.

Beaches are very important land features. They serve to protect coastlines from storms and erosion. In addition, they are ecologically important, serving as homes, in intertidal flats and tidepools, for useful and sometimes rare flora and fauna.

—*Sanford S. Singer*

See also: Ecosystems; Food chains and food webs; Habitats and biomes; Lakes and rivers; Marine biology.

Bibliography

Bascom, Willard. *Waves and Beaches: The Dynamics of the Ocean Surface*. Garden City, N.Y.: Anchor Press, 1980. Discusses waves, tides, surf, beaches, and storms.

Braun, Ernest. *Tideline*. New York: Viking Press, 1975. Describes tidepool and seashore ecology thoroughly.

Carson, Rachel, and Bob Hines. *The Edge of the Sea*. Boston: Houghton Mifflin, 1955. A timeless book which fascinatingly and thoroughly describes beaches, tidepools, and aspects of seashore life.

Lencek, Lena, and Gideon Bosker. *The Beach: The History of Paradise on Earth*. New York: Viking, 1998. Holds a lot of useful information on beaches, their formation, and their use by people.

Levinton, Jeffrey S. *Marine Biology: Function, Biodiversity, Ecology*. New York: Oxford University Press, 1995. Covers many issues thoroughly, including oceanography, marine organisms, coastal environments, and seabed processes.

TIGERS

Type of animal science: Classification
Fields of study: Anatomy, conservation biology, genetics, zoology

Tigers are carnivorous mammals which are the largest members of the Felidae family. They are an endangered species in all their habitats, and three species became extinct in the twentieth century.

Principal Terms

ALLELES: alternate forms of genes
DIGITIGRADE: walks on toes
FLEMEN: lip movement when cats detect an unusual scent
PUGMARKS: pawprints
SAGGITAL: bony skull top

The first tigers were members of the Felidae family of big cats that lived in northern Asia during the Late Pleistocene epoch. They migrated south and east and evolved into specific types of larger tigers according to their habitats. The ancient saber-tooth tigers were not ancestors of modern tigers. Three of the eight tiger subspecies, Bali, Caspian, and Javan, became extinct during the twentieth century.

Anatomy

Tigers range in weight according to their subspecies and gender. Sumatran male tigers are the lightest, weighing 110 kilograms (250 pounds). Females weigh approximately twenty kilograms less. Siberian male tigers weigh as much as 225 kilograms (500 pounds). The heaviest known tiger weighed 465 kilograms (1,025 pounds). From head to the base of the tail, tigers measure from 1.4 to 2.8 meters (4.5 to 9 feet) long, and their tails are from 90 to 120 centimeters (3 to 4 feet). They use their tails to balance and to communicate.

Tiger skulls have a big saggital crest which anchors a large jaw muscle. Tigers' vertebra and joints are flexible. Their hind legs are longer than their front legs, providing impulsion and assisting leaping when they are chasing game. They have five toes on their front paws and four toes on their hind paws. Each toe has a retractable claw which is 80 to 100 millimeters (3 to 4 inches) long and helps them restrain prey and climb trees.

Tigers stalk their prey, then leap to make the kill. (Corbis)

Mature tigers have thirty teeth. The canine teeth are 75 to 90 millimeters (2.5 to 3 inches) long. Tigers have triangular, erect ears set atop broad skulls, with their eyes positioned on the front of their face. Tigers' eyes have reflecting retinas which enable excellent night vision. Tigers' sense of smell is also acute; they can distinguish different animals by smell and exhibit the flemen response.

Tigers have nineteen pairs of chromosomes, which determine genetic patterns. Their coats are colored shades of orange, with black or brown stripes of varying widths and lengths and white accents around the eyes, ruffs, and other body parts. Rarely, Bengal tigers with the two necessary alleles are born with a white foundation coat and blue eyes. They are not albinos or a separate subspecies. Tigers' stripes vary according to subspecies, with Sumatran tigers having the most and Siberian tigers having the fewest. Each tiger's stripes are unique and function as camouflage. Fur thickness varies with seasonal changes and geography. Siberian tigers have almost twice the number of hairs per square centimeter than Sumatran tigers.

Tiger Facts

Classification:
Kingdom: Animalia
Subkingdom: Metazoa
Phylum: Chordata
Subphylum: Vertebrata
Class: Mammalia
Subclass: Eutheria
Order: Carnivora
Suborder: Feloidea
Family: Felidae
Genus and species: Panthera tigris
Subspecies: P. t. tigris (Bengal), *P. t. altaica* (Siberian), *P. t. amoyensis* (South Chinese), *P. t. balica* (Balinese), *P. t. corbetti* (Indochinese), *P. t. sondaica* (Javan), *P. t. sumatrae* (Sumatran), *P. t. virgata* (Caspian)

Geographical location: Asia, specifically India, Thailand, Manchuria, China, and Indonesia

Habitat: Jungles, forests, tundra, mountains, and swamps

Gestational period: 3.5 to 4 months

Life span: Up to fifteen years in the wild, up to twenty years in captivity

Special anatomy: Sharp canine teeth, saggital crest

Behavior

Tigers are solitary, preferring to hunt alone. A male tiger's territory averages twenty-six to seventy-eight square kilometers (ten to thirty square miles), depending on the availability of prey. Some Siberian tigers roam territories of 1,036 square kilometers (400 square miles). Tigers' territories often overlap, with several females sharing territorial space with one male. Tigers scratch on trees, leave fecal droppings, and spray urine to mark their territory. Male tigers occasionally fight. Tigers have several vocalizations to communicate aggression and receptiveness to other tigers.

Females attain sexual maturity at age three and males at age four. After a four-month gestation, females have litters of two to five cubs which are born blind and are vulnerable to predators such as pythons. The cubs drink their mother's milk for two months, then feed at her kills until they are about two to three years old and capable of hunting alone.

Tigers can catch and kill prey as large as 160 to 900 kilograms (440 to 2,000 pounds). They stalk and ambush ungulates, knocking prey to the ground and biting the neck or throat to sever the spinal cord or suffocate the animal. Tigers can consume twenty to twenty-five kilograms (sixty to seventy pounds) of meat daily. They drag carcasses into vegetated areas and gorge on a kill, then fast. Tigers also eat termites and snakes. Some tigers, especially in the Sunderbans river delta of India and Bangladesh, have attacked and killed humans.

Conservation

Adult tigers are hunted by poachers for their hides, bones, teeth, and body parts or for sale to exotic pet traders. Much of their jungle habitat has been destroyed during wars or for agricultural use. As a result, only about five thousand to seven thousand tigers are alive in the wild. Authorities estimate that an equivalent number are kept as exotic pets in North America and in zoos, sanctuaries, and circuses. Tiger censuses have been taken by counting pugmarks in known tiger habitats.

In captivity, tiger hybrids include ligers, the hybrid of lion fathers and tiger mothers, and tigons, produced by tiger fathers and lioness mothers. Conservation breeding programs are aspiring to preserve and increase the tiger population.

—Elizabeth D. Schafer

See also: Breeding programs; Cats; Cheetahs; Endangered species; Fauna: Asia; Jaguars; Leopards; Lions; Mountain lions; Wildlife management.

Bibliography

Hornocker, Maurice, ed. *Track of the Tiger: Legend and Lore of the Great Cat*. San Francisco: Sierra Club, 1997. Essays advocating the preservation of tigers.

Meacham, Cory J. *How the Tiger Lost Its Stripes: An Exploration into the Endangerment of a Species*. New York: Harcourt Brace, 1997. A journalistic account of tigers' roles in modern societies.

Nichols, Michael, and Geoffrey C. Ward. *The Year of the Tiger.* Washington, D.C.: National Geographic Society, 1998. Excellent source comparing wild tigers in India's Bandhavgarh National Park and captive tigers living in zoos and commercial breeding facilities or performing in circuses and Las Vegas clubs.

Seidensticker, John, Sarah Christie, and Peter Jackson, eds. *Riding the Tiger: Tiger Conservation in Human-Dominated Landscapes*. New York: Cambridge University Press, 1999. Scholarly papers addressing the threats endangering tigers' survival.

Tilson, Ronald L., and Ulysses S. Seal, eds. *Tigers of the World: The Biology, Biopolitics, Management, and Conservation of an Endangered Species*. Park Ridge, N.J.: Noyes, 1987. A valuable collection of scientific papers written by the world's leading tiger researchers.

TOOL USE

Type of animal science: Behavior
Field of study: Ethology

Tools extend an animal's ability to interact with or modify its environment. Most of these interactions involve obtaining food, but animals are known to use tools in many different ways.

Principal Terms

ECHOLOCATION: the ability of animals to locate objects at a distance by emitting sound waves which bounce off an object and then return to the animal for analysis

ECTOPARASITE: a parasite, such as a tick, that lives on the external surface of the host

ETHOLOGY: the study of an animal's behavior in its natural habitat

INSIGHT LEARNING: using past experiences to adapt and to solve new problems

PHEROMONE: a hormone produced by an animal and then released into the environment

PREDATOR: an organism that kills and eats another organism, generally of a different species

PRIMATES: a group of mammals including apes, chimpanzees, monkeys, humans, lemurs, and tarsiers

In general, a tool is considered to be something which is not an integral part of an animal's body but is used by the animal to accomplish a specific task. For example, a lobster may use its claw to crack open shells; however, since the claw is a normal appendage of the lobster, it is not considered to be a tool. When humans use a similar object, a nutcracker, to open shells, the nutcracker serves as a tool. It is difficult to define tools accurately. Examples of tools acceptable under the definition of one scientist may not meet the criteria set down by another investigator. Some scientists expand the definition of tool use to include specialized structures some animals use to extend their capability to locate and capture prey. These capabilities might include echolocation or sonar, electromagnetic fields, and specialized cells used for feeding such as the cnidocytes used by jelly fish. Other scientists consider products produced by an organism to be used to capture food as tools. Under this definition, a spider's web can be considered to be a tool.

Quite often, objects taken directly from the environment, such as stones or sticks, are used as tools without further modification by the animal. Other times, the object may be modified by actions such as stripping the leaves from a stick prior to use. Tools allow the user to complete a task more easily or to accomplish a task that may not have been possible without the advantage provided by the tool. The size, shape, and even texture of tools varies across the animal kingdom. Some animals use trees as tools and others use grains of sand. Some fish use spurts of water as tools. In addition to capturing or obtaining food, tools are also used in grooming, for defense, or even as protection from the elements. Thus, animals that use tools are actively interacting with and even modifying their environment.

Sticks and Stones Used as Tools

Many different species of animals, including insects, fish, birds, mammals, and primates, are known to use tools in some way during their everyday activities. While many different types of tools are used in the animal kingdom, the stick is a common and readily available tool. The use of sticks as tools has been well documented in nonhuman primates, such as chimpanzees, apes, and orangutans. Primates often use insight to solve a problem using tools and the young learn to use tools from either observing or being taught by the adults. A classic example of insight learning leading to multiple tool use in chimpanzees was shown by Wolfgang Köhler, an early twentieth century psychologist. Chimpanzees held in captivity were offered food that had been placed beyond their normal reach. When boxes and sticks were added in the enclosure, the chimpanzees stacked the boxes, climbed them, and then used the sticks to knock down bananas that were hanging overhead. If one stick was not long enough, they would connect them together.

Orangutans and chimpanzees will strip the leaves from a stick and then use it to probe into the nest of insects such as ants or termites. When the stick is removed from the nest, the insects crawling over it can be eaten. Leaves themselves have been used by chimpanzees to gather water for drinking. Birds, too, use sticks to probe for insects and to remove them from crevices in the bark of trees. Some birds, the Galápagos woodpecker finch for example, will use their bill to trim and modify the twig before using it as a probe. Pacific island crows use their beaks to modify sticks as well as leaves before using them as probes. In the absence of sticks or leaves, some animals will use cactus spines as probes. Elephants use trees and sticks in various ways. They will rub against a tree or they may pick up a stick with their trunk to scratch. They have been observed to use tree trunks as levers and to use sticks to remove ectoparasites. When monkeys throw sticks and rocks, they are using these objects as tools for defense.

Stones are another common tool. Sea otters use stones in two different ways. Some otters will carry stones with them when they dive and use the stone as a hammer to free a tightly adhered abalone from a rock. While floating along the surface on their backs, otters use stones to crack open the shells of abalone or of bivalves such as clams, mussels, or oysters, which they also pluck from under the water. Otters may use bottles floating in the water to crack shells. Birds use stones in a similar way. Egyptian vultures pick up stones in their beaks and use them in a pecking fashion, like a hammer, to crack open an ostrich egg. If this method fails, they will fly at the egg while clasping the stone in their talons. Mongooses also use rocks to crack eggs. Other birds, such as eagles, gulls, and crows, drop shelled animals such as turtles onto the rocks to crack their shells. Vultures are known to drop bones of prey onto rocks to crack them open and expose the marrow. Chimpanzees use stones to crack open nuts, analogous to humans using a hammer and anvil. Even spiders use stones as tools. The trap-door spider, *Stanwellia nebulosa*, uses a stone as a defensive tool. If forced to retreat when being attacked, the spider uses a stone to close off its burrow behind it.

Using Traps to Catch Prey

Traps are one example of animals using a tool to ambush and capture prey. When a predator is an ambusher, it lies in wait for another animal to happen upon its territory, and then the predator strikes. This technique has been especially perfected by the ant lion. The ant lion is the larval form of hundreds of species of insects in the order Neuroptera, family Myrmeleontidae. Using a series of circular and backward body movements combined with a quick side-to-side motion of the head, the ant lion digs into sandy soil, forming an inverted, cone-shaped impression. When an ant or a small insect crawls along the margin of the cone or happens to fall over the edge, the ant lion vigorously begins to throw grains of sand out of the bottom of the pit. This causes the prey to fall deeper into the pit and into the grasp of the ant lion's two large mandibles. The ant lion thus has used the method of tossing sand grains as a tool to capture food. Natural selection will favor the gene pool of those individuals with the greatest ability to move and throw quantities of sand quickly.

Sea otters use stones as tools to open tasty clams and mussels. (PhotoDisc)

Other Tools

In Japan, one species of crow uses a very different tool, a car. It has been reported that these crows use cars as nut crackers by placing the nut on the road and, after a car has run over it, retrieving the nut meat. If the car should miss hitting the shell, the crow may try again.

Humans are not the only species to use tools for fishing. Some green herons are known to drop objects into the water to attract fish looking for food. The herons then consume the curious fish. The archer fish uses jets of water shot from its mouth to knock insects off overhanging branches and into the water. Some scientists do not view this as a tool because the water passes along a specialized region of the mouth. However, it is similar to using a bow and arrow to subdue prey from a distance. Octopuses use water shot from their siphon system as a broom to clean the exoskeletons of eaten invertebrates from its den. An octopus may also use the jet of water to modify the size of the den. Another group of animals that uses a form of liquid tool belongs to the spider family, Scytodidae. These spiders shoot sticky material from modified venom glands to entangle their prey.

Spiders use their webs as tools in various ways. Those species of spiders that construct webs make them with silk produced from modified appendages called spinnerets. Webs are used to ambush animals that happen to enter into them. Some spiders strum their webs and use them as tools for communicating. Others may spin a long single strand of silk that they use as a drag line to find their way back or as a safety line to catch themselves. In some species, young spiders make silk parachutes which trap the air currents and allow them to be dispersed far from the nest. Spiders of the genus *Mastophora* spin a single thread, on the end of which is a sticky globule. By suspending the thread from one leg, the spider uses the web to "fish" for male moths, which are attracted to the sticky globule containing chemicals similar to the pheromones produced by female moths to lure males for mating.

The jellyfish and the hydra, two members of the phylum Cnidaria, have specialized cells, cnidocytes, concentrated on the surface of their tentacles. Inside these cells is an organelle, the nematocyst, which contains a thread. The nematocyst is stimulated to discharge when prey are near to it. This thread may have a barb on its tip that will penetrate the body surface of the prey, or it may be a lasso that wraps around the prey. The prey is then pulled into the digestive cavity of the cnidarian.

Sonar

Bats and dolphins are two good examples of animals that use echolocation to locate prey. Since sound waves can travel over great distances, the prey can be well beyond the predator's immediate area. The objects do not need to be large in order to be detected. Bats are able to locate mosquitoes. By analyzing the sound waves returning after bouncing off an object, the bat knows which objects are moving and which are stationary. The moving objects represent potential prey. Some potential prey, moths, have evolved a way to detect that they are being tracked by a bat. Thus, they are able to take evasive action and seek shelter near a stationary object such as a tree, or by landing on the ground. In a similar manner, dolphins use a series of high-frequency clicks to track fish. However, the fish, unlike the moths, are often not aware that they are being followed.

—Robert W. Yost

See also: Beaks and bills; Displays; Ethology; Grooming; Intelligence; Learning.

Bibliography

McFarland, David. *Animal Behavior.* 3d ed. Boston: Longman Science and Technology, 1998. An upper-level textbook on animal behavior. Chapter 27, "Intelligence, Tool Use, and Culture," discusses tool use among animals. This is a good book for readers with some previous background on the subject.

McGrew, W. C. *Chimpanzee Material Culture: Implications for Human Evolution.* New York: Cambridge University Press, 1992. Describes and analyzes the use of tools by chimpanzees in their native habitats through field studies conducted across Africa.

Maier, Richard. *Comparative Animal Behavior: An Evolutionary and Ecological Approach.* Boston: Allyn & Bacon, 1998. An excellent general-audience textbook on various aspects of animal behavior, with sections on tool use in animals.

Sherman, Paul W., and John Alcock, eds. *Exploring Animal Behavior: Readings from "American Scientist."* 2d ed. Sunderland, Mass.: Sinauer Associates, 1998. A good collection of essays on various aspects of animal behavior.

TRICERATOPS

Types of animal science: Classification, evolution
Fields of study: Anatomy, ecology, evolutionary science, paleontology, systematics (taxonomy)

Triceratops was a quadrupedal, three-horned herbivore that lived in the western United States at the end of the Cretaceous period and had an ecological role similar to that of the rhinoceros today.

Principal Terms

DENTAL BATTERY: unit of teeth in the upper and lower jaws consisting of the cutting teeth and the rows of replacements below them

FRILL: elaborate crest at the back of the skull that was used for visual display but not for protection

OCCIPITAL CONDYLE: ball-shaped bone that connected the back of the skull to the fused upper vertebrae of the spine

PREDENTARY BONE: keeled and pointed bone that terminated the lower jaw

ROSTRAL BONE: keeled and pointed bone that terminated the upper jaw

STEGOSAURS: quadrupedal, herbivorous dinosaurs with vertical bony plates arranged along their backbones

Triceratops became the first genus of horned dinosaur known to science when its skull was described by Othniel C. Marsh in 1889. The remains of its horns were originally attributed to the high-horned bison (*Bison alticornis*), and its occipetal condyle was originally named *Ceratops montanus*. In his preliminary description of the skull, Marsh named its owner *Ceratops horridus* and felt it was related to the stegosaurs. After the skull had been cleaned, Marsh changed the name to *Triceratops horridus*. Thirteen species of *Triceratops* have been described, but only one (or possibly two) species actually occurred in nature. *Triceratops* lived in western North America at the end of the Cretaceous, between 68 and 65 million years ago.

Characteristics

The most characteristic feature of the animal was its large, V-shaped head which terminated in an elongate frill. The skull can be more than 6 feet long (2.2 meters). Only whales have larger skulls. The frill allowed an animal to recognize members of the same species as well as members of the opposite sex. Since *Triceratops* had color vision, the frill was probably pigmented, and its ornamentation was designed for visual display and not for protection or to serve as a point of attachment for the jaw muscles. The head bore three horns that functioned in display, ritual combat, and protection from predators. One short horn arose over the nose, and two others, the longest, arose over the eyes. Males had large, erect horns while females had smaller, somewhat forward-pointing horns. The large number of skulls that have been found indicates that *Triceratops* was an abundant, gregarious species.

No complete skeletons are known. A composite, presumably female, skeleton on display at the Science Museum of Minnesota is 26 feet (7.9 meters) long and 9 feet, 7 inches (2.9 meters) high. With a weight of 8.5 metric tons (9.4 tons), *Triceratops* was three times heavier than a rhinoceros. The shin bone (fibia) was notably shorter than the thigh bone (femur). The size relationship between these two bones is the reverse of what is seen in animals that are fast runners. Evidence from ceratopsian trackways and the anatomy of its shoulder (the hind legs were located directly below the hips while the forelimbs sprawled out-

Triceratops *and other large-headed, frilled dinosaurs such as its ancestor* Anchiceratops, *pictured here, probably had color vision. This suggests that the characteristic frill around the head was colored for display, rather than serving a defensive function.* (©John Sibbick)

ward and were not located below the shoulders) also indicates that *Triceratops* was rather slow. Its running speed has been estimated at about 4.2 kilometers per hour (2.6 miles per hour).

—*Gary E. Dolph*

See also: *Allosaurus*; *Apatosaurus*; *Archaeopteryx*; Dinosaurs; Evolution: Animal life; Extinction; Fossils; Hadrosaurs; Ichthyosaurs; Paleoecology; Paleontology; Prehistoric animals; Pterosaurs; Sauropods; Stegosaurs; *Tyrannosaurus*; Velociraptors.

Bibliography

Dodson, Peter. *The Horned Dinosaurs: A Natural History*. Princeton, N.J.: Princeton University Press, 1996. This book must be read by anyone interested in the horned dinosaurs.

Dodson, Peter, and Philip J. Currie. "Neoceratopsia." In *The Dinosauria*, edited by David B. Weishampel, Peter Dodson, and Halszka Osmólska. Berkeley: University of California Press, 1990. A highly technical review of the bone structure of *Triceratops* and its relatives.

Triceratops Facts

Classification:
Kingdom: Animalia
Subkingdom: Bilateria
Phylum: Chordata
Subphylum: Vertebrata
Class: Reptilia
Subclass: Dinosauria
Order: Ornithischia (bird-hipped dinosaurs)
Suborder: Ceratopsia (beaked dinosaurs)
Family: Ceratopsidae (horned dinosaurs)
Subfamily: Chasmosaurinae (horned dinosaurs with long frills)
Genus and species: Triceratops horridus

Note: A number of competing classification schemes exist and will probably continue to do so in the future.

Geographical location: All large, horned, frilled dinosaurs were confined to western North America from present-day Denver, Colorado, to southern Alberta, Canada, between the western interior seaway to the east and the forming Rockies to the west

Habitat: Restricted to the arid, coastal, lowland plain

Gestational period: Although no eggs have been found, *Triceratops* must have been an egg layer; the frequency of egg laying, the time it took for the eggs to hatch, and the reproductive life span of the adults are unknown

Life span: Based on mammalian models, sexual maturity would be reached after ten years and the life span was probably in excess of one hundred years

Special anatomy: Rostral and predentary bones combined to give the snout its parrotlike appearance in side view; the occipital condyle projected off the back of the skull and gave the head a high degree of movement; each tooth had two roots, with the crown of each lower tooth in the dental battery fitting into the notch formed by the two roots of the tooth above it

Fastovsky, David E., and David B. Weishampel. *The Evolution and Extinction of the Dinosaurs*. New York: Cambridge University Press, 1996. Presenting a bare minimum of anatomical detail on the Ceratopsia, the authors concentrate on the analysis of their behavior.

Forster, Catherine A., and Paul C. Sereno. "Marginocephalians." In *The Complete Dinosaur*, edited by James O. Farlow and M. K. Brett-Surman. Bloomington: Indiana University Press, 1997. This article is a very brief overview of *Triceratops* and a number of its relatives.

Lucas, Spencer G. *Dinosaurs: The Textbook*. 3d ed. Boston: McGraw-Hill, 2000. A well-written text that would serve as a good supplement to the other references listed.

TUNDRA

Types of animal science: Ecology, geography
Fields of study: Ecology, environmental science, ethology, ornithology, wildlife ecology, zoology

The tundra, a characteristically treeless area of shrubs and patchy grass, lies above 60 degrees north latitude and covers about 10 percent of the earth's surface. It is inhabited by large numbers of a small variety of animals.

Principal Terms

CONIFEROUS FOREST: great northern forests, mostly evergreen, that end where the tundra begins
HERBIVORES: plant-eating animals
LICHENS: organisms formed by algae and fungi that are a source of food for tundra animals
MIGRATORY ANIMALS: animals that move from one place to another for feeding or breeding
PERMAFROST: a permanently frozen layer below the earth's surface
TEMPERATE REGION: a mild climatic area between the Tropic of Cancer and the Arctic Circle

From March to September, the tundra, which covers about one-tenth of the earth's surface, is warmed by sunshine, with the sun visible for days or weeks. Over much of the tundra, from September to March the sun does not rise at all, with temperatures of –40 to –50 degrees Fahrenheit common during the dark months.

Basically, two kinds of tundra exist: the Arctic tundra, which follows the coniferous forest belt and covers the northernmost landmass of Europe, Asia, North America, Greenland, and Iceland; and alpine tundra, found on mountain slopes above the tree line in temperate areas.

Permafrost is a consistent feature of the Arctic tundra, as well as of the poorly developed tundra of the Antarctic. The summer thaw in the Arctic tundra extends to a depth of six to twelve inches, with most plant root depth and burrowing of animals limited to this thawable area. The tundra permafrost retards drainage, thereby causing boggy, saturated lowland during the summer thaw. Alpine tundra is drier because of drainage.

Year-Round Inhabitants

Some hardy animals have learned to cope with the cold. Musk oxen, bulky animals with shaggy protective coats of hair, live along the shores of the Arctic Ocean and in other desolate areas of the tundra, where darkness continues for months and the temperature never rises above zero. These herbivores do not seek shelter even in the coldest weather, but search for frozen twigs or grass beneath the snow. Also capable of enduring the cold winter is the arctic fox, whose thick coat of fur turns white in winter and provides camouflage from his enemies. Foxes feed on arctic hares, birds, and lemmings, small rodents that dig beneath the snow. Lemmings remain active all winter, living under the snow and feeding upon grass and roots. Frequently, they reproduce under the snow.

Ptarmigans spend winters on the tundra, having feathers on their feet to aid traveling in the snow. Living in the driest areas of the tundra, they eat berries and tender leaves in the summer and rely on frozen vegetation during the winter. Their feathers turn white in winter, helping camouflage them from their fiercest enemy, the snowy owl. The owl, a bird of prey, moves southward into the forest in winter when the food supply becomes scarce in the tundra. Owls feed upon lemmings, small birds, insects, and arctic hares that are twice their size.

Some alpine mammals, such as marmots and ground squirrels, hibernate, eating large amounts of vegetation in summer and early fall before hibernation begins. Other small animals, including rabbits, forage as they can for winter feeding. Foxes range over alpine tundra in winter.

Migratory Inhabitants

As summer approaches and the days slowly grow longer and warmer, the frozen tundra begins to sprout grass, leaves, and wildflowers. The plants which have adapted to the tundra's short growing season and shallow topsoil above the permafrost are unusually small and provide food for animals. Huge herds of caribou migrate north from the forest to roam across the tundra, feasting on tender, young vegetation. The tundra wolf, or arctic wolf, a strong, fast animal endowed with exceptional hearing, vision, sense of smell, and endurance, follows the caribou into the tundra. Living and hunting in family groups or packs, the wolves feed off young, sick, and old members of the caribou herd.

After hibernating in the forest for the winter, grizzly bears move north to enjoy the berries and plants of the summer. These blond-colored bears also eat fish, lemmings, and carrion (remains left over by other animals). Polar bears, who prefer to eat meat, live near the Arctic Ocean to enjoy the seals, walruses, and fish. Occasionally, they move inland and devour berries, carrion, and other tundra animals.

Many birds also move north from the warm southern areas to feed, nest, and raise their young. Swans, ducks, and geese converge on the ponds, while gulls and terns occupy the tundra's beaches, and falcons and eagles soar at great heights above the tundra seeking their prey. In summer on the tundra, birds hatch as food becomes plentiful. As new vegetation is abundant and fish are plentiful in ponds and lakes, there are large numbers of insects on the tundra. Butterflies, moths, and bees abound, and flies and mosquitoes swarm after the caribou herds, providing food for the birds and fish.

Alpine animals, such as mountain sheep, ibex, wildcats, and many birds, who are not equipped for year-round alpine life, migrate south into more temperate forest environments in winter, returning to higher regions in the summer.

—*Mary Hurd*

See also: Chaparral; Ecosystems; Food chains and food webs; Forests, coniferous; Forests, deciduous; Grasslands and prairies; Habitats and biomes; Lakes and rivers; Marine biology; Mountains; Rain forests; Savannas; Tidepools and beaches.

Bibliography

Johnson, Rebecca L. *A Walk in the Tundra.* Minneapolis: Carolrhoda Books, 2000. Newest entry in *Biomes of North America* series, including a map and a resource section. Includes much information about the habitat, the flora and fauna, and color photography. Intended for juveniles.

Sayre, April Pulley. *Tundra*. Frederick, Md.: Twenty-first Century Books, 1995. Part of the series *Exploring Earth's Biomes.* Aimed at young readers. Presents an overview and variety of facts and activities about the ecological community of the tundra. Encourages young readers to read further and communicate with pertinent ecological organizations.

Shepherd, Donna W. *Tundra*. New York: Franklin Watts, 1997. Clear, succinct introduction to the tundra that surveys climate, landforms, seasonal change, plants, animals, and people; includes color photographs and maps. Intended for juvenile readers.

Walker, Tom. *Caribou: Wanderer of the Tundra*. Portland, Oreg.: Graphic Arts Center, 2000. Using vivid photography, Alaskan wildlife photographer and author follows large herds of elusive caribou through their life cycle; includes natural history information and personal observation.

Zwinger, Ann H., and Beatrice E. Willard. *Land Above the Trees: A Guide to American Alpine Tundra*. Boulder, Colo.: Johnson Books, 1996. Interesting, well-written coverage of Colorado Rockies, California Sierra Nevada, Olympia Mountains, White Mountains, and Cascades. Discusses natural science and ecology.

TURTLES AND TORTOISES

Types of animal science: Classification, evolution
Fields of study: Evolutionary science, herpetology, systematics (taxonomy)

Turtles are reptiles with shells that enclose the major body as well as the girdles to which limbs are attached. Shells not only define turtles, but determine by their structure what lifestyle is available to a particular species.

Principal Terms

AMNIOTE: animals with eggs in which embryos develop within fluid-filled membranes (the amnion), allowing eggs to be laid on land; amniotes include reptiles, birds, and mammals

BRIDGE: the portion of the shell that connects the carapace to the plastron

CARAPACE: the portion of the shell that covers a turtle's back (dorsum)

CLADE: a group of animals and their common ancestor

PLASTRON: the portion of the shell that protects a turtle's belly (venter)

Turtles, tortoises, and terrapins are all turtles. The term "tortoise" is used for terrestrial turtles with high-domed shells and elephantine hindlimbs, whereas the term "terrapin" is used properly for some highly aquatic turtles (genus *Malaclemys*) of eastern North America, although it frequently is used in error for American box turtles in the genus *Terrapene*.

Turtles are easily recognized and distinguished from all other vertebrates by their shells. Shells are composed of a dorsal carapace and a ventral plastron. These are usually rigidly connected on the sides by bridges. Shells are composed of bony plates that form within the skin. These are fused to underlying vertebrae and ribs. Most shells have a covering of horny plates made of keratin, a protein which, in other vertebrates, forms scales, hair, nails, claws, or horns. In some turtles, the plates of bone and keratin are reduced or absent, and the shell is covered by leathery skin. Many turtles have one or more hinges in their shells, usually in the plastron. These allow the shell to completely enclose the withdrawn head, limbs, and tail. The plastron of males in many species is indented to accommodate the female's shell during mating.

Turtle Lifestyles

Shell shape largely determines the lifestyle of its owner. Terrestrial (land-dwelling) turtles such as box turtles (*Testudo*) and tortoises (*Geochelone*) have high-domed shells. These reduce surface area through which water is lost and also are difficult for predators to grasp and break. Most aquatic and all marine turtles have relatively flat, streamlined shells for ease in swimming. However, African pancake tortoises (*Malacochersus*) have flat shells that allow them to hide in rocky crevices, and some bottom-dwelling aquatic turtles, such as the mud turtles (*Kinosternon*) of the southeastern United States, have high-domed shells. Snapping turtles in the genera *Chelydra* and *Macrochelys* have rough shells on which algae grow. This camouflages these turtles as they wait to ambush prey.

Limbs also provide clues to lifestyles. Aquatic turtles have webbed feet, and sea turtles have forelimbs modified into flippers that allow them to "fly" through water. In contrast, terrestrial turtles often have spadelike feet for digging and/or columnlike limbs to support them as they walk. Regardless of shape or function, the girdles to which the limbs attach are enclosed by the ribs and shell. Turtles are the only vertebrates with this skeletal arrangement.

Other anatomical modifications include nostrils on top of the snout or at the very tip of a long proboscis; these allow aquatic turtles to breathe at the surface with minimal exposure. Modern turtles, like modern birds, lack teeth. Instead, they have horny beaks of keratin variously shaped to cut leaves, tear flesh, or crush the shells of snails or clams. Because the shell prevents expansion and contraction of the thorax when breathing, turtles compress or expand the lungs by altering the location of other internal organs to which the lungs are attached. Shells limit mobility to a great extent; consequently, turtles have long and flexible necks. These allow them to reach up to browse or down to graze, or to quickly extend their necks in order to ambush quicker prey. In addition, neck vertebrae are modified to allow the head to be withdrawn into the shell, either by pulling it straight back while the neck assumes an S-shape (cryptodiran turtles) or by laying it to the side under the overhanging lip of the carapace (pleurodiran or sideneck turtles).

Turtle Facts

Classification:
Kingdom: Animalia
Subkingdom: Bilateria
Phylum: Chordata
Subphylum: Vertebrata (vertebral column and braincase)
Class: Reptilia
Order: Chelonia (turtles), consisting of twelve families, with fifty-two genera and about 260 species

Geographical location: Worldwide

Habitat: Terrestrial, aquatic, and marine habitats, except at high latitudes

Gestational period: Varies according to temperature

Life span: Turtles live longer than other vertebrates, with some small species living fifty years and larger forms up to two hundred years

Special anatomy: The only vertebrates with shells; lack teeth and instead have a horny beak; neck vertebrae are modified to allow the head to be extended or retracted from the shell; lungs are compressed and expanded by altering the position of other, attached internal organs

Some Types of Turtles

ALLIGATOR SNAPPING TURTLES (*Macrochelys temminckii*), from the southern United States, may exceed three hundred pounds. These bottom-dwellers rest motionless with their mouths open, waving a wormlike projection on the tongue to lure unsuspecting fish.

SIDENECK TURTLES (families Chelidae and Pelomedusidae), of the Southern Hemisphere, include the matamata (*Chelus fimbriatus*), a bottom-dweller that uses its long neck as a snorkel and to ambush small fish attracted to flaps of skin that act as lures, and the large Amazonian river turtles (*Podocnemis expansa*), whose nesting behavior is much like that of sea turtles.

SEA TURTLES (families Cheloniidae and Dermochelyidae) nest on beaches and flee to the sea immediately after hatching; males never return to land and females do so only to lay eggs. The leatherback (*Dermochelys coriacea*) is the largest living turtle; specimens with shells eight feet long and weighing over a ton have been taken.

GIANT TORTOISES (*Geochelone elephantopus*) from the Galápagos Islands were collected by the thousands by nineteenth century whalers heading for Antarctic waters; stored upside down, they provided fresh meat for months. Populations on many islands were extirpated.

Turtles are the only vertebrates that have shells. (Digital Stock)

Origins and Future of Turtles

Fossil turtles are known from the Jurassic. Most systematists (biologists who study evolutionary relationships) group turtles with some extinct relatives in a clade called the Parareptilia. Although turtles traditionally have been considered reptiles, many experts now place them a separate vertebrate class. Regardless, the ancestors of turtles arose from the first amniotes before the ancestors of other reptiles. This and their many unique features justify placing turtles into their own class.

Unlike many reptiles, turtles are perceived positively by most people. Nevertheless, many species are threatened or endangered. Habitat destruction and alteration are responsible in most cases. Aquatic habitats are drained or polluted and nesting sites, especially beaches, are developed, rendering them unusable by turtles. Many species are exploited as food, either as eggs or adults, and others are killed for their shells or body parts, which are thought by some cultures to have medicinal or aphrodisiac qualities. Exotic predators, such as rats and dogs, dig up nests and kill adults. Hundreds of thousands of wild-caught turtles die each year in the pet trade, much of it illegal. Many species become roadkills when they migrate to new habitats or breeding sites. Only a few species are formally protected in at least some parts of their ranges, and several, including the sea turtles, may be nearing extinction in spite of efforts to conserve them.

—*Robert Powell*

See also: Beaks and bills; Endangered species; Fins and flippers; Lakes and rivers; Lungs, gills, and tracheas; Marine animals; Reptiles; Shells.

Bibliography

Alderton, David. *Turtles and Tortoises of the World*. New York: Facts on File, 1988. This informative book was written with the nontechnical reader in mind.

Ernst, Carl H., and Roger W. Barbour. *Turtles of the World*. Washington, D.C.: Smithsonian Institution Press, 1989. This book provides information on turtles from throughout the world.

Ernst, Carl H., Roger W. Barbour, and Jeffrey E. Lovich. *Turtles of the United States and Canada*. Washington, D.C.: Smithsonian Institution Press, 1994. This extremely well researched volume provides photographs and extensive information on North American species.

Laurin, Michel, and Robert R. Reisz. "A Reevaluation of Early Amniote Phylogeny." *Zoological Journal of the Linnean Society* 113 (1995): 105–225. This article presents a modern view of relationships among ancestors of turtles and reptiles.

Pough, F. Harvey, Robin M. Andrews, John E. Cadle, Martha L. Crump, Alan H. Savitsky, and Kentwood D. Wells. *Herpetology*. 2d ed. Upper Saddle River, N.J.: Prentice Hall, 2001. This textbook attempts to address most aspects of herpetology; of particular interest are chapters dealing with vertebrate origins and classification of reptiles.

TYRANNOSAURUS

Type of animal science: Classification
Fields of study: Archaeology, evolutionary science, paleontology

The largest terrestrial carnivore, Tyrannosaurus *appeared late in the Cretaceous era that ended some sixty-five million years ago.*

Principal Terms

CARNIVORES: meat eaters
FIBULA: the smaller of two bones between the knee and ankle
JURASSIC ERA: dating from 135 to 190 million years ago
PALEONTOLOGIST: a scientist who studies fossils
TRIASSIC ERA: dating from 190 to 225 million years ago

Dinosaurs have a 175 million-year history beginning in the Permian period and extending through the Triassic and Jurassic periods to the Cretaceous, the last period in the Mesozoic era, which lasted 160 million years and ended 65 million years ago. *Tyrannosaurus* dates to the latter part of the Cretaceous period, which ceased when a great deal of life on earth disappeared for reasons that are not fully known, although it is speculated that a huge meteor crashed into earth, causing a heavy cloud to hang over the planet long enough to kill most vegetation.

No land animal weighing over about fifty-five pounds survived whatever catastrophe caused the sudden end of an era when huge animals roved the earth. *Tyrannosaurus*, the largest terrestrial carnivore, was almost forty feet long. *Tyrannosaurus* had a huge head, a large mouth, and menacing teeth.

Physical Characteristics and Habitat

An adult *Tyrannosaurus* standing upright would have been as tall as a four-story building, but *Tyrannosaurus* did not stand erect. Its hind legs provided sturdy underpinnings, whereas its arms were short and weaker than its legs. They could be used defensively when necessary. Fossil footprint evidence substantiates that *Tyrannosaurus* has feet over three feet long.

This dinosaur depended on its hind legs for most of its locomotion, although it used its arms minimally when it walked. Bulky in the midsection, its long tail aided its balance. Its long neck supported a huge head with a large mouth and seven-inch-long serrated teeth. Adults weighed about seven tons.

Most of the extant remains of *Tyrannosaurus* have been found in the United States, mostly in the South Dakota, Montana, and Wyoming Badlands. The first three *Tyrannosaurus rex* remains were found in Montana and Wyoming in the early 1900's. This area was also inhabited by duck-billed dinosaurs, much smaller animals than *Tyrannosaurus*, who were often eaten by their larger counterparts.

During the late Cretaceous period, the area in which *Tyrannosaurus* remains were found was warmer than it currently is. It is known from fossil remains that its climate resembled the current climate of the southern states. The area was rich in such plant life as ferns, palm trees, redwoods, and flowering plants, which contributed to the diet of dinosaurs. The preserved contents of *Tyrannosaurus* stomachs reveal that they ate many of these plants.

Most of the animals that coexisted with dinosaurs were small, seldom exceeding the size of a domestic cat. Birds were abundant, as were such insects as spiders and beetles. Opossums existed in large numbers, and the waterways of the ancient landscape were filled with fish and turtles, all of which became part of the *Tyrannosaurus* diet. One thing is clear: For a period of 150 million years, dinosaurs ruled the earth. They were the largest, most complex organisms in existence, and *Tyrannosaurus* was preeminent among dinosaurs.

Tyrannosaurus rex lived closer to the beginnings of human existence than it did to the time when the earliest dinosaurs roved the earth. Some paleontologists believe that it was descended from a species of carnivores in Mongolia that migrated from Asia to North America over a formation that once bridged the Bering Straits, but has since disappeared. The Badlands are the richest depository in the United States discovered to date of dinosaur remains.

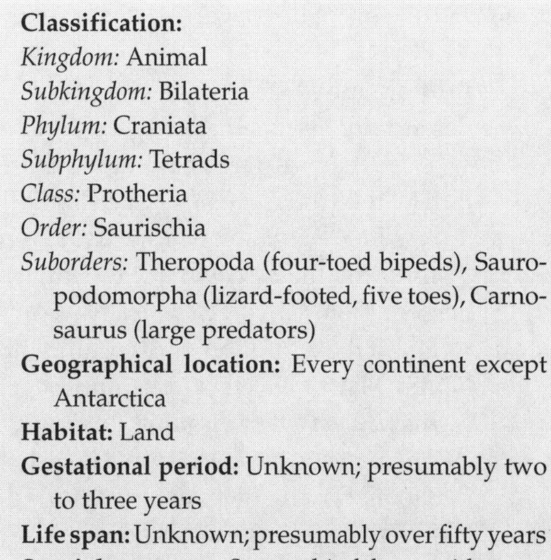

Tyrannosaurus Facts

Classification:
Kingdom: Animal
Subkingdom: Bilateria
Phylum: Craniata
Subphylum: Tetrads
Class: Protheria
Order: Saurischia
Suborders: Theropoda (four-toed bipeds), Sauropodomorpha (lizard-footed, five toes), Carnosaurus (large predators)
Geographical location: Every continent except Antarctica
Habitat: Land
Gestational period: Unknown; presumably two to three years
Life span: Unknown; presumably over fifty years
Special anatomy: Strong hind legs, with arms weaker than the legs; two fingers on each hand; large head; sharp, serrated teeth

The Largest *Tyrannosaurus rex* Ever Found

Peter Larson, an independent collector of fossils who knew a great deal about paleontology, unearthed the skeleton of Sue, the largest and most perfect *Tyrannosaurus* skeleton ever found. Painstakingly cleaned and reassembled, it is dramatically displayed in Chicago's Field Museum.

Tyrannosaurus *was the largest terrestrial carnivore, with an average length equal to the height of a four-story building.* (©John Sibbick)

Sue was discovered by Sue Hendrickson in August of 1990, when she noticed three large dinosaur vertebrae and a femur protruding from a cliff in the Badlands of South Dakota. These items obviously belonged to *Tyrannosaurus* because the vertebrae were concave from the disk, unlike the straight vertebrae of duck-billed dinosaurs.

With Larson's help, Hendrickson determined that because part of its skull was damaged, Sue died in some sort of conflict. Sue had suffered a fractured fibula that healed but that must have left her defenseless for some time. Contents of its stomach indicated that Sue's last meal was a duck-billed dinosaur.

—R. Baird Shuman

See also: *Allosaurus*; *Apatosaurus*; *Archaeopteryx*; Dinosaurs; Evolution: Animal life; Extinction; Fossils; Hadrosaurs; Ichthyosaurs; Paleoecology; Paleontology; Prehistoric animals; Pterosaurs; Sauropods; Stegosaurs; *Triceratops*; Velociraptors.

Bibliography

Berenstain, Michael. *King of the Dinosaurs: Tyrannosaurus Rex*. Racine, Wis.: Western, 1989. A good starting point for those unacquainted with the field.

Fiffer, Steve. *Tyrannosaurus Sue: The Extraordinary Saga of the Largest, Most Fought Over T. Rex Ever Found*. New York: W. H. Freeman, 2000. Account of the controversies that followed the discovery of the largest *Tyrannosaurus* skeleton ever unearthed.

Lessem, Don. *Kings of Creation*. New York: Simon & Schuster, 1992. Detailed information about the evolution of dinosaurs through a two-hundred-million-year period.

Norell, Mark, Lowell Dingus, and Eugene Gaffney. *Discovering Dinosaurs: Evolution, Extinction, and the Lessons of Prehistory*. Berkeley: University of California Press, 2000. The best source to date on the history of dinosaurs.

Peterson, David. *Tyrannosaurus Rex*. Chicago: Children's Press, 1989. Directed toward juvenile audiences, this book is clear, concise, and accurate.

Stein, Wendy. *Dinosaurs: Opposing Viewpoints*. San Diego, Calif.: Greenhaven Press, 1994. Brief, accurate overview of the history of dinosaurs.

UNGULATES

Types of animal science: Anatomy, classification, ecology
Fields of study: Anatomy, zoology

Ungulates are herbivorous, hoofed mammals. Those having an odd number of toes are perissodactyls (such as horses), while those with an even number of toes are artiodactyls (such as cattle). Elephants (proboscids) and rabbitlike hyraxes are also ungulates.

Principal Terms

ANTLERS: branched, temporary horns made of solid bone, shed and regrown yearly
CARNIVORE: any animal that eats only the flesh of other animals
GESTATION: the term of pregnancy
HERBIVORE: an animal that eats only plants
NOCTURNAL: active at night
OMNIVORE: an animal that eats both plants and other animals
TRUE HORNS: straight, permanent, hollow bone horns

Ungulates are the hoofed mammals, belonging to the phylum Chordata. The word "ungulate" comes from Latin *ungula*, meaning "hoof." Ungulates are a large group of dissimilar vertebrate animals, grouped together because their outermost toe joints are encased in hooves. There are four ungulate orders. Those ungulates having an odd number of toes belong to the order Perissodactyla. This includes horses (one-toed), rhinoceroses (three-toed), and tapirs (four-toed on the front feet and three-toed on the back feet). Entirely even-toed ungulates belong to the order Artiodactyla. This includes pigs (four-toed) and two-toed ruminants such as camels, giraffes, antelope, deer, cattle, sheep, and goats.

The two other orders are Proboscidea (elephants) and Hyracoidea (rabbitlike hyraxes). The size extremes among ungulates range from the seven-ton male African elephant to the rabbit-sized dik-dik antelope. Most ungulates are herbivores. Ungulates are also the only mammals with horns or antlers, although not all of them have this bony headgear. They are native to all earth's continents except Australia.

Wild Ungulates

Antelope, elephants, hippopotamuses (hippos), and deer are some common wild ungulates. Horses, sheep, goats, cattle, and pigs are mostly domestic ungulates. Ungulate appearance varies widely, but there are common physical and digestive characteristics. Most are artiodactyls, which walk on two toes. Their ancestors had five toes, but evolution deleted the first toe and made the second and fifth toes vestigial. The third and fourth toes provide support, and end in protective hoofs. Hippopotamuses, unique among artiodactyls, have four toes of equal dimensions.

Most ungulates are herbivorous ruminants. They eat only plants, and have specialized digestive tracts with three or four chambers in their stomachs. They chew and swallow vegetation, which, after partial digestion, is regurgitated, chewed again, and reenters the stomach for more digestion. This leads to maximum nutrient uptake from vegetable food. Ungulates usually lack upper incisor and canine teeth. They have hard pads in their upper jaws, which help the lower teeth to grind food.

Deer and antelope are swift-running, hoofed ruminants. Male deer have solid, branched antlers (temporary horns) made of bone, which are shed and regrown yearly. Antelope of both genders have unbranched, permanent, hollow bone horns (true horns). Deer inhabit Asia, Europe, the Americas, and North Africa. Antelope inhabit Africa, Asia, and Europe. Both deer and antelope live in woods, prairies, marshes, mountains, and tundra. Their sizes range from huge moose and elands to rabbit-sized species. Deer and antelope eat twigs, leaves, bark, and grass. The largest antelope are ox-sized.

Giraffes and hippos are very unusual African ungulates. Giraffes live south of the Sahara desert. They have very long legs and necks. Males are over sixteen feet tall and both sexes have short horns. Their flexible tongues and upper lips pull leaves—their main food source—from trees. The two-ton animals can go months without drinking, getting most of their water from the leaves they eat. The three- to four-ton hippos walk on all four toes of each foot. They have short legs, large heads, no horns, small eyes and ears, and nostrils that close underwater. Hippos have long, sharp incisors and canines in both jaws. A hippo can be fifteen feet long and five feet high at the shoulder. They spend most of their time submerged, eating aquatic plants.

Brazilian tapirs inhabit South American forests from Colombia and Venezuela to Paraguay and Brazil. They look a bit like elephants and pigs, but are related to horses. The tapirs have stout bodies and short necks and legs, well adapted for pushing through dense forests, and have short, rigid manes, which protect them from predators. Each tapir has a short trunk with a flexible "finger" at its tip. Like elephants, they use the finger to pull leaves into the mouth. They are dark brown to reddish colored, about 6 feet long, 2.5 feet tall at shoulder height, and weigh up to six hundred pounds. Tapirs are nocturnal herbivores, spending much of the night eating grass, grasses, aquatic vegetation, leaves, buds, soft twigs, fruits,

Zebras are perissodactyl ungulates, walking on only one hoofed toe. (Corbis)

and plant shoots. The tapirs roam the forest and can climb river banks and mountains. Excellent swimmers, they spend a lot of time in the water, eating and cooling off.

Domesticated Ungulates

Bovids—cattle—are domesticated ungulates. Most have true horns. Bovid horns are spiral, straight, tall, or grow from the sides of the head and then up. Most are herbivorous ruminants. Cattle are raised to provide meat, milk, and leather. Modern cattle come from European, African, and Asian imports. Breeding modern cattle began in mid-eighteenth century Europe; today there are three hundred breeds. Dairy cattle such as Holsteins make milk, and beef cattle such as Angus yield meat.

Sheep and goats are also domesticated, ruminant ungulates. Sheep were domesticated eleven thousand years ago from Asiatic mouflons. They have paired, spiral true horns, largest in males. Adults reach lengths of five feet and weights from 250 to 450 pounds. The eight hundred domesticated breeds provide wool for clothing, meat, and milk. Goats, closely related to sheep, have shorter tails, different horn shape, and beards. They eat grass, leaves, and branches. Numerous breeds are domesticated for meat and milk. Angora goats yield silky mohair. Goat milk is as nutritious as cow milk.

The horse, donkey, zebra (HDZ) family are perissodactyls. They live in habitats ranging from grassland to desert. They eat grasses, bark, leaves, buds, fruits, and roots, spending most of their waking hours foraging and eating. Wild specimens inhabit East Africa and the Near East. Domesticated horses and donkeys are used for food, meat, and leather. Zebras are too savage to domesticate.

Members of the HDZ family lack horns. They have long heads and necks, slender legs, manes on their necks, and long tails. They have good wide-angle day and night vision and a keen sense of smell. The smallest family members are African wild donkeys, 4 feet tall, 6.5 feet long, and weighing nearly five hundred pounds. The largest, Grevy's zebras, are five feet tall, nine feet long, and weigh up to nine hundred pounds. Zebras have black or brown and white, vertically striped coats. The other HDZ family members are brown, black, gray, white, and mixtures of these colors.

The Lifestyles of Ungulates

The lifestyles of ungulates are very different. Many of them are very sociable and live in large herds, including many bovids, horses, deer, antelope, and zebras. In other cases, the animals live in smaller family groups, or are solitary, coming together only to breed.

Wild donkeys and horses live in herds made up of a male and his mates. Young stay in the herd until two or four years old for females and males, respectively. Males then join other bachelors until winning a herd. Females join other herds. Goats and sheep are also herd animals. Young goats, sheep, and cattle join herds or live in solitary fashion after they are weaned.

Moose are quite different. Males are solitary until they fight for mates and breed in the fall. A successful male often leads several females and babies all winter. In the spring he returns to the solitary life. Giraffes and male elephants are solitary. Female elephants and young form herds whose members breed with visiting males, protect each other, and raise young.

Ungulates are of great importance to humans and to the world. First of all, in the wild state they are food for many carnivores and omnivores. Domesticated, they provide meat, milk, hides, and sinew for human use. Furthermore, elephants, horses, and reindeer have long been used as beasts of burden. In addition, ungulates are biologically important because as herbivores, they prevent overgrowth of all sorts of plants by eating them.

—*Sanford S. Singer*

See also: Antelope; Cattle, buffalo, and bison; Deer; Donkeys and mules; Elephants; Elk; Giraffes; Goats; Hippopotamuses; Horns and antlers; Horses and zebras; Moose; Pigs; Reindeer; Ruminants; Sheep.

Bibliography

Arnold, Caroline, and Richard Hewitt. *Zebras*. New York: William Morrow, 1987. This brief book describes characteristics, habitats, and lives of zebras.

Gerlach, Duane, Sally Atwater, and Judith Schnell. *Deer.* Mechanicsburg, Pa.: Stackpole, 1994. Discusses deer habits, habitats, species and much more. It has a useful bibliography.

Rath, Sara. *The Complete Cow.* Stillwater, Minn.: Voyageur Press, 1990. Covers cattle, their history, and their breeds.

Sherr, Lynn. *Tall Blondes: A Book About Giraffes*. Kansas City, Mo.: Andrews and McMeel, 1997. This useful book contains a great deal of information on giraffes.

Shoshani, Jeheskel, and Frank Knight. *Elephants: Majestic Creatures of the Wild*. Emmaus, Pa.: Rodale Press, 1992. Covers many elephant topics, including natural history.

Walker, Sally M., and Gerry Ellis. *Hippos*. Minneapolis: Carolrhoda Books, 1998. This book, written for juvenile readers, explores the natural history of the two hippo species.

URBAN AND SUBURBAN WILDLIFE

Types of animal science: Behavior, ecology
Fields of study: Ecology, ethology, wildlife ecology

The global increase in the human population growth and density has seen a simultaneous increase in the growth of urban and suburban areas throughout the world. As urban habitats and their suburban extensions have become more common, the wildlife of these human landscapes has become the focus of attention and study.

Principal Terms

ANTHROPOGENIC: originating from human sources, such as aerosols and other pollutants

BIODIVERSITY: variety of life found in a community or ecosystem; includes both species richness and the relative number of individuals of each species

EXOTICS: organisms, usually animals, that have been deliberately or inadvertently introduced into a new habitat, such as monk parakeets in New England, or brown snakes on Guam

FERAL ANIMALS: domestic animals that have reverted to a wild or semiwild condition, such as cats, dogs, or caged birds that have been released or escaped and now survive in the wild

MORPHOLOGY: development, structure, and function of form in organisms

OPEN SPACE: natural or partly natural areas in and around cities and suburbs, such as woodlots, greenbelts, parks, and cemeteries

URBAN WILDLIFE: generally the nondomestic invertebrates and vertebrates of urban, suburban, and urbanizing areas; may include domestic animals that have escaped and are subsequently feral

Animals and plants of cities and suburbs are categorized as urban wildlife. As cities and suburbs grow ever larger and displace natural habitats, many city and suburban landscapes have become more attractive for certain kinds of wildlife, or at least urban wildlife has become more noticeable. Urban wildlife consists of an eclectic and unlikely mix of escaped pets (mostly exotics and caged birds), feral animals, furtive and temporary intruders from adjacent natural habitats, and species whose natural ecology and behavior enables them to fit within human-modified landscapes and tolerate living in close proximity to humans.

Urban landscapes present a seemingly stark and forbidding environment for wildlife. The horizontal pavement of streets and sidewalks is punctuated by rising angles and arches of concrete and steel which in turn are topped by wood and metal rooftops. Overhead, a maze of telephone, power, and cable lines limits vertical movement, while vehicle and foot traffic pose a constant threat to surface movement. All of these edifices and connecting corridors and lines result in a complex, vertically structured environment within which some animals find difficult to maneuver yet to which other animals quickly adapt. In addition to monotonous and often dangerous structural diversity, urban wildlife is subject to elevated and often almost continuous noise and disturbance and is constantly exposed to an enormous variety of residential wastes (garbage, litter, excess water, salts, sewage), vehicular pollutants (lubricants, greases, gasoline, hydrocarbons, nitrogen oxides), and chemical wastes (pesticides, paints, lead, mercury, contaminants).

Despite the forbidding features offered by urban habitats, a surprising variety of wildlife manages to exist on a more-or-less permanent status. In fact, some kinds of wildlife can be found even in the midst of the most degraded forms of urban blight. *Ailthanthus*, which is also commonly called tree-of-heaven, is but one of many opportunistic trees and shrubs that can take root and grow given a bare minimum of soil and nutrients. A simple linear crack in the pavement of a sidewalk, a little-used roadway, an unused parking area or vacant lot can trap enough windswept dirt to offer a growing substrate for *Ailanthus* and similar hardy plants. Each *Ailanthus*, in turn, provides food and shelter for equally tough and adaptable wildlife, ranging from the variety of invertebrates that colonize and feed upon *Ailanthus* to birds and mammals that take shelter or find food in its branches and foliage. Similarly, every invading sprig of grass, wildflower, shrub, or tree, however large or small, creates its own suite of microhabitats which, in turn, offer colonization opportunities for other plants and animals, the whole ultimately contributing to an overall increase in urban biodiversity.

Characteristics of Urban Wildlife

Ailthanthus is an example of those plants and animals able to tolerate the most extreme urban conditions, but in reality most urban wildlife derives a number of benefits by living within the confines of cities and suburbs. Far from being homogenous expanses of concrete, most urban centers are a patchwork of different habitats—residential, commercial, and industrial buildings, warehouses, power stations, vacant lots, detached gardens, rooftop gardens, and alleyways—that each offers innumerable opportunities for wildlife. Many urban areas also have a number of limited access areas that animals are quick to adopt for shelter and breeding places; these include fenced-in lots and boarded-up buildings, along with a rabbit warren of underground tunnels, ducts, steam and water pipes, basements, and access ways.

City lights extend foraging time and opportunities, allowing wildlife to hunt for food not only throughout the day but also during much of the night, as needed. Urban nooks and crannies offer an extensive variety of microhabitats that differ fundamentally in size, microclimate, and other structural features. These microhabitats serve primarily as shelters and breeding sites for city wildlife. Many birds, such as house sparrows (*Passer domesticus*) and Eurasian starlings (*Sturnis vulgaris*) nest in innumerable crevices, cracks, nooks, niches, and sheltered rooftops. Pigeons (*Columbia livia*) and starlings hide in sheltered enclaves offered by bridge abutments and supports, archways, and other edifices.

The most adaptable wildlife are quick to find and take advantage of subtle advantages offered in urban habitats: Many birds cluster around chimneys and roof reflectors or in shelters afforded by lee sides of rooftops during harsh cold and windstorms. Others are equally quick to obtain warmth by sitting on poles, rooftops, or other elevated perches to orient toward sunlight, while at ground level animals gather near gratings, vents, and underground heating pipes.

Urban wildlife quickly concentrates in areas where potential food is made available, for instance, during trash pickup, then just as quickly disperses to find new food sources. Most urban wildlife forage opportunistically as scavengers, specializing in finding and consuming all bits of discarded food, raiding trash cans, and concentrating at waste collection and disposal centers. Thus, the rubbish dumps, found in or immediately adjacent to every city of the world, attract an amazing diversity of small mammals and birds. Feeding on the scavenged food of urban areas and bird feeders is much more efficient because it requires less energy to find or catch and is usually available throughout the year.

Because of the need to find and exploit temporary food resources, some of the most successful urban animals forage in loose groupings or flocks—the more eyes, the more searching, and the more feeding opportunities can be identified and exploited. Solitary and nonsocial species often do less well in urban environments simply because they lack the collective power of the group to find food and shelter, and avoid enemies.

The availability of a year-round food supply—however tenuous and temporary—along with the presence of an enormous variety of safe shelters and breeding sites promotes a higher life expectancy, which partly or mostly balances the higher vehicle-related death rates to which urban wildlife is continuously subject.

Parks and open space provide the only true refuges of natural habitats set deep within urban and suburban landscapes. Such open-space habitats function as ecological islands in a sea of urbanism. Most are necessarily managed habitats rather than entirely natural and, like the urban environment that surrounds them, are usually subject to constant disturbance from adjacent traffic, noise, and other forms of pollution. Economically, since most open-space parks are set aside and maintained for a variety of recreational purposes rather than as natural habitats, the wildlife that colonizes these unnatural natural habitats must have an unusually high tolerance for human presence and recreational activities of all kinds.

Sources and Types of Urban Wildlife

For some urban wildlife, the urban landscape is merely a manmade version of their natural environment. Thus, for pigeons the ledges, cracks, and crevices of buildings and bridges represent an urban version of the cracks and crevices of cliffs and rock outcrops that they use for roosting and nesting in the native habitats. Similarly, the short-eared owls (*Asio flammeus*) and snowy owls (*Nyctea scandiaca*) that show up in winter to stand as silent sentinels at airports, golf courses, and other open areas are simply substituting these managed short-grass habitats for the tundra habitats preferred by snowy owls and the coastal marshes hunted by short-eared owls. Their summer replacements include a host of grassland nesting species such as grasshopper sparrows, kildeer, and upland sandpipers, which all find these managed habitats to be ideal substitutes for the native grasslands which they displaced or replaced.

Many bird inhabitants of urban and suburban environments are exotics which were deliberately or inadvertently introduced into urban areas. Certainly the three birds with the widest urban distribution in North America, the pigeon or rock dove, European starling, and house sparrow or English sparrow, all fit within this category. The introduction of the European starling into North American cities and suburbs resulted from the dedicated efforts of the American Acclimitization Society of the late 1800's. The goal of this society was the successful introduction of all birds mentioned in the works of Shakespeare into North America. Unfortunately for North Americans, the character of Hotspur in *Henry IV* makes brief note of the starling, so the society repeatedly attempted to introduce the starling into Central Park until they were finally successful. Since then, the starling has become the scourge of cities and suburbs throughout much of North America and the rest of the civilized world. Starlings damage and despoil crops, and dirty buildings with their droppings.

The association between house sparrows and urban centers is apparently very old. Evidence suggests that they abandoned their migratory ways to become permanent occupants of some of the earliest settlements along the Nile and Fertile Crescent, a trend that has continued to this day. Sparrows and starlings both share certain characteristics that enable them effectively to exploit urban and suburban habitats; both are aggressive colonizers and competitors, able to feed opportunistically on grains, crops, discarded bits of garbage, and other food supplies.

Avian occupants also include an increasing diversity of released caged pets. Thus, urban locales in Florida, Southern California, and along the Gulf Coast support an ever increasing diversity of parakeets, parrots, finches, and lovebirds, all stemming from caged pet birds either deliberately released or lost as escapees.

Feral animals, mostly dogs (Canidae) and cats (Felidae), represent another important source and component of urban wildlife. Feral dogs revert to primal adaptive behaviors, gathering in loose packs that usually forage and take shelter together, but have limited success because almost all cities in developed countries have ongoing measures to control and remove them whenever found. Feral cats are often more successful because they are secretive, mostly nocturnal, and can clearly better exploit available urban food sources. The role of other feral animals as urban wildlife, mostly escaped pets, is not well known.

Humans and Urban Wildlife

The attitude of urban dwellers toward urban wildlife varies greatly. For many humans, urban wildlife offers a welcome respite from their otherwise dreary and mundane surroundings. Urban wildlife in all of its forms and colors can be aesthetically attractive, even beautiful, and is also compellingly interesting. For example, the nesting of a pair of red-tailed hawks (*Buteo jamaicensis*) in New York City's Central Park sparked a remarkable interest in birdwatching in the city and a heightened awareness of exactly how exciting wildlife watching can actually be. All facets of the pair's courtship and nesting were watched and reported in newsprint, novellas, and even a book, *Red-Tails in Love*. Other animals, while not nearly as large, conspicuous, and glamorous in their color and disposition, also elicit interest. Urban wildlife adds lively color and contrast to the otherwise monotonous gray and grime of streets and sidewalks. Part of the attraction is that urban birds are usually already sufficiently tolerant to be semitame in spirit, easily seen and observed, and, in some instances, easily attracted. Strategically placed bird feeders and bird houses also attract these birds.

Public attitude toward urban predators varies considerably. Some people find them attractive and interesting and even put out food for them. Others consider them pests or potentially dangerous and avoid them. During rabies outbreaks or public scares, most urban wildlife is targeted by various control programs to remove unwanted animals.

Suburban Wildlife Habitats

The vast sprawl of suburbs across the landscape offers many types of wildlife yet another habitat opportunity to exploit, either as residences or as temporary components of the search for food or shelter. Like urban areas, suburbs offer a range of differing habitats. The simplest suburbs are merely extensions of urban row houses with minimal yardscapes, but there is an increasing progression toward more open and natural yards in outlying suburbs that merge with rural areas and natural habitats. The larger and more diverse yards at the edges of suburbs often help blur the distinction and diversity between human landscapes and natural landscapes.

Ornamental trees, shrubs, flowers, gardens, and lawns that characterize almost all suburban habitats provide a series of artificial habitats that can actually increase wildlife diversity. Again, the chief wildlife benefactors are species that can best ecologically exploit the unnatural blend of woodland, edge, and meadow that suburban landscapes offer. It is no accident that some of the most common components of suburban wildlife include thrushes such as robins, finches, and cardinals, titmice, blue jays, crows, and many other similar birds. All of these species are actually responding to the structural components of the suburban landscape, which provide suitable substitutes for their natural landscapes.

The blend of ornamental and garden vegetation offered by most suburban landscapes offers food for a diversity of what were once considered less tolerant wildlife. Deer, wild turkey, grouse, and a host of other animals, large and small, make periodic forays into suburbs in search of foods. Crepuscular and nocturnal wildlife is much more likely effectively to exploit food sources offered by suburban landscapes than diurnal wildlife, which is more at risk because of its high visibility during daylight hours.

Well-wooded suburban habitats that attract a variety of wildlife also attract an increasing number of predators. American kestrels (*Falco sparverius*), Cooper's hawks (*Accipiter cooperi*), barn owls (*Tyto alba*), screech owls (*Otus* spp.), and little owls (*Athene noctua*) provide but a small sampling of birds of prey that nest deep within urban and suburban environments, taking advantage of open-space habitats deep within cities and quickly exploiting unused areas within most suburbs. Terrestrial predators are almost equally common, but most are nocturnal or nearly so; consequently, their contacts with humans are quite limited. Many urban predators are, in fact, mistaken for neighborhood pets and left alone or avoided: Coyotes (*Canis latrans*) are often mistaken for dogs, especially when seen in twilight. The wily coyote is equally at home in the suburbs of Los Angeles, California, and the urban parks of New Haven, Connecticut, joining a host of small and medium-sized mammal predators such as foxes (*Vulpes* spp.) and scavengers such as opossums (*Didelphis marsupalis*), raccoons (*Procyon lotor*), and skunks (*Mephitis mephitis*). These urban predators have many behavioral attributes in common. All are omnivorous and able to feed on a wide variety of natural foods such as fruit, small birds and mammals, insects, and invertebrates such as beetles, grasshoppers, and earthworms.

Foraging and food habits of urban predators sometimes conflict with human concerns. Urban foxes hunt and kill cats, especially kittens, if given the opportunity, while the larger and stronger urban coyote will often not hesitate to kill and eat cats and dogs, to the pet owners' dismay.

Conservation and Management of Urban and Suburban Wildlife

Urban wildlife must be much more closely managed than wildlife of natural environments because urban and suburban habitats attract an enormous number of pest species as well as interesting and beneficial species. Introduced species such as starlings may also transmit histoplasmosis, a fungal disease that attacks human lungs. Other birds may also be harbingers, carriers, and vectors of various diseases, the most notable of which are the parrots and parakeets, which transmit parrot fever or psittocosis. Rats and mice (Rodentia) carry and spread disease and despoil both residential and public buildings and other structures.

The growing interest in urban wildlife has stimulated innumerable programs to promote beneficial wildlife. Both public and private organizations and agencies have embarked on a variety of programs aimed at remodeling existing habitats and even creating new habitats for urban wildlife.

Programs aimed at creating new or modifying existing urban habitats come in a variety of categories, such as linear parks, greenways, urban wildlife acres programs, backyard gardens, and treescaping streets and roadways, all of which create biodiversity, which in turn provides attractive habitats for colonization by additional animals and plants. Modification of existing habitats to increase animal biodiversity includes "critter crossings," roadside habitats, backyard gardens, and arbor plantings, all of which provide refuges, shelters, breeding sites, connecting corridors, and safe havens that promote the welfare of urban and suburban wildlife.

Many existing open-space habitats are also being modified. Many urban renewal commissions have placed new and more restrictive regulations on the use of pesticides and fertilizers on golf courses, which not only reduces the incidence and intensity of nonpoint pollution from the golf courses but also reduces the incidence of wildlife poisoning. These steps cannot help but increase the biotic potential of golf courses for supporting local biodiversity.

—Dwight G. Smith

See also: Birds; Dogs, wolves, and coyotes; Ecological niches; Ecosystems; Foxes; Habitats and biomes; Hawks; Mice and rats; Opossums; Owls; Raccoons and related mammals; Scavengers; Skunks; Sparrows and finches; Weasels and related mammals.

Bibliography

Adams, Lowell W. *Urban Wildlife Habitats: A Landscape Perspective*. Minneapolis: University of Minnesota Press, 1994. One of the few recent works to focus specifically on urban wildlife. Separate chapters on city wildlife, characteristics of urban wildlife, and planning and management of urban wildlife.

Bird, David, Daniel Varland, and Juan Josè Negro, eds. *Raptors in Human Landscapes*. San Diego, Calif.: Academic Press, 1996. A good overview of the natural history of raptors in a variety of human settings: urban, suburban, industrial, and farming communities. Some of the topics treated in this book include the impact of raptors in human landscapes and the impact of nest box programs and other management programs to ensure the presence of these forms of wildlife in a variety of urban habitats.

Forman, Richard, and Michel Godron. *Landscape Ecology*. New York: John Wiley & Sons, 1986. A broad-brush overview of the role of landscape ecology with respect to sustaining wildlife populations in urban and suburban habitats. The roles of corridors, habitat islands, patches, and other structural components of landscapes for wildlife are thoroughly presented.

Gill, Don, and Penelope Bonnett. *Nature in the Urban Landscape: A Study of City Ecosystems*. Baltimore: York Press, 1973. One of the earliest books to summarize urban habitats for wildlife. It is comprehensive in treatment of all major groups of animals that occur in city landscapes. Appendices include landscaping for wildlife and glossary of useful terms.

McDonnell, Mark J., and Steward T. A. Pickett, eds. *Humans as Components of Ecosystems*. New York: Springer-Verlag, 1993. The subtitle of this book is "the ecology of subtle human effects and populated areas," which really sums up the general aims and intentions of the authors. Much more of a theoretical approach to the role and utility of wildlife in urban and other highly modified environments.

VELOCIRAPTORS

Types of animal science: Classification, evolution
Fields of study: Anatomy, evolutionary science, paleontology, systematics (taxonomy)

Velociraptors are a group of human-sized theropods (carnivorous dinosaurs) that existed during the Cretaceous era in North America and Asia.

Principal Terms

CLADISTICS: a method of analyzing biological relationships in which advanced characters of organisms are used to indicate closeness of origin

CRETACEOUS: a period of time that lasted from about 146 to 65 million years ago, the end of which was marked by the extinction of the dinosaurs

ECTOTHERMY: a form of metabolism in which internal temperature is regulated by the ambient temperature

JURASSIC: a period of geological time that lasted from about 208 to 146 million years ago

PUBIS: one of the three bones that make up the pelvis (the others are the ischium and ilium)

SAURISCHIA: one of the two main dinosaur groups, characterized by a pelvis in which the pubis points forward

Velociraptors are represented by the genus *Velociraptor* ("speedy predator"), which is one of a group of human-sized theropods or carnivorous dinosaurs that existed during the Cretaceous in North America and Asia. Theropods include all dinosaurian carnivores and are allied to the sauropods, very large, long-necked herbivores, in a major group of dinosaurs called the Saurischia. These dinosaurs are all characterized by the forward projection of the pubis, which separates them from the other major group, the Ornithischia, in which the pubis is directed backward. Velociraptors are most closely related to a group of similar predators that includes the dromaeosaur *Deinonychus*, which is extremely well known from articulated skeletons found in Montana and Wyoming, described by Yale University's John Ostrom in the late 1960's and early 1970's. One result of cladistic analysis of the relationships of this group is that they have been shown to be the closest relatives of *Archaeopteryx*, the earliest known bird. *Archaeopteryx* is known only from the latest Jurassic, and the dromaeosaurs are entirely Cretaceous, so it has been suggested that either dromaeosaurs represent a group of secondarily flightless birds,

Velociraptor Facts

Classification:
Kingdom: Animalia
Subkingdom: Bilateria
Phylum: Chordata
Subphylum: Vertebrata
Class: Reptilia
Order: Saurischia
Suborder: Dromaeosaurida (medium-sized carnivores with an enormous claw on the second toe of the hind foot)
Geographical location: North America and Asia
Habitat: Terrestrial habitats
Gestational period: Unknown
Life span: Unknown
Special anatomy: Medium-sized bipedal carnivores with a well-developed, sickle-shaped claw on the hind foot

The velociraptor (right) used its forelimbs to grasp prey and its long hind limbs and sickle-shaped claws to kick and slash at the prey's sides and belly. (©John Sibbick)

or alternatively that ancestral forms in the Jurassic await discovery.

Anatomy and Life Habits

The velociraptor *Deinonychus* ("terrible claw") is the best-known genus of this group and is thus often used as a typical example. It has an elongated, lightly built skull with numerous backward-curving, serrated teeth, and a relatively large brain. There are large openings in the side of the skull for the eyes and for jaw muscles, which suggest a sharp-eyed predator with a fearsome bite. The neck was quite slender and flexible in contrast to the back and tail, which were fairly stiff due to the presence of ligaments and (in the tail) bony rods that provided support to the vertebral column. The arms were unusually long, and the three-fingered hands bore long, sharp claws. The hind leg is particularly interesting, as the relatively short femur (upper leg bone) indicates a fast-running animal while the second toe is modified to form a large, sickle-shaped claw that must have been held above the ground during locomotion.

These anatomical features suggest that *Deinonychus* (and other velociraptors) was a nimble predator that was able to grasp its prey and dispatch it with lethal kicks from its sickle-claw. The femur has a special process on it for the attachment of a muscle that would have allowed a very powerful backward and downward kick, enabling effective use of the claw. Additionally, the tail is flexible near the body but stiffened by bony rods more distally, which would have made it ideal as a balancing organ. The only actual evidence of velociraptor predation comes from a specimen of *Velociraptor* from Mongolia, in which the individual is interlocked with the skeleton of a small *Protoceratops* (a herbivore), suggesting that they died in mutual combat. The most common herbivores found with *Deinonychus* remains are ornithopods, large, bipedal animals that would have been too large for attack by a solitary individual. This has been used as the basis for the hypothesis that velociraptors might have attacked in packs, which in turn suggests a level of organization not usually present in reptiles. The active and agile lifestyle of velociraptors, together with the possibility of group behavior in hunting, have been advanced as evidence that they had an endothermic metabolism, similar to that of modern mammals, in which internal temperature was controlled by food intake. As dinosaurs are reptiles, they had been thought to have an ectothermic metabolism similar to that of modern reptiles. Velociraptors have become well-known dinosaurs due to their role in the film *Jurassic Park* (1993), where they are depicted as intelligent and warm-blooded, in contrast to the long-held perception of dinosaurs as ponderous and slow-witted animals.

—David K. Elliott

See also: *Allosaurus*; *Apatosaurus*; *Archaeopteryx*; Dinosaurs; Evolution: Animal life; Extinction; Fossils; Hadrosaurs; Ichthyosaurs; Paleoecology; Paleontology; Prehistoric animals; Pterosaurs; Sauropods; Stegosaurs; *Triceratops*; *Tyrannosaurus*.

Bibliography

Alexander, R. McNeill. *Dynamics of Dinosaurs and Other Extinct Giants*. New York: Columbia University Press, 1989. Informative coverage of how dinosaurs operated, including a discussion of balance and locomotion in dromaeosaurs.

Benton, Michael J. *Vertebrate Palaeontology*. 2d ed. London: Chapman and Hall, 1997. General vertebrate paleontology text that devotes part of one chapter to the dromaeosaurs.

Currie, Philip J., and Kevin Padian. *Encyclopedia of Dinosaurs*. San Diego, Calif.: Academic Press, 1997. Excellent coverage of all aspects of dinosaur biology.

Norman, David. *The Illustrated Encyclopedia of Dinosaurs*. New York: Crescent Books, 1985. Although dated, this book has wonderful illustrations and an excellent text with good coverage of dromaeosaurs, including *Velociraptor*.

Paul, Gregory S. *Predatory Dinosaurs of the World*. New York: Simon & Schuster, 1988. The best and most comprehensive coverage of predatory dinosaurs, with beautiful illustrations by the author.

VERTEBRATES

Types of animal science: Anatomy, classification, reproduction
Fields of study: Anatomy, marine biology, ornithology, zoology

Vertebrates, chordate animals with backbones, include fish, amphibians, reptiles, birds, and mammals. Although numerically smaller than invertebrates in total population, vertebrates are the largest animals on earth, due to the increased weight-carrying capacity allowed by the spine and endoskeletal system.

Principal Terms

ARTICULATE: interconnect by joints

BONE: dense, semirigid, calcified connective tissue, the main component of skeletons of adult vertebrates

CARTILAGE: elastic, fibrous connective tissue, the main component of fetal vertebrate skeletons, it turns into bone

COLLAGEN: a fibrous substance plentiful in bone, cartilage, and other connective tissue

CONNECTIVE TISSUE: fibrous tissue that connects or supports organs

NOTOCHORD: a flexible, rodlike structure in lower chordates and vertebrate embryos, it supports like the backbone of a vertebrate

The Vertebrata are a subphylum of the phylum Chordata. The six vertebrate classes are lampreys, true fish, frogs and toads, reptiles, birds, and mammals. A vertebrate animal has a spinal column (backbone) made of bone or cartilage, and a brain case (skull). Vertebrates also have two pairs of limbs (though some have lost limbs through evolution), and a bilaterally paired muscular system. The backbone that gives the subphylum its name is a group of small bones or cartilage pieces with articulating surfaces (vertebrae). Ribs and bones that support the limbs are attached to the backbone. The ribs protect the heart, lungs, and other internal organs, and can expand and contract. The earliest vertebrate fossils occur in rock from the Paleozoic era.

The vertebrae serve to encase and protect the spinal cord, a major part of the vertebrate nervous system. The central nervous system also has an enlarged, highly differentiated upper portion, the brain. The bony skull of a vertebrate also serves to encase and protect that brain, as well as providing a base for the vulnerable sensory organs of eyes, ears, nose, and mouth, which are thus efficiently located close to the brain.

The more primitive members of the phylum Chordata have notochords, solid or segmented columns that cover the nervous system. Vertebrates resemble chordates in having notochords in their embryonic state. The vertebrae develop around the notochord. The trunk of a vertebrate is a hollow cavity in which the heart, lungs, and digestive tract are suspended. The central nervous system branches out from the spinal column to reach all internal organs, muscles, and skin. The support offered to the brain and nervous system by the vertebrae has allowed vertebrates to evolve increasingly large brains, which in turn has allowed vertebrates to become increasingly intelligent and responsive to their environment.

Bone and Bones

Bone, the material that form the vertebrae, first evolved 500,000 years ago. Bone is the hard supportive framework of all vertebrates. The framework, a skeleton, has hundreds of separate parts; for instance, there are 206 bones in a human skeleton. Bones protect delicate organs, such as the brain and lungs. Muscles, attached to bones, enable walking, flying, swimming, and all other means of motion. Bones provide body calcium needs and contain sites for making blood cells.

Much bone in adult vertebrates arises from cartilage, an elastic, fibrous connective tissue, and the main component of fetal vertebrate skeletons. Such bone is cartilage bone. Cartilage is an extracellular matrix made by chondrocyte cells. It is firm and elastic, due to its collagen fibrils. The fibrils provide mechanical stability and high tensile strength, while allowing nutrients to enter chondrocytes. Blood vessels around cartilage supply nutrients and remove wastes.

Cartilage-containing skeletons of newborn vertebrates become bone by a process of calcification, chondrocyte destruction, and replacement by bone cells. In young vertebrates, cartilage is the site of growth and calcification that lengthens bone to attain adult size.

The Common Characteristics of Vertebrates

- Complex, paired eyes
- Muscular mouth and pharynx
- Epidermis and dermis, often modified into protective coverings such as scales, feathers, or hair
- Blood containing red (hemoglobin-carrying) and white blood corpuscles
- Large body cavity for holding internal organs
- Digestive system consisting of digestive glands, liver, and pancreas
- Two- to four-chambered ventral heart
- Paired, ducted kidneys
- Male and female genders (with appropriate reproductive organs)

Vertebrae

In higher vertebrates, each vertebra consists of a lower part, called the centrum, and an upper, Y-shaped part, called the neural arch. The arch has a downward and backward projection, which can be felt as the bumps along a vertebrate's back, and two sideways projections, where muscles and ligaments can attach. The space between the arms of the Y on the neural arch and the centrum create an opening called the vertebral foramen, through which the spinal cord passes. Intervertebral disks, made of cartilage, separate the centrums and serve as shock absorbers.

The vertebral column has five regions: the cervical region (neck), thoracic (chest), lumbar (lower back), sacral (pelvic girdle), and caudal (tail). The top two cervical vertebrae, called the atlas and axis vertebrae, make a joint to attach with the skull. The number of vertebrae varies by species.

Vertebrate Evolution

Vertebrates, all designed on the same general plan, flourish on land, in the air, and in both fresh and salt water. Vertebrates first evolved in the Silurian period, around 438 to 408 million years ago. In the intervening millennia, according to need, evolution produced valuable morphological changes to optimize vertebrate biofunctions. For example, whales, once land dwellers, evolved into ocean dwellers with lungs and sonar to help them navigate and find food. Birds developed wings to ride the air, and mammals proliferated in forms that fit varied habitats worldwide. Fish were the first vertebrates to appear, around 480 million years ago. Amphibians and reptiles appeared around 360 million years ago, while birds and mammals begin to appear around 205 million years ago.

Morphological change led to a balance of nature, where herbivores ate plants, preventing their overgrowth, and carnivores ate herbivores, preventing their superabundance. Then the ultimate vertebrates, humans, developed civilization. Humans domesticated animals for food, clothing, transportation, and pets. In so doing, many species have been eradicated, endangered, or put at risk. For example, blue whales were endangered because their blubber was useful. Wolves and tigers have been eradicated or endangered by the quest for hunting trophies and to keep them from eating livestock. Mustelids were treated similarly because their pelts made attractive fur garments. Fortunately, human tolerance for this treatment of animals is decreasing, lest many species be seen only in zoos or faded photographs.

—*Sanford S. Singer*

See also: Amphibians; Animal kingdom; Birds; Bone and cartilage; Brain; Circulatory systems of vertebrates; Endocrine systems of vertebrates; Endoskeletons; Fish; Fur and hair; Hormones in mammals; Invertebrates; Mammals; Muscles in vertebrates; Nervous systems of vertebrates; Reproduction; Reproductive systems of female mammals; Reproductive systems of male mammals; Reptiles; Water balance in vertebrates; Zoology.

Bibliography

Chatterjee, Sankar. *The Rise of Birds: 225 Million Years of Evolution*. Baltimore: The Johns Hopkins University Press, 1997. The book describes evolution of birds, describes bird fossils, and covers flight.

Crispens, Charles G. *The Vertebrates: Their Forms and Functions*. Springfield, Ill.: Charles C Thomas, 1978. An interesting book with information on vertebrates, including anatomy, development, and natural history.

Pough, F. Harvey, Christine M. Janis, and John B. Heiser. *Vertebrate Life*. Upper Saddle River, N.J.: Prentice Hall, 1999. An excellent text covering vertebrate diversity, function, and evolution in general, aquatic vertebrates, reptiles, birds, mammals, and humans.

Rosen, Vicki, and R. Scott Theis. *The Molecular Basis of Bone Formation and Repair*. Austin, Tex.: R. G. Landes, 1995. This nice book covers many topics related to bone growth, other remodeling, and growth factors.

VETERINARY MEDICINE

Types of animal science: Anatomy, behavior, fields of study, physiology, reproduction, scientific methods

Fields of study: Anatomy, biochemistry, cell biology, embryology, genetics, histology, marine biology, neurobiology, ornithology, pathology, physiology, reproduction science, zoology

Veterinary medicine enables professionals to treat and prevent diseases and injuries in animals to better their quality of life and as a public health measure to protect humans from transmissible diseases through contact with infected animals and contaminated animal by-products.

Principal Terms

BRUCELLOSIS: cattle disease caused by brucellae bacteria

CHOLERA: gastrointestinal diseases

DISTEMPER: viral disease affecting respiration

LEPTOSPIROSIS: bacterial infection of spirochetes spread by urine

RABIES: viral disease of the nervous system

TUBERCULOSIS: communicable disease caused by the tubercle bacillus

ZOONOSES: diseases transmissible from animals to humans

The treatment and prevention of animal diseases is crucial to maintain animal vigor and to secure public health. The term "veterinary" is derived from the Latin word *veterinarius*, which refers to beasts of burden and was used to describe people who were animal healers. Before veterinary science became professionalized in the late nineteenth century, animal owners tended to their livestock's health needs. Gradually, universities worldwide included courses which focused on specific animal ailments and wounds, and professional organizations such as the American and Canadian Veterinary Medical Associations were established. Because horses were essential for transportation, and cattle, sheep, and goats provided necessary meat, skins, wool, and milk, early veterinary practitioners focused on tending to the health concerns of those domesticated animals.

Veterinary medicine gained public acceptance because of campaigns to eradicate and control zoonoses that threatened humans. Legislation affirmed veterinarians' status as public health professionals responsible for inspecting and vaccinating animals. Licensing procedures established educational and professional standards for qualified veterinary practitioners. Globally, cultural beliefs regarding animals and health care determined the degree to which veterinary medicine was incorporated in societies.

Veterinarians pursue many roles, such as practicing at private hospitals or zoos, teaching, researching, consulting with pharmaceutical industries, and working for national, state, and local governments as food inspectors, military veterinarians, and other such positions. Veterinary medicine is crucial for disease control in domestic animals through the examination of animals for parasites and contagious diseases and the quarantine of imported animals until they pass health inspections.

Prevention

Veterinarians strive to protect both animals and humans from animal-transmissible diseases such as rabies. Bacterial, viral, fungal, and parasitical diseases can weaken animals already genetically predisposed to certain conditions. The livestock industry relies on veterinarians to inoculate swine, poultry, cattle, goats, sheep, and other animals raised and sold as meat, milk, and hide and fur producers. Ill animals can detrimentally affect agriculturists' profits and the overall farm economy due to the necessity of destroying valuable animals exposed to hazardous diseases, especially in confinement units where animals live closely together.

During the twentieth century, veterinarians achieved some control of hog cholera, brucellosis, and tuberculosis. With serums, medicated dips, and quarantines, they continue efforts to eradicate these diseases and prevent further outbreaks and transmission to additional herds. Through mandatory regulated meat and dairy inspections, veterinarians prevent contaminated foodstuffs from reaching human consumers, although some people in several countries have encountered tainted meat from bovines suffering from mad cow disease.

By the mid twentieth century, the popularity of keeping small animals and some large livestock, such as horses and ponies, as pets led many veterinarians to begin specializing in practices especially for those animals. Owners relied on veterinarians' preventative measures to ensure the healthy well-being of their pets. Dogs and cats were the most frequently seen patients, but veterinarians also attempted to prevent diseases in more exotic pets, including birds, rodents, fish, reptiles, and amphibians.

Routine Examinations

Veterinarians initiate preventive measures by taking a medical history of each animal, preferably first examining them when they are several weeks old. The condition of the animal is assessed and basic diagnostic tests for parasites and infectious diseases such as distemper are administered and evaluated. If animals test positive, appropriate treatment is begun. For both cats and dogs, a heartworm test is crucial. When a mosquito that has bitten an infected animal bites another dog or cat, these spaghetti-like parasites are injected as

Most veterinary medicine has focused on the care of domesticated animals, such as these cows whose hooves are being treated to help prevent foot and mouth disease. (AP/Wide World Photos)

Emergency Veterinary Medicine

Veterinarians dress wounds, inject snake antitoxin and antibiotics, splint broken limbs, and euthanize animals as needed after accidents. Educating clients about possible dangers, veterinarians provide safety information for preemergency preparations for pets and livestock. Veterinarians distribute guides advising how animal owners should respond when faced with natural disasters such as fires, floods, earthquakes, hurricanes, blizzards, and tornados. Other emergencies that can victimize animals include hazardous materials spills, evacuations, gas leaks, and airplane, train, and car crashes.

Anticipating critical care needs, veterinarians instruct clients how to prepare animal evacuation and first aid kits with copies of veterinary records, proof of ownership, pharmaceuticals, and instructions appropriate for their pets. They also provide owners emergency supplies such as animal identification tags with veterinary contact information and waivers for veterinary care in case owners are separated from their animals.

After an emergency, veterinarians open their animal hospitals to treat victims or travel to accident and disaster sites where both domestic and wild animals need assistance. When areas have been evacuated during a storm or emergency, veterinarians help locate, rescue, restrain, shelter, and identify animals left behind and administer first aid and medication to panicked or shocked animals unable to be moved.

After disasters, Veterinary Medical Assistance Teams (VMAT) assist the Red Cross, humane societies, and local animal control officials. To protect public health, they examine carcasses and food and water sources for possible contamination. They also vaccinate animals with tetanus and rabies boosters to prevent postemergency epidemics. Veterinarians assist in the reunion of owners and missing pets.

larvae into the skin, then migrate to the bloodstream. Adult heartworms can grow as long as fourteen inches inside host animals' hearts, impeding their circulation and potentially causing death. Mosquito control and regular doses of heartworm medicine help prevent the spread of heartworms. Cats are also tested for feline leukemia virus, a fatal, contagious disease. Horses are given a Coggins test to detect equine infectious anemia.

A series of vaccinations prevent diseases from occurring in young animals and are given as immunity boosters in mature animals. Some diseases can be combated with combination injections which combine vaccines for distemper, hepatitis, leptospirosis, parvovirus, and parainfluenza. Rabies vaccinations are required by law in many countries to prevent the disease spreading between animals and from animals to people.

Veterinarians recommend parasite control through flea and tick repellents, many of which can be applied topically to the skin or swallowed in a pill, as well as nutritional diets to prevent many diseases from occurring. Some veterinarians offer other preventive measures in the forms of tattoos and implanted microchips to prove ownership in case animals are stolen or lost.

Treatment

Veterinarians treat animals infested with parasites with deworming medicines. Tapeworms are prevalent in dogs and cats because they often ingest fleas or raw fish and meat which carry worm eggs. Some animals are born with roundworms. These parasites can cause serious health problems for both humans and animals, and infestations must be carefully treated to rid animals of the parasite without killing the host. Veterinarians also treat animals who have been exposed to poisons and toxins from plants, insects, and reptiles or that have suffered extremes of heat or cold. They set fractured limbs and suture skin tears.

Animal reproduction is another aspect of veterinary treatment. Veterinarians advise pet and livestock breeders before conception, then monitor the mother animals during their pregnancies and often assist in the births. Evaluating the newborn animals, some veterinarians perform cosmetic surgery such as docking tails and cropping ears to meet societal expectations for appearance. Others observe animals' development in order to fix anatomical defects. Veterinarians also routinely spay and neuter animals to prevent unwanted litters from being born, and also to assure good reproductive health by eliminating prostate and uterine cancers.

Research and Specialties

Many veterinarians choose to research a specific veterinary medicine topic. These professionals often earn advanced degrees and certification in veterinary science or related fields such as biochemistry. Usually affiliated with veterinary colleges or universities, research veterinarians provide expertise for the specialized diagnostic testing or treatment that some animals require. Pharmaceutical research concerning the control of parasitic heartworms and the parvovirus were two of the most outstanding twentieth century developments in veterinary medicine. The parvovirus, an acute, contagious intestinal disease, was first identified in 1978 in the United States and is the most common infectious canine disorder.

Many veterinary experiments explore the role of genetics in health and also examine alternative methods for reproduction, such as artificial insemination and embryo transplantation. Veterinarians also devise methods to use technology such as ultrasound to diagnose animals' ailments.

Most veterinary medicine specialties address physiological aspects such as dentistry, dermatology, radiology, cardiology, orthopedics, nutrition, toxicology, anesthesiology, pharmacology, and ophthalmology. Other specialties concern animals' intellectual and emotional capabilities; veterinarians specializing in behavior seek new ways to measure how animals react to stimuli and how to utilize this information to train animals. Some veterinarians also address legal and ethical issues related to veterinary medicine. Others seek alternative veterinary medical techniques through holistic healing, massage, and acupuncture.

Military Veterinary Medicine

Veterinary medicine has played a crucial role during wars throughout history because armies rely on animals for transportation, food, and communication. Before warfare was mechanized during World War II, cattle and poultry were driven behind troops to assure fresh meat supplies, and veterinarians often inspected these animals prior to slaughter to protect troops from transmissible diseases and local populations from epidemics.

Horses were vital as cavalry mounts to carry troops. Both horses and mules pulled artillery and supply wagons. Veterinarians treated these animals' wounds and injuries in the field and at evacuation hospitals, where they determined whether the animals should be quarantined, rehabilitated, or destroyed.

In modern wars, veterinarians inspect and evaluate military animals such as pigeons and dogs. They vaccinate war dogs at training camps before canines are sent to warfronts and provide care throughout their service.

Many veterinarians have enlisted as soldiers during wars and were assigned animal-related duties. These veterinarians often lacked adequate and suitable supplies in the field and innovated and adapted surgical procedures to save animals' lives from wounds, parasites, and diseases. For example, war dogs serving in the World War II Pacific theater received blood collected from local donor dogs who were immune to indigenous parasites. Veteran veterinarians returned home and applied useful wartime methods to their peacetime practices.

Veterinary corps were established to maintain a sufficient quantity of enlisted veterinarians to meet military demands. During peacetime, military veterinarians monitor the health of working and ceremonial military animals and conduct laboratory research essential to combat biological warfare.

Because advances in veterinary medicine have resulted in animals aging beyond previously expected life spans, geriatric practices have gained demand. Veterinarians are exploring new ways to treat conditions associated with old age, such as arthritis, cancers, organ failures, and reduced senses. Some veterinarians participate in ventures such as Operation Arctic Care to vaccinate and examine animals in remote regions who would otherwise receive no veterinary attention.

—Elizabeth D. Schafer

See also: Aging; Biology; Birth; Breeding programs; Diseases; Domestication; Embryology; Ethology; Fertilization; Life spans; Nutrient requirements; Physiology; Pregnancy and prenatal development; Zoos.

Bibliography

Beaver, Bonnie V. G. *The Veterinarian's Encyclopedia of Animal Behavior.* Ames: Iowa State University Press, 1994. Discusses veterinary concerns and techniques regarding domestic animals' behavior.

McCorkle, Constance M., Evelyn Mathias, and Tjaart W. Schillhorn van Veen, eds. *Ethnoveterinary Research and Development.* London: Intermediate Technology Publications, 1996. Examines the global role of veterinary science and cultural factors influencing research.

The Merck Veterinary Manual. 8th ed. Rahway, N.J.: Merck, 1998. A valuable reference work which provides technical information about diseases, diagnoses, and treatments.

Putney, William W. *Always Faithful: A Memoir of the Marine Dogs of World War II.* New York: Free Press, 2001. A veteran military veterinary officer explains how veterinary techniques were adapted to battlefield conditions.

Rollin, Bernard E. *An Introduction to Veterinary Medical Ethics: Theory and Cases.* Ames: Iowa State University Press, 1999. Explores the moral and ethical aspects of professional veterinarians.

Smithcors, J. F. *The Veterinarian in America, 1625-1975.* Santa Barbara, Calif.: American Veterinary Publications, 1975. Provides historical information about both international and North American veterinary science.

Swabe, Joanna. *Animals, Disease, and Human Society: Human-Animal Relations and the Rise of Veterinary Medicine.* London: Routledge, 1999. Explores the relationship of disease and animal domestication with the professionalization of veterinary science.

VISION

Types of animal science: Anatomy, behavior
Fields of study: Anatomy, biochemistry, cell biology, evolutionary science, neurobiology, physiology

Vision is the ability of animals to analyze light information in their environment. Given the many unique visual worlds in the animal kingdom, there are many different types of eyes and light gathering mechanisms in them.

Principal Terms

ACCOMMODATION: changing the shape of the lens in order to keep objects at different distances focused on the retina

FOVEA: area, often a pit in the retina, of maximal acuity, where each photoreceptor has its own nerve cell, as opposed to many receptors converging on one nerve cell

OPSIN: a membrane-bound protein or pigment, which absorbs light

PHOTON: a unit used to describe light intensity

PHOTORECEPTOR: cell containing membranes which house light-sensitive pigments

RETINA: the light-sensitive film at the back of the eye

Despite the fact that it is often taken for granted, vision is one of the more interesting and complex sense systems. Different animals have, over the course of their evolution, independently fine-tuned their visual systems to adapt to their unique environments and needs. The possession of a good visual system can be the factor dictating a species' survival. Vision is essential for many animal behaviors, such as foraging for food, prey avoidance, and mate choice. Upon considering the many different species in the world, it is evident that there must be many different types of vision; for example, a fish in its unique underwater environment would have a vastly different visual world than an insect in the rainforest.

When animals forage for food, vision is used along with many other senses, especially the senses of smell and hearing, to make a good food choice. Many animals scan a visual field before deciding to forage; thus, the visual system must be acute, enabling the animal to understand what is in its field of view, often in a very short time. The brain must be able to determine such things as shape, form, and color of objects. Many animals utilize color vision when foraging. Cues as to the suitability of a food choice are often indicated by color, for example, the difference between a poisonous berry and an innocuous one. Eye placement (the exact positioning of eyes on the head) is crucial; appropriate positioning of the eyes from each other, as in humans, makes binocular vision possible. Binocular vision gives the viewer a sense of depth in the field of view, critical for many animals when catching prey. Species with laterally spaced eyes have a smaller amount of binocular vision, and some of their visual field has to be viewed monocularly.

Vision is essential in many animals' communication with each other. Body markings and displays in many species are often used as mating signals. Birds provide a striking example of visual communication in animals, where often specific body markings are used to attract mates. Some birds attract mates by elaborate nest construction; the bird with the most elegant nest attracts a mate and ensures reproduction and the passing of its genes to the next generation.

The Physiology of Vision

In simple terms, the visual system takes a signal in the form of light and translates it into a chemical change and later a nervous impulse in the brain; this nervous impulse is what the animal perceives as sight. The two main characteristics of eyes, no matter how complex or simple, are light-sensitive receptors (photoreceptors), and a mechanism to control light. In simple eyes, the light-sensitive receptors make up a layer known as the retina. The nature of a photoreceptor is dictated by opsin, the photosensitive proteinaceous pigment present within the membranes in the photoreceptor. The photoreceptors are variable in size, shape, and content. There are two types in vertebrates: rods and cones. Rods are especially sensitive to light and as such make up the majority of the retina of nocturnal species, who require as sensitive a system as possible. The photopigment in rods is called rhodopsin. Cones are less sensitive to light, but there are different types which are sensitive to light of different wavelengths (or colors), and can give animals color vision. There are many different types of cones, and hence many types of color vision; possession of these cone types and their specific positioning within the retina is an evolutionary adaptation particular to animals who benefit from color vision.

The eagle's keen vision, necessary for its raptor lifestyle, has led to the term "eagle-eyed" for exceptional sight. (Adobe)

Polarized Light Vision

Polarized light plays an important role in the visual world of many animals, including many species of fish and insects. These animals are able to use their visual systems to decode one particular aspect of polarized light, namely the angle of polarization. Light which is polarized has an electric vector which has a specific angle, or orientation. Natural light is unpolarized, but becomes polarized by scattering in air, water, or by reflection off surfaces. Scattering of the ultraviolet (UV) light in sunlight is quite predictable and produces obvious patterns in the sky, invisible to humans without a polarizing filter. These patterns are analyzed by some insects and are used as a kind of map for navigation, for example, to get back to the nesting area after a foraging trip. Similarly, the polarization pattern aids in orientation of the animal. Analysis of the skylight polarization patterns is possible because of the unique positioning of the microvillar photoreceptors within the eye. Pairs of microvillar photoreceptor populations are positioned perpendicular to each other. The light-sensitive molecules within the photoreceptor lie parallel to the axes of the microvilli. Each member of the pair of photoreceptors is maximally sensitive to a different angle of polarized light, and the combined response to a given angle of both members allows the animal to compute the exact angle of polarization. Enhancement of contrast is an advantage of polarized light vision used by fish, although the mechanism of detection of the light is obscure.

There are two main types of eye design in the animal kingdom: simple eyes and compound eyes. Simple eyes have a single layer of photoreceptors, which, in the least complicated case, form a cup of photosensitive material. The human eye, with its complex light-focusing apparatus, is still a simple eye. Compound eyes, which are present in most insects, have many separate optical units, called ommatidia. Each ommatidium has a rhabdom, containing a group of up to nine tubular rhabdomeres, with ciliary or microvillar (finger-shaped) photoreceptors. The orientation of groups of photosensitive cilia is structured, often with pairs of rhabdomeres organized at right angles to each other. This is especially key in analysis of polarized light.

Many visual systems have mechanisms to control light. Restriction of the amount of light entering the eye is useful; the opening through which light enters is referred to as an aperture, as in a camera. Many animals have a contractible iris which constricts and dilates to control light entry through the pupil. For example, in dim light conditions the iris can dilate and let in as much light as possible. Some eyes have lenses which enable light to be focused on the retina, allowing for better resolution of objects in the visual field. Many are able to change the shape of the lens in order to bring an image to focus on the retina, a process called accommodation. Other animals use a cornea to bend light onto the retina, although the cornea is rigid and cannot change shape. On the other hand, there are many species with much simpler eyes, which do not possess any kind of light control apparatus.

Upon reception of light by a photoreceptor, a biochemical cascade of events occurs within the photoreceptor itself, which amplifies the original signal received. The result of this cascade is a nervous signal which proceeds through many neural layers to the brain. Throughout most of the retina in the simple eyes of vertebrates, several photoreceptors connect to one neuron (convergence), but there is often an area of the retina where one photoreceptor connects to one neuron. This area is called the fovea and is the part of the retina which has best acuity. The area within the retina which comprises the fovea is variable. Fish possess what is known as a "visual streak" fovea, which gives excellent vision along a horizontal slice of the visual field. This is an ideal adaptation for fish given their particular habitat.

Information is passed through the nervous system in layers of neurons. The retina is, in fact, an extension of the brain, and contains many nerve cells. The exact arrangement and mechanism of action of neural cell types, and the precise pathway to and within the brain, differ greatly from species to species. Phototransduction in vertebrates is different from that of invertebrates, from the arrangement of the retina to the biochemical cascade and the types of neurons involved.

Color Vision

Reflected natural light has a unique property that may be exploited by the visual system of an animal, namely its wavelength, or color. Color vision can be vital for a species in regard to mate choice and

Ultraviolet (UV) Light Vision

Sunlight is the natural source of ultraviolet light, whose wavelength runs from less than 280 to 400 nanometers. For the sake of vision, however, one need only be concerned with UV light between 320 and 400 nanometers, since any light below these wavelengths becomes absorbed by air before it reaches earth. Many members of the animal kingdom are able to see in the ultraviolet region of the spectrum, due to their possession of a particular class of opsin molecule which is maximally sensitive to UV light. Humans also possess a UV-sensitive opsin, but are unable to actually see UV because the lens, which is yellow, absorbs UV light. There are certain behavioral advantages for the species who do possess UV vision. For example, fish use UV vision extensively in signaling and communication with members of the same species; many fish and birds have UV markings on their bodies which make it easier for other individuals to see them and which may also be involved in mate choice. There are also very strong UV markings on flower petals, which act as guides for pollinators such as bees.

foraging. Photoreceptors can be sensitive to different wavelengths of light, or colors; in vertebrates the color-sensitive receptors are called cones because of their shape. The maximum sensitivity of a photoreceptor is dictated by the nature of the photo-sensitive protein (opsin) within the receptor. Opsins can be classified according to the approximate wavelength of light that stimulates them maximally. Opsins have been studied that are sensitive to light from the ultraviolet region of the visible spectrum all the way to the far red region.

To have the possibility of color vision, an animal must possess at least two photoreceptors with differing sensitivities. The brain must then be able to compare the outputs of both these receptors and discriminate color. Often more than two types of photopigment type are present, as in the fish retina, which results in very complex color vision, including sensitivity to ultraviolet light. Color vision has been shown to exist in many animals within the animal kingdom. Positioning and distribution of the different types of photoreceptor within the retina are also key to the ability to discriminate color. In vertebrates, this aspect of retinal structure is called the cone mosaic, and its nature is often closely related to some behavioral aspect of the animal in question. Color vision also requires a central processing system that can decode the various light signals and turn them into a brain output which is useful to the animal.

—Lucy A. Newman

See also: Anatomy; Brain; Ears; Eyes; Hearing; Nervous systems of vertebrates; Noses; Physiology; Sense organs; Vocalizations.

Bibliography

Baylor, D. "How Photons Start Vision." *Proceedings of the National Academy of Science* 93 (January, 1996): 560-565. This review explores the biochemistry of phototransduction and is fairly detailed.

Jacobs, G. H. "The Distribution and Nature of Color Vision Among the Mammals." *Biological Reviews* 68 (1993): 413-471. This review covers the topic of color vision and addresses the different types of color vision in many mammalian species.

Lodish, H., A. Berk, S. L. Zipursky, P. Matsudaira, D. Baltimore, and J. Darnell. *Molecular Cell Biology*. 4th ed. New York: W. H. Freeman, 2000. Chapter 21 offers in-depth coverage of phototransduction from a biochemical point of view.

Marshall, J. "Visual Function: How Spiders Find the Right Rock to Crawl Under." *Current Biology* 9, no. 24 (December, 1999): 918-921. An interesting article on polarization vision in spiders.

Solomon, E., L. Berg, D. Martin, and C. Villee. *Biology*. 3d ed. Fort Worth, Tex.: Saunders College Publishing, 1993. Chapter 41 covers image formation and, briefly, phototransduction.

VOCALIZATIONS

Types of animal science: Anatomy, behavior, physiology
Fields of study: Anatomy, physiology

Vocalization is any sound produced through the action of an animal's respiratory system and used in communication. In the animal kingdom, vocal sounds are limited to frogs, crocodilians, geckos, birds, and mammals. Animal calls are used to locate mates, define territory, or indicate the location of food. Mammals also vocalize to express emotion, while primates utilize more complex cries to indicate danger, show aggression, and exhibit love. The complex auditory communication system of humans is a distinguishing mark of the species.

Principal Terms

LARYNX: the vocal mechanism of mammals, consisting of a structure of cartilage at the upper end of the trachea containing the vocal folds

PHARYNX: lower part of vocal tract, connecting the mouth and nasal cavities to the larynx

SYRINX: the vocal mechanism of birds, consisting of one or more membranous structures at the lower end of the trachea, where the windpipe divides into two bronchial tubes leading to the lungs; the membranes vibrate due to pressure differences when air streams across their surfaces

TRACHEA: a cartilaginous tube that transports air from the lungs to the pharynx

VOCAL FOLDS: small, laminated sheets of muscle which meet at the front of the larynx; they are open for breathing, and are brought together to vibrate for voiced sounds

A vocal mechanism typically consists of lungs to provide an air stream, a trachea to conduct the air to a larynx or syrinx, a pharynx, and the associated oral and nasal cavities. Vocalization, any sound produced by the respiratory system, implies that air flowing from the lungs has been converted into an oscillating air stream. The sound can be melodious or noisy; when the vocal folds vibrate the sound is termed phonation.

Animal vocalizations evolved with hearing because of the many evolutionary advantages of sound communication. Sound can be varied in pitch, duration, tonality, and repetition rate, making it possible to communicate considerable amounts of detailed information quickly. Animals may vocalize while keeping their limbs free or while hiding. Because sound waves pass readily through vegetation and around obstacles, vocalization is used among insects, frogs, and birds to indicate sexual receptivity.

Vocalizations are also an important component in the behavioral displays of reptiles, birds, and mammals. Although animals typically employ body language and nonvocal noises in their displays, vocalization provides an impressive elaboration impossible to achieve otherwise. The fearsome sight of a gorilla beating its chest and stomping the ground is enhanced considerably by its bloodcurdling roar.

There are two different types of sound-generating mechanism used for animal vocalizations. The first requires a vibrating structure, such as the voiced sounds of human speech produced by vibrating vocal folds. The second is aerodynamically excited, such as for unvoiced sounds or whistled tones. The vibrating element of voiced sounds is a flow valve, which vibrates when pressure is applied from the lungs. Two different systems occur in the animal kingdom. The first is the larynx, common among mammals; when pressure is applied from the lungs, the vocal folds swing outward, stretching muscular ligaments which tend to restore them. These alternating forces induce vibration in the form of air pulses, which propagate through the trachea to be emitted by the mouth or nose. The second is the avian syrinx, which is blown open by excess pressure on either side of the mechanism and restored by forces supplied by pressurized air sacs surrounding the syrinx. In both cases, pitch can be varied by the muscles associated with the valve mechanism. Airflow through the valve is nonlinear, generating a complete set of harmonics in the radiated sound.

Whistled sounds are created when an air jet impinges on a sharp edge or an aperture, creating a sinuous instability in the air stream. The sound is considerably louder when the air stream is acoustically coupled to a resonating oral or nasal cavity, the size and shape of which determines the resulting pitch.

In any acoustic communication system, signals must compete with noise in the environment. The environment has a preponderance of natural low-frequency noise, but high frequencies are produced more easily by small animals. Most vertebrates, however, do not communicate with high frequency sounds, because they are more readily attenuated and thus do not travel far. The optimum frequencies for auditory communication thus depend on the desired communication range. High frequencies are best for small animals communicating over short distances, while low frequencies better serve large animals communicating over longer distances.

Reptiles and Frogs

Although the vocal apparatus of alligators is rather primitive, they can produce noise-excited roaring and hissing vocalizations when provoked by low-frequency sounds, such as a horn or a cannon. Crocodilians live as lone individuals and establish individual territories defined by their loud vibrant roars. To roar, crocodiles tense their body

Wolves vocalize at dawn and dusk to establish and maintain their territories. (Corbis)

muscles and raise their heads and tails high above the water. The emitted sound vibrates the animal's flanks so violently that water is sprayed into the air. Crocodiles are also capable of deep grunting sounds, used during courtship. Geckos are small nocturnal lizards with soft skin. Their voice varies, by species, from faint chirps to loud squawks.

Frog vocalizations encode several pieces of information; the species, the sex, and whether or not it has mated. Male mating calls attract females and indicate the number of other males nearby, critical information for females who wish to deposit their eggs in the most receptive habitat. The vocalizations are produced by primitive vocal cords, consisting of a pair of slits at the throat opening in the floor of the mouth. When the frog forces air from the lungs, the cords vibrate to produce sound. Many species also have a vocal sac, an inflatable chamber located in the throat region of males, which swells to a large size when calls are produced. Air vibrates the vocal cords while passing back and forth between the lungs and the sac, while the vocal pouch acts as a resonating chamber which amplifies the sound. The frog's mouth remains closed while vocalizing, so it can call even while under water.

Birds

In birds, the voice is well developed, having such distinctive sound patterns that many species are named onomatopoeically, such as the whippoorwill. Although birds use different calls for different purposes, each species has a primary song, often repeated incessantly, used for species recognition. Male birds also use vocalizations to mark their territory and to attract mates. Some species can even identify their mates by sound. During the breeding season, the male emperor penguin leaves for several days to forage for food; when he returns he is able to locate his mate, out of a pack of hundreds of birds, from the calls emitted.

The green-backed sparrow utters a hoarse scream when attack or escape is likely to occur, and medium hoarse notes when the bird's indecision between the two courses of action make it unlikely that either will occur. To a family of migrating geese, the sounds of other geese on the ground conveys the information that there is probably food and a safe shelter.

Cuckoos are a highly vocal species; they use a variety of contact calls, alarm notes, and melodious songs used to define territory or attract mates. The male's song is characterized by a repetition of loud, short notes on a descending scale. The common cuckoo found throughout Europe, Asia, and Africa emits the well-known two-note call imitated in cuckoo clocks.

Owls produce a variety of vocalizations with a pitch, timbre, and rhythm unique to each species. Most vocalize at dusk and dawn before beginning to hunt. Their songs vary from the deep hoots of large species to the chirps and warbles of small owls. When its nest is threatened, the nestling burrowing owls emit a buzzing noise, resembling the warning sound of the rattlesnakes that frequently inhabit rodent burrows. North American screech owls begin mating when a special song, commenced by the male, is answered by a distant female. After fifteen minutes of antiphonal singing while gradually approaching, the couple meets, sings a duet with a different song pattern, and mates. Other calls of the screech owl include sounds to prompt the young to reveal their location, a food-soliciting call by the young, and barking calls used to eject the matured young from the parents' territory.

The "voicebox" used by birds to produce birdsong is the syrinx, located where the windpipe divides into the two bronchial tubes leading to the lungs. The syrinx varies considerably among different birds. In the ordinary chicken it is quite simple, consisting of four uncomplicated membranes, which produce the characteristic clucking sounds when activated. An asymmetric chamber at the base of the ducks' trachea adds a noise component to its vocalization, which humans hear as "quacking." The trachea of trumpeter swans enters the sternum, flexes twice into bony pockets, then coils back to the lungs, somewhat analogous to bass orchestral wind instruments. This long resonator implements the production of its clarion, trumpetlike call.

The human brain can perceive speech in sounds having only the remotest resemblance to speech if the rhythm and intonation matches that of a simple sentence. Mynah birds use this phenomenon to deceive us into believing they can speak. They have a syrinx valve on each bronchial tube which can be independently controlled to produce two simultaneous wavering tones, which we perceive as speech when they mimic the rhythm of a sentence.

Mammals

Among mammals, vocalization is used first for survival. Infants vocalize to express hunger or pain, or to be located when lost. Other cries, such as the lion's roar or the trumpeting of an elephant, mandate caution. Animal vocalizations of this type, often accompanied by an offensive posture, are used to startle or intimidate an opponent. There is a direct correlation between vocal anatomy and behavior among mammals. Social animals that readily vocalize have larynges which open less widely when they breathe, thus reducing breathing efficiency. The vocal folds must close to start phonation (wailing of cats, howling of wolves). For breathing, they must open wide so as to not obstruct air flow to the lungs. Horses and animals whose survival depends on running long distances while breathing aerobically have simple vocal folds that can open wide to offer an unobstructed air passage, but which consequently cannot be effective phonators. The giraffe's vocal folds are so poorly developed that the animal was long thought to be mute. In actuality, giraffes can phonate to a limited extent; they groan when injured and call their young when they stray. The more highly developed vocal folds of primates enhance phonation, but at the expense of a more constricted airway.

Vocalization is an important aspect of mammal communication. When the wild dogs of India (dholes) hunt, the leader coordinates the pack's motions with a series of sharp yelps. The black-tailed prairie dog combines a visual and vocal display consisting of jumping into the air with its nose straight up while emitting an abrupt two-part vocalization. This display indicates that some behavior is about to be interrupted or prevented by fleeing, which usually occurs immediately thereafter. The display is only employed when an alternative to flight also exists. Hyenas have no organized social behavior but often cooperate while

Evolution of Human Language

Both the anatomy and the brain mechanism involved in human speech resulted from the evolutionary development of organs originally used for other purposes. The mouth, throat, and larynx were designed for swallowing food and breathing, but evolutionary modification rendered them capable of producing a variety of easily understood sounds. Unfortunately, a flexible speech-producing organ was achieved by so rearranging the throat anatomy that humans are particularly susceptible to choking on food that lodges in the larynx. Even our closest living relatives, the great apes, do not possess a vocal system capable of producing human speech.

In conjunction with a flexible organ of speech, the brain must evolve to allow the fine motor control necessary to control the speech mechanism precisely and to decipher the semantic content of vocal utterances. This process began about two million years ago, when the brains of tool-using hominids began to grow rapidly in proportion to the body. This was probably when human language began, and it is highly plausible that human thought and ingenuity evolved concurrently with articulate speech. As the new hominid societies evolved, groups of males hunted with tools in order to provide meat for females and children. Language was necessary to keep these societies together, and accurate communication was essential for the hunt. The development of language encouraged the brain to grow even faster; by 150,000 years ago the brain and basic anatomy of hominids had attained their modern structures.

hunting. Their cries suggest human laughter—a low-pitched, hysterical chuckling that rises to higher tones. The female deer emits a sharp, staccato bark to warn its young when it senses danger, and lions coordinate a hunt by grunting while stalking prey.

Elephants use their trunks for communication by trumpeting, humming, roaring, piping, purring, and rumbling. At least three dozen distinct elephant vocalizations have been documented, including an assortment of trumpeting sounds ranging from outright blasts to a low groan that males use to indicate that a jousting session is finished. Elephant screams range from expressions of social agitation to the pulsating bellow emitted by a female pursued by an unwanted suitor. Babies scream when they want milk; the scream gets progressively louder until their hunger is satisfied.

Elephants also communicate with infrasonic frequencies (below the range of human hearing), which humans detect as an air pulsation accompanied by low-frequency rumblings. Rumbles constitute the majority of elephant vocalizations and explain the uncanny ability of widely separated groups to coordinate their activities. There are rumbles of reassurance, rumbles to say "Let's go," rumbles to maintain contact, rumbles to cry "I'm lost," courtship and mating rumbles, and a humming rumble produced by mothers for newborn calves. Rumbles also coordinate activities within a given group when preparing to fight a dominance battle with another group, and mothers use a special rumble to reassemble the younger members of her family. About fifteen of the known rumbles have an infrasonic component, which enables elephants to maintain contact over long distances. Because low frequencies dissipate less rapidly in air, they can travel up to five miles. Elephants also emit infrasound to alert others to listen carefully for faint, higher-frequency sounds containing more detailed information.

Highly territorial mammals, such as lions, coyotes, and wolves, vocalize extensively at dawn and dusk to establish and maintain territory. Some species vocalize to attract mates and to intimidate rivals. Male moose give hoarse, bellowing cries during mating season to locate cows; the cows respond with a softer, somewhat longer lowing sound. The male elk, in an effort to collect as large a harem as possible, challenges competitors by emitting a buglelike sound. This vibrant call begins in the low register, ascends to high pitch, then abruptly drops in a scream. Bull seals, arriving at breeding grounds before the females, attempt to obtain as many cows as possible for their harems by frightening away competitors with loud roaring.

Bats and dolphins utilize high-frequency ultrasonic vocalization, or echolocation, as a type of animal sonar to find prey and to navigate in the dark. High frequencies are desirable to locate small targets, as high frequency waves are more directional, and a wave cannot "see" an object smaller than its wavelength.

Primates

Primates communicate by various vocal sounds as well as by facial expressions. Apes and monkeys use growls, grunts, twitterings, chirpings, whispers, barks, screams, and cries to warn of danger, indicate alarm or distress, and keep the members of a clan together. Velvet monkeys are known to have three different alarm calls: One warns of eagles, one of snakes, one of leopards. Howling monkeys emit loud, disconcerting, barking roars. The sound is produced by air passing through a resonating chamber in the throat. Rival groups fighting over territory engage in a duel of roaring until one group retreats. While roaring, all other activity, such as feeding, playing, or exploring comes to a halt. A female howler may also wail in distress when one of her young falls from a tree, while the youngster emits diminutive cries to indicate its position.

Primates lower the natural resonant frequencies of their vocal tracts when faced by danger in order to project the sonic aura of a larger animal. Apes and monkeys achieve this by protruding and partly closing their lips to generate low-pitched, aggressive sounds.

Among the great apes, gibbons are the most vociferous. Their raucous cries, especially boisterous at sunrise and sunset, can carry more than a mile. Apes only vocalize to express an emotional state. The ability to produce vocal sounds not linked to instinct or emotion is the primary difference between human speech and ape calls. The changing sound patterns of human speech represent abstract concepts, while apes produce simple melodies tied to their mood. The seat of human language is the cerebral cortex, while ape vocalizations are controlled by the subcortical neural structures involved in emotion. Emotionally based human vocalizations, such as sobbing, crying, laughing, giggling, or shouting in pain, are also controlled subcortically.

—George R. Plitnik

See also: Brain; Communication; Defense mechanisms; Ears; Emotions; Eyes; Groups; Hearing; Mammalian social systems; Mating; Noses; Sense organs; Smell; Vision.

Bibliography

Brackenbury, J. H. "The Structural Basis of Voice Production and Its Relationship to Sound Characteristics." In *Acoustic Communication in Birds*, edited by D. E. Kroodsma and E. H. Miller. New York: Academic Press, 1982. A detailed discussion of avian anatomy and physiology, applied to understanding birdsong.

Chadwick, Douglas. *The Fate of the Elephant*. San Francisco: Sierra Club, 1992. Includes descriptions of elephant vocalization and their elaborate communication system.

Fletcher, Neville. *Acoustic Systems in Biology*. New York: Oxford University Press, 1992. A detailed exposition of the physical principles involved in the various animal communication systems found in nature.

Lewis, B., ed. *Bioacoustics: A Comparative Approach*. New York: Academic Press, 1983. A detailed treatise on the comparative anatomy and functioning of animal vocal and hearing systems.

Stebbins, W. C. *The Acoustic Sense of Animals*. Cambridge, Mass.: Harvard University Press, 1983. A complete dissertation on animal acoustics which includes discussions of vocal mechanisms.

Strong, William, and George Plitnik. *Music, Speech, Audio*. 3d ed. Provo, Utah: Soundprint, 1992. A general text which details the anatomy, physiology, and psychoacoustics of human speech.

VULTURES

Types of animal science: Anatomy, classification, ecology, reproduction
Fields of study: Anatomy, reproductive biology, zoology

Vultures are large, carrion-eating birds, useful ecologically because decomposing carrion might otherwise endanger the health of humans and animals. Vultures live in Europe, Asia, the Americas, and Africa.

Principal Terms

CARRION: corpses of dead animals
NEW WORLD VULTURES: storklike vultures of the Americas
OLD WORLD VULTURES: hawklike vultures of Europe, Asia, and Africa
THERMAL UPDRAFT: rising air currents, encountered where the sun warms the air

Vultures comprise two groups of carrion-eating birds. They are useful because they eat carrion, which otherwise might decay and endanger the health of other animals. The twenty-one vulture species inhabit temperate to tropical regions of the Americas, Europe, Asia, and Africa.

All vultures exhibit similar eating habits, behavior, and appearance, including bare heads and necks. Many also have somber-colored feathers. Vultures of Europe, Asia, and Africa (Old World vultures) arise from eaglelike birds. Vultures of the Americas (New World vultures), similar in appearance to Old World vultures, are anatomically related more closely to storks.

Some Characteristics of Vultures

Vultures have bare heads and necks and hooked bills. Carrion is their main food, and on some occasions they attack newborn or wounded animals. Most hunt by long-distance soaring to scavenge with their keen sight. New World vultures differ from Old World vultures in their lack of the ability to vocalize. Six species compose New World vultures. Three live in North America: turkey vultures of the southern United States and northern Mexico; black vultures of the southwestern United States and Central America; endangered California condors; king vultures; Andean condors; and yellow-headed turkey vultures of South America.

There are fourteen Old World vulture species. Among the most interesting are the cinereous (with a color resembling ashes) vultures of southern Europe, northwest Africa, and Asia; the similar griffon vultures; white (Egyptian) vultures found from the Mediterranean to India; and the bearded vultures (lammergeiers) of Europe, Asia, and Africa.

Vultures lack feathers on their heads and necks, which keeps them free of gore from carrion. Among New World vultures, several have interesting appearances. Black vultures have black heads and plumage, with white feathers under the wings. King vultures, in contrast, have feathered neck ruffs and yellow, red, white, and blue heads.

California condors, the largest North American land birds, average four feet in length, with wingspans up to eleven feet. They have black neck ruffs, bald, orange to yellow heads, and black plumage except for white feathers under wings. Andean condors are similar. South American yellow-headed turkey vultures resemble North American turkey buzzards.

Notable among Old World vultures are cinereous vultures, about four feet long with bare, pinkish heads and black feathers. They inhabit Europe, northwest Africa, and Central Asia. Griffon vultures are similar in size and appearance.

Egyptian vultures, two feet long, have yellow heads and white feathers except for black wings. They inhabit Mediterranean areas and are found as far east as India.

Bearded vultures (lammergeiers) are especially interesting. They live on Asian, African, and European mountains. They have tan plumage on the chest and stomach and dark brown wing and tail plumage. Lammergeiers have red eyes in white heads. Conspicuous black feathers surrounding the eyes end in beardlike tufts and led to the name "bearded." These vultures average four feet long and weigh up to twenty-four pounds. Their huge wings allow soaring for hours on thermal updrafts. Lammergeiers are unusual in building large, conical nests on or in rock ledges or caves. A mated, monogamous pair uses the nests many times.

Life Cycles of Vultures

Most vultures nest on bare ground underneath mountain overhangs, or in caves. They build no nests, and females lay eggs on bare rock. After hatching, both parents feed the chicks partly digested carrion regurgitated into their mouths. For example, Andean (great) condors live in mountain caves, and females lay one or two greenish-white to bluish-white eggs on the cave floor. Both parents incubate the eggs until they hatch. The scarcity of the California condor is partly due to the fact that it lays only one egg at two- to three-year intervals. Young condors fly in six months, but parents feed them for another eighteen months. Andean condors first mate at seven years old, and at two-year intervals after that. They are monogamous and may live for forty-five to fifty years. Lammergeiers, as noted, are unusual in building several nests used over and over. The female lays her eggs, incubates them, and feeds chicks with the help of the male.

Vulture Facts

Classification:
Kingdom: Animalia
Phylum: Chordata
Subphylum: Vertebrata
Class: Aves
Order: Falconiformes
Family: Accipitridae (Old World vultures), with subfamily Aegypiinae (ten genera, twenty species); Cathartidae (New World vultures, five genera, five species)

Geographical location: Temperate to tropical regions of the Americas, Europe, Asia, and Africa

Habitat: Mountains, deserts, and other regions where carrion is available

Gestational period: Up to four months

Life span: Fifteen to fifty years, depending on species

Special anatomy: No feathers on head or neck; weak and blunt claws; New World vultures lack larynxes and thus lack voices

Vultures are identifiable by their curved bills, and their naked heads and necks. (Rob and Ann Simpson/Photo Agora)

Marabous: Storks or Vultures?

Marabou storks (marabous) combine stork and vulture anatomy and occur throughout Africa. Adults are five feet tall. They have long, storklike legs and sharp, straight bills. Their heads and necks are vulturelike. Most marabou food is deer, antelope, and zebra carrion.

Marabou plumage is gray on the back and wings, with white bellies and ruffs encircling red necks. Most inhabit African wetlands, rivers, and lakes. Pairs build nests in trees or on rocky terrain. Usually, three eggs are laid and incubated by both parents. Chicks hatch during dry season when carrion is plentiful. They stay with their parents for six months. Marabous live for over twenty years.

Vultures consume carrion, preventing decay and danger to health. This activity is one of their main ecological functions. Some vultures (such as condors) eat live food, giving them another ecological function, killing injured or weak members of other species. This helps the species that are eaten to select for individuals which enhance long-term survival.

—Sanford S. Singer

See also: Beaks and bills; Birds; Feathers; Flight; Molting and shedding; Nesting; Respiration in birds; Scavengers; Wildlife management; Wings.

Bibliography

Grady, Wayne. *Vulture: Nature's Ghastly Gourmet*. San Francisco: Sierra Club Books, 1997. This interesting book provides much information on vultures.

Mundy, Peter, Duncan Butchart, John Ledger, and Steven Piper. *The Vultures of Africa*. San Diego, Calif.: Academic Press, 1992. A comprehensive illustrated guide to the eleven African vulture species.

Stone, Lynn M. *Vultures*. Minneapolis: Carolrhoda Books, 1993. This nice book describes the lives, habitats, and other aspects of Old and New World vultures.

Wilbur, Sanford R., and Jerome A. Jackson. *Vulture Biology and Management*. Berkeley: University of California Press, 1983. This professional text describes vultures.

WARM-BLOODED ANIMALS

Types of animal science: Anatomy, behavior, physiology
Fields of study: Anatomy, environmental science, physiology, zoology

Warm-blooded animals—mammals and birds—maintain a consistent body temperature by adjustments in behavior, insulation, and metabolism when exposed to differing environmental temperatures; however, their body temperature can vary through a wide range depending on the time of day, season of the year, and the nutritional state of the animal.

Principal Terms

CORE TEMPERATURE: the internal body temperature around the heart, brain, and spinal cord; warm-blooded animals maintain a consistent core temperature; however, the skin or peripheral temperature varies with their surroundings

ECTOTHERMIC: heat from without; cold-blooded, animals that depend on environmental heat sources, usually solar radiation, to maintain body temperature

ENDOTHERMIC: heat from within; the preferred term for warm-blooded animals because it describes how they maintain their body temperature

HIBERNATION: a sustained period of torpor (lack of activity) where an animal reduces its metabolic rate

HOMEOTHERM: an animal with a constant, steady body temperature

METABOLIC RATE: the rate (expressed in calories per minute) at which an animal produces and consumes energy; can be indirectly measured by the amount of oxygen consumed per minute

METABOLISM: the conversion of carbohydrates, proteins, or fats into chemical energy that can be used to accomplish work and generate heat

POIKILOTHERM: an animal with a changing body temperature

Warm-blooded animals (birds and mammals) were so named because they are warm to the touch. However, a cold-blooded animal lying in the sun would also feel warm, so additional terms have been created. Homeotherm (same temperature) and poikilotherm (changing temperature) have been used for warm- and cold-blooded animals, but hibernating warm-blooded animals reduce their body temperature to a few degrees above their surroundings, and some cold-blooded deep-sea fish live in a constant temperature environment, thus maintaining a constant body temperature.

More accurate terms have been developed that focus on the heat source maintaining body temperature. Warm-blooded animals use the heat generated from their body's metabolism to maintain body temperature, whereas cold-blooded animals depend more on outside sources of heat, such as solar radiation, to maintain body temperature. Thus, warm-blooded animals are called endotherms (heat from within) and cold-blooded animals are called ectotherms (heat from outside). Warm-blooded animals also use solar radiation to help warm themselves on a cold day, and cold-blooded animals have some internal metabolic heat to maintain their body temperature, but endotherm and ectotherm best describe the primary difference between warm- and cold-blooded animals.

The Metabolic Furnace and Body Temperature

Warm-blooded animals (endotherms) are characterized by using metabolism to keep their body temperature relatively constant and independent of the environmental temperature. The overall metabolic process is about 25 percent efficient, which means 25 percent of the food energy accomplishes work and the other 75 percent is lost as heat. The internal organs such as the heart, kidneys, brain, and lungs generate the majority of the heat, which is then distributed throughout the body by the circulatory system.

Because warm-blooded animals produce heat, they must also lose heat to maintain a steady body temperature. The rate of heat loss from the body is dependent on the difference between the skin temperature and the surroundings. The colder the surroundings, the faster heat is lost. To maintain a steady body temperature, warm-blooded animals can adjust their behavior, insulation, and metabolic rate.

Small rodents and insectivores seek shelter in burrows to reduce heat loss, while larger animals, such as elk and deer, graze on warm, sunny slopes on cold winter days to gain solar radiant heat. Penguins huddle together to reduce heat loss in cold Antarctic winds, otherwise the metabolic heat required to keep warm would deplete their body fat stores before the winter's end.

The arctic fox reduces its heat loss by growing a thick, insulating fur coat. In contrast, diving mammals such as the Weddell seal use a thick layer of fat (blubber) under the skin to insulate and reduce heat loss to cold ocean water.

If behavioral adjustments and increased insulation cannot prevent excess heat loss, then warm-blooded animals increase their metabolic rate to generate additional heat to maintain body temperature. Shivering produces an increase in muscle metabolism, and the heat produced adds additional warmth. These metabolic adjustments are costly and indicate that heat loss is greater than metabolic heat production. In this situation, an animal will have difficulty in maintaining a steady body temperature in its surroundings.

Warm-blooded animals that live in cold environments, such as bears, have thick fur coats to help retain their natural body heat. (PhotoDisc)

Cooling the Body

During periods of cold weather and reduced food supply, small warm-blooded animals will decrease their body temperature to a few degrees above their surrounding temperature to reduce heat loss and conserve bodily energy stores. This can be a hibernation lasting throughout the winter months or a period of torpor lasting for the night. This lower body temperature, however, is not a decrease in body temperature as observed in cold-blooded animals, but a new, highly regulated body temperature attained by lowering the animal's metabolic rate. It is a very important strategy for energy conservation and survival.

To maintain a steady body temperature, warm-blooded animals depend on a balance between heat produced by metabolism and heat lost to their surroundings. When the ambient temperature exceeds body temperature, an animal will gain heat instead of losing heat to the surroundings. This situation can then elevate body temperature. Body temperature, however, is held constant by heat loss through evaporation of water from the animal. Water, in changing phases from liquid to vapor, requires energy (heat), and this represents the only mechanism available to an animal if the surrounding temperature is greater than body temperature.

Two mechanisms for cooling the body by increasing water loss are panting and sweating. Zebras, gazelles, and bison are examples of closed-mouth panters. These animals rapidly move air through the nostrils, which cools the tissues lining their upper airways (nasal cavities). Dogs are the best example of open-mouth panters; they move air rapidly over the moist tongue to remove heat. In contrast to panting, sweat glands secret a hypotonic salt water over the skin surface. The sweat evaporates and cools the body surface. Both panting and sweating can reduce an animal's bodily water reserves, resulting in dehydration. Thus, increasing water intake during periods of hot weather is especially important.

Camels are known for their tolerance of hot environments. Although they are closed-mouth panters, they also allow their body temperature to rise during the day by 3 to 4 degrees Celsius. This helps the camel tolerate a hot environment in two ways. By increasing its body temperature, the camel will gain less heat from the environment because there is less of a difference between the surroundings and the camel's temperature. Second, the camel can reduce the amount of panting and water loss needed to keep a steady body temperature. A camel stores body heat during the hot day and loses this heat during the cool desert night. The camel's humps are not stores of water, but fat. This makes the camel's skin thinner, which allows for better heat loss at night.

—Robert C. Tyler

See also: Cold-blooded animals; Fur and hair; Hibernation; Mammals; Thermoregulation.

Bibliography

Pough, F. H., C. M. Janis, and J. B. Heiser. *Vertebrate Life*. 5th ed. Upper Saddle River, N.J.: Prentice Hall, 1999. Emphasis on the differences in animals based on the sequences of evolution.

Schmidt-Nielson, K. *Animal Physiology: Adaptation and Environment*. 5th ed. New York: Cambridge University Press, 1997. This classic textbook in animal physiology is a standard for many high school and college courses and focuses on physiologic function and adaptation to different environmental conditions.

Willmer, P., G. Stone, and I. Johnston. *Environmental Physiology of Animals*. Malden, Mass.: Blackwell Science, 2000. This textbook has simple-to-understand diagrams and explanations of animal organ systems and their adaptation to different environments.

Withers, P. C. *Comparative Animal Physiology*. Fort Worth, Tex.: W. B. Saunders, 1992. Gives a broad overview of animal physiology comparing the characteristics of each species.

WASPS AND HORNETS

Types of animal science: Anatomy, classification, reproduction
Fields of study: Anatomy, entomology, zoology

Wasps, stinging insects of the order Hymenoptera, are hornets, which are social species, or mason, digger, carpenter, and mud-dauber wasps, which are solitary species. They help humans by killing insect pests.

Principal Terms

DRONE: a male wasp
HORNET: a social wasp
MANDIBLE: a wasp jaw
SOCIAL WASP: a wasp living in a large colony

Wasps are stinging insects of the order Hymenoptera. Many live in large colonies which have a queen, males, and sterile female workers. Such social wasps are hornets. Wasps are called solitary if they do not live in communities, but build small brood nests to hold their young.

Social wasps (hornets) make paper nests. One example is white-faced hornets, found all over North America. These wasps, 1.25 inches long and black with white markings, build nests, up to half-bushel size, in tree limbs. Yellow jacket species live in colonies of many thousands, close to or under the ground. Giant European hornets, in the United States since the 1850's, are brown with yellow streaks and nest in hollow trees. In some wasp species, no workers are born and females lay eggs in the nests of other wasps. Wasp size varies from parasitic wasps, that can develop in insect eggs, to species attaining body lengths of over two inches.

Physical Characteristics of Wasps and Hornets

Wasp bodies, which are covered by coarse hairs, have a head, thorax (midbody), and abdomen (hind body) segments. Thoraxes hold four wings and six legs. The bodies are steel blue, black, yellow, or red, with abdominal rings. Reproductive, digestive, and excretory systems are in abdomen and thorax. Females have stingers at abdomen ends. Parasitic wasps use stingers to insert eggs into hosts. Female nonparasite wasps (which are most wasps) use stingers to paralyze their prey and inject venom. The stings are painful, because the venom contains histamine and a factor that

Hornets, such as this giant hornet, are social wasps that live in colonies with clearly defined castes of workers, males, and queens. (Rob and Ann Simpson/Photo Agora)

dissolves red blood cells. Wasp stings, especially by hornets, can kill allergic humans.

Wasp heads contain sharp, strong mandibles (jaws), designed to chew hard things, tear up food, dig burrows, and pulp wood and earth. Wasp mouths can also lap liquids. Above the mandibles, their heads contain paired, keen, compound eyes and paired sensory antennae. Queen wasps in social species are reproductive females who use sperm obtained in mating flights to fertilize eggs that become females. Unfertilized eggs become drones (males). Production of queens-to-be or female workers depends on diet.

Wasp Nests and Life Cycles

Nests of social wasps range from combs without protecting covers, to round nests up to ten inches in diameter, having paper tiered combs and waterproof outside covers. Social wasps nest wherever possible. Small combs occur under porch roof and rafters or in trees. There are two yellow jacket types, the long-faced and short-faced species. Long-faced yellow jackets nest in trees, bushes, and roofs, while the short-faced type nests in the ground.

A wasp colony lasts one year. Wasps store no food, and in fall the whole colony dies except for the future queens. They hibernate in crannies over the winter and become queens of new colonies in spring. A colony starts after a queen makes a few cells, lays an egg in each, and feeds larvae with chewed-up insects. Next, larvae spin cocoons and pupate for several weeks, emerging as workers. After this, a queen does nothing except lay eggs. The eggs yield worker wasps until late summer, when the queen lays eggs that will become males and queens. Workers tend the young and enlarge the nests. A hornet nest may have thousands of males, females, workers, and young.

Solitary wasps live alone except for breeding. Afterward, females build flat, one-comb nests. Instead of being papermaker wasps, they are mason, carpenter and digger wasps. Among mason wasps are potters and stoneworkers. Potters wasps make mortar of mud and saliva and place brood nests in trees. Mud dauber wasps mix mud with saliva and build nests under porch roofs. Stone worker wasps mix pebbles with mud and nest on rocks. Carpenter wasps tunnel into trees and digger wasps tunnel into the ground.

All adult wasps eat caterpillars, spiders, beetles, flies, other insects, and nectar. Solitary wasp species feed their larvae with specific live insects. Mothers set up nurseries, paralyze prey by piercing nerve centers with their stings, and take the live food to nests. Then they lay an egg on each body. Larvae feed on the insects until they begin

Wasp and Hornet Facts

Classification:
Kingdom: Animalia
Subkingdom: Bilateria
Phylum: Uniramia
Class: Insecta
Order: Hymenoptera (wasps, ants, and bees)
Suborders: Apocrita, Symphyta (wood wasps)
Superfamilies: Vespoidea, Sphecoidea, Bethyloidea, Scolioidea, Chalcidoidea
Families: Include Vespidae (hornets), Sphecidae, Chrysididae (cuckoo wasps), Pompilidae (spider wasps), Tiphiidae, Scoliidae, Mutillidae
Genus and species: Include *Vespula*, such as *V. vulgaris* (common wasp, or yellow jacket); *V. maculata* (bald-faced or white-faced hornet), *V. crabro* (European hornet); *Polistes*, such as *P. fuscatus* (golden paper wasp); *Eumenes iturbide pedalis* and *E. fraternus* (potter or mason wasp); *Sceliphron caementarium* (black-and-yellow mud-dauber wasp); *Pepsis limbata* (tarantula hawk); *Sphecius speciosus* (cicada killer)

Geographical location: Europe, Asia, and the Americas

Habitat: Trees in forests, woods, and plains; on or under rocks; in the ground; around human habitations

Gestational period: About one month for hatching and pupation

Life span: Varies; queens live up to ten months, while drones or workers live for a few weeks

Special anatomy: Six legs, three on each side of thorax; two pairs of wings

spinning cocoons to pupate, emerging after pupation as adult wasps.

Yellow Jackets

North American short-faced yellow jacket wasps (hornets) are 0.75 inches long, with yellow and black head, thorax, and abdomen markings that give them their name. They nest below grass level near decaying wood. Their nests are paper, made from saliva and wood.

Each nest has a queen, who lays all eggs. Fertilized eggs become females and unfertilized eggs become males. Reproductive females are produced when the colony is ending its one-year life span. Sterile females tend the nest and larvae. Reproductive females eventually become queens, and males mate with queens-to-be.

The yellow jacket diet is insects, fruit, and nectar. Only worker yellow-jackets hunt food, which they eat by tearing it with their mandibles. Queens live for ten months, while drones or workers only live for a few weeks.

Helpful Wasps

Most wasps help humans and the environment. They damage some fruit, but they destroy myriad caterpillars, beetles, flies and other harmful insects. Thus, they do far more good than harm. Several species pollinate farm crops. Furthermore, the parasitic varieties lay their eggs in the bodies and eggs of pests such as aphids, thereby reducing their numbers.

—Sanford S. Singer

See also: Bees; Insect societies; Insects; Termites.

Bibliography

Crompton, John. *The Hunting Wasp*. Reprint. New York: Lyons Books, 1987. An interesting book on wasp natural history and predation.

Gullan, P. J., and P. G. Cranston. *Insects: An Outline of Entomology*. 2d ed. Malden, Mass.: Blackwell Science, 2000. A fine text on entomology, including wasps and hornets.

Johnson, Sylvia A., and Hiroshi Ogawa. *Wasps*. Minneapolis: Lerner, 1984. A book for juveniles that describes social wasp nests, egg laying, and life.

Spoczynska, Joy O. I., and Melchior Spoczynski. *The World of the Wasp*. New York: Crane, Russak, 1975. A solid, illustrated book on wasps.

WATER BALANCE IN VERTEBRATES

Type of animal science: Physiology
Fields of study: Cell biology, physiology

Water balance in vertebrates must be maintained within narrow limits so that electrolytes and enzymes are in the correct concentrations for proper functioning. Since sources of water are limited and the losses occur from systems with very different functions, the integration of retention and excretion is complex but essential.

Principal Terms

APOCRINE GLAND: a type of sweat gland that becomes active at puberty and responds to emotional stress; the glands are found at the armpits, groin, and nipples

ECCRINE GLAND: a type of sweat gland that helps maintain body temperature; the glands are located on the palms and soles, forehead, neck, and back

EXTRACELLULAR FLUID: the fluid outside cell membranes, including fluid in spaces between cells (interstitial), in blood vessels (plasma), in lymph vessels (lymph), and in the central nervous system (cerebrospinal)

HOMEOSTASIS: the dynamic balance between body functions, needs, and environmental factors which results in internal constancy

HYPEROSMOTIC: a solution with a higher osmotic pressure and more osmotically active particles than the solution with which it is being compared

HYPOOSMOTIC: a solution with a lower osmotic pressure and fewer osmotically active particles than another solution with which it is being compared

ISOSMOTIC: a solution with the same osmotic pressure and the same number of osmotically active particles as the solution with which it is being compared

Water, which makes up about 60 percent of the body weight of vertebrates, may be the most neglected nutrient. Drinking and eating are the obvious ways of obtaining water, but metabolism, the processes of synthesis and breakdown within the body's cells, also provides water for organisms. In fact, diet and metabolism provide 40 percent of the water necessary for human life.

Drinking water is limited by the environment. Areas such as deserts, which have little rainfall, have little potable groundwater available; even plants have to develop some means of conserving the little water their roots can find or the dew that settles on exposed surfaces during the cool desert night. With no surface water and few plants as sources of water, some desert mammals, such as the kangaroo rat, get most of their water from metabolism. They often do not drink even when a supply of water is nearby.

Water balance varies both daily and seasonally as environmental factors such as temperature, humidity, and wind vary and as activity levels change. The body must maintain nearly constant volume and composition of the extracellular fluid despite fluctuations caused by drinking, eating, metabolism, activity, and environment.

Water Loss from Kidneys and Lungs

There are four sites of water loss: kidneys, lungs, skin, and intestines. Control of water balance depends on the efficiency of water retention compared with the necessity of water loss in the normal functioning of both terrestrial and marine vertebrates. For freshwater vertebrates, for example, excretion of excess water without losing salts is vital.

The kidney can produce urine that can be highly concentrated or very dilute. For humans, urine can be four times more concentrated than body fluids and contain 1,200 milliosmoles of solute. Other animals, particularly some desert species, can produce urine five times more concentrated than that. The more concentrated the urine, the more water is retained during the excretion of waste materials.

Generally, increasing length of the loop of Henle in the nephron is associated with increased concentrating ability of the kidney in mammals. Although some of their nephrons contain loops of Henle, birds cannot match the mammals' concentrating ability. The maximum urine-to-plasma concentration ratio in birds is only a little more than five. The reason for this is that mammals excrete osmotically active urea, whereas birds excrete precipitates of uric acid and uric acid salts that do not contribute to osmotic pressure. The osmotic pressure of birds' urine primarily comes from sodium chloride. Birds also allow their plasma to become twice as concentrated as that of mammals during dehydration.

When water must be conserved, urine is concentrated to a greater degree than when water is plentiful. As water becomes scarce, the concentrations of solutes in the body fluids increase. Osmoreceptors in the hypothalamus sense the increase and stimulate the release of antidiuretic hormone (ADH). Antidiuretic hormone is a small peptide that varies somewhat in composition among vertebrate classes. It affects kidney cells, increasing the permeability of the distal tubule of the nephron and the collecting duct to water. More water is reabsorbed and, therefore, retained by the body. In some species, ADH also affects the numbers of nephrons filtering.

The earliest water-balance problem that the vertebrates faced in their evolutionary history was an excess of fresh water and a scarcity of sodium. The hormone aldosterone evolved in the fishes to cope with this. Aldosterone is secreted by the adrenal gland and increases the reabsorption of sodium by the kidney. Some potassium is lost from the body in exchange for sodium, but since plants are rich in potassium, its loss could be made up in the diet.

Water is also lost or gained when respiring; all gas-exchange surfaces must be moist. Aquatic organisms take in water through their gills in fresh water and lose water through the gills in marine environments. Terrestrial organisms have lungs, and the moist exchange surface is inside the body. Even if the atmosphere is dry, air entering the lungs is moistened to nearly 100 percent relative humidity during its passage through the airways. Expired air loses water to those same airway walls as it leaves the body. Some moisture, however, remains in the air and is lost to the body. The cloud of vapor seen as animals exhale in cold weather is this lost moisture condensing in the cold air.

Breathing rate and volume influence the amount of water lost from the lungs. As breathing rate increases, air moves out of the body more quickly, and there is less time for moisture to condense on the cool airway walls before it is exhaled. The movement of greater volumes of air also increases loss of water vapor from the respiratory tract, because less comes in contact with the airway walls. Since heavy exertion often involves breathing through the mouth rather than through the moisture-conserving nasal cavity, further water is lost. As much as 25 percent of the moisture present in expired air may be lost in a dry environment.

Water Loss from Skin and Intestines

The skin is another site through which water is lost. The sweating which occurs in warm weather is an obvious example. Even during the winter, when the air is cool, "insensible perspiration" occurs as water diffuses through the skin. Insensible perspiration occurs at all times. The perspiration that can be sensed comes from eccrine glands. They secrete a watery fluid that cools the skin by using body heat to evaporate the liquid. The amount of salts and nitrogenous wastes in perspiration is not large; however, when one is working in a hot environment, the loss may be significant.

Primarily, though, it is water that is lost and must be replaced.

The apocrine sweat glands are located in the armpits and groin. They become active during emotional stimulation. These are the glands associated with the musky odor that some animals exude. These glands are not important in regulating body temperature, and their evaporation of water is not a major component of water balance mechanisms. The losses of water from the respiratory tract and the skin are obligatory, usually amounting to about 850 milliliters a day.

The intestinal tract is a source of water gain, as it ingests both liquids and food (with its associated water). Some water, however, is also lost because the copious intestinal secretions contain water. In fact, one day's intestinal secretions may amount to twice the body's plasma volume (from which it is derived). Not all water is reabsorbed in the passage through the stomach, small intestine, and large intestine: About one hundred milliliters are lost in the feces each day.

In a normal human diet, ingested fluid tends to exceed the minimum required by about one liter. Whatever excess is not used in evaporative cooling and lost from expired air or in feces is excreted by the kidney. The minimum water uptake required for balance is defined as that required to provide minimum urine volume without weight loss. The stomach and the small intestine reabsorb most of the ingested and secreted water. Only 35 percent reaches the large intestine. The large intestine is specialized to absorb water and produce semisolid feces for excretion. The maximum rate of water absorption by the intestines lies well above what is normally required.

The body fluid compartments provide an excellent example of the steady-state system characteristic of living things. Intracellular fluid must maintain a composition that promotes chemical reactions and diffusion despite the changes that those reactions bring. Extracellular compartments must retain their individual characteristics even though they communicate with one another. Hormones and nerves coordinate the interactions of the digestive, respiratory, integumentary, and urinary systems, which contribute to the constant conditions of volume and composition of the body fluid compartments.

Studying Water Balance

The study of water balance is often difficult because so many body systems and physical factors are involved. Gross methods include measuring moisture in respired air, feces, and urine, as well as in ingested foods. In other studies, the weight of water in bodies or organs may be obtained by drying: The amount of water is the difference between the wet and dry weights. These methods are crude and give rough estimates of the fluid volume or fluid balance in the body or an individual compartment or organ.

To obtain more precise information, dilution techniques are used. One method is the injection or infusion of an indicator or test substance. This substance must be distributed only in the fluid compartment being measured. It must be safe for the organism, while not being metabolized or synthesized in the body. If the substance can be excreted, it must be measurable in the excreta. Radioactive tracers are also used to determine dilution of an injected or infused sample. The isotope chosen must not influence the mechanisms governing the size and composition of the compartment being measured. Moreover, the body must not be able to distinguish between the radioactive molecule and the unlabeled isotope.

The concentration of the test substance in a sample of the blood, or lymph, or cerebrospinal fluid gives an indication of the dilution caused by the volume of the fluid within the compartment. There is no perfect test substance; each is associated with problems affecting the accuracy of the measurement. Inulin is used to determine the volume of interstitial fluid, but inulin diffuses slowly through dense connective tissue. Radioactive sodium enters most compartments, but it binds to the crystalline structure of bone. The dye Evans blue, which binds to plasma proteins, and radioiodinated serum albumin are used to measure plasma volume, but these substances move out of capillaries.

All the fluid compartments communicate, but some, such as bone and cartilage, communicate more slowly than others, such as lymph and plasma. The resistance between compartments is often supposed to be at the interface between compartments. There is evidence, however, that the rate-limiting factors may not only be the permeability of the cell membrane alone but also the physical state of some of the water within the compartment. For example, in red blood cells, some water is bound to protein and is not accessible to solutes. A portion of the water in mitochondria is not free to participate in osmotic processes. For these reasons, dilution techniques may underestimate the amount of water in the compartment being measured. Although dilution techniques use sophisticated technology, the measurements are often extrapolations and not exact. These techniques do provide a general picture of the distribution of body fluids, but since water balance is a continuing, dynamic process, the values are not stable.

Regulating Water Loss

Loss of water through the skin and respiratory tract is obligatory. All respiratory surfaces must be moist so that gases can pass through them. Amphibians are limited in their geographical distribution primarily because their skin is a respiratory surface and, therefore, water loss from it cannot be curtailed. The water losses vary with environmental factors such as temperature, humidity, and wind. Because of this, amphibians must remain near a source of water which their skin can imbibe.

Loss of body water through "insensible perspiration" is not controllable; it is obligatory. The sweating that helps regulate body temperature is facultative, and it varies with weather and exercise. If sweating is prevented when the ambient temperature is high, the body temperature can rise explosively. This will cause death as surely as the dehydration which was prevented by not sweating would have.

Mammals and birds can minimize water loss by modifying the depth and rate of breathing. On exertion, the rate and depth increases, and correspondingly, the loss of water through the airways and across the skin surface increases. Unless the organism intends to quit breathing and allows its body temperature to rise, some water must be lost in this way.

Losses through the digestive tract are often involuntary. Diarrhea and vomiting accompany many illnesses, and since these uncontrolled losses are from the digestive system, which secretes a volume twice that of the plasma each day, their unreclaimed losses are massive. Dehydration and electrolyte imbalance follow quickly if these losses are not made up. This is particularly crucial for infants, for whom the daily diet makes up 25 percent of the total body water. With dehydration, the volume of the circulatory system decreases and circulatory failure results. In addition, since the extracellular fluid compartments are continuous with the intracellular compartments, the fluid inside the cells becomes hyperosmotic, and metabolic reactions cannot take place.

The retention of water depends upon several conditions. The most important is the sources of water available. If fresh water is not available, a human will die after eleven to twenty days, depending upon the rapidity of onset of dehydration. By that time, the person will have lost 15 to 20 percent of initial body weight. The excretion of wastes, the act of breathing, and insensible perspiration (even in moderate temperatures with shade available) are accompanied by obligatory water losses that cannot be reduced.

For other organisms, water is present in excess and becomes a problem. Freshwater fish take in water through their respiratory surface, the gills. They must release great quantities of urine without losing the salts necessary to maintain proper internal osmotic conditions. The hormone aldosterone promotes that salt absorption from the nephrons.

For marine creatures, on the other hand, the entry of salt is a problem. They must eliminate the excess salt without losing too much of their precious body water. Because kidney function always involves water loss as well as the loss of ions, and because fish and reptiles do not concentrate urine efficiently, many of these vertebrates have evolved salt glands. The salt glands use metabolic energy to excrete sodium chloride with very little water. Each environment presents its own water-balance problems to an organism. Yet even in the world's harshest, driest conditions—in the Antarctic—tiny mites and spiders, penguins, and predatory birds have found ways to live and obtain all the water they need in a land where water is solid most of the time.

—Judith O. Rebach

See also: Kidneys and other excretory structures; Osmoregulation; pH Maintenance; Thermoregulation.

Bibliography

Deetjen, Peter, John W. Boylan, and Kurt Kramer. *Physiology of the Kidney and of Water Balance*. New York: Springer-Verlag, 1975. An easily understandable and brief analysis of the kidney and its functions, along with discussion of the importance of water and its regulation in the body. The effects of environmental factors on water balance are also considered.

Raven, Peter H., and George B. Johnson. *Understanding Biology*. 3d ed. Dubuque, Iowa: Wm. C. Brown, 1995. A textbook for college-level students in introductory biology.

Slonim, N. Balfour, and Lyle H. Hamilton. *Respiratory Physiology*. 5th ed. St. Louis: C. V. Mosby, 1987. This brief text presents information about the normal functioning of the lungs, their development, and the influences of environmental, pathological, and physiological factors on gas exchange and water balance. A helpful appendix explains abbreviations and equations and presents charts used in analyzing pulmonary function. The glossary and references are detailed and complete.

Weisburd, Stefi. "Death-Defying Dehydration: Sugars Sweeten Survival for Dried-Out Animals, Membranes, and Cells." *Science News* 133 (February 13, 1988): 107-110. An interesting article on research which has shown that the insertion of trehalose molecules into membranes and enzymes allows some animals to survive drought.

Widdicombe, John, and Andrew Davies. *Respiratory Physiology*. 2d ed. London: Edward Arnold, 1991. This small book contains chapters on the structure and function of the lungs, physical factors involved in their functioning, and the dynamics of ventilation. It considers pulmonary relationships and controls. Each chapter includes learning objectives and references.

Withers, Philip C. *Comparative Animal Physiology*. Fort Worth, Tex.: W. B. Saunders, 1992. A broad overview textbook for upper-level college students, taking a comparative approach to animal physiology and zoology.

WEASELS AND RELATED MAMMALS

Types of animal science: Behavior, classification, reproduction
Fields of study: Anatomy, classification

Weasel family members live throughout the world, in many different habitats. They include badgers, martens, mink, otters, skunks, weasels, and wolverines. Most are carnivores, and all are hunted for their fur.

Principal Terms

CARRION: dead, partly decomposed animal bodies

ERMINE: valuable white fur made from weasel pelts taken in winter

GESTATION: time period for carrying developing mammalian offspring in the uterus

MUSK: bad-smelling liquids animals make to mark territory or for self-defense

OMNIVOROUS: able to eat both plants and animals

PERINEAL: located between scrotum and anus in males or the equivalent region in females

About seventy species belong to the weasel family (Mustelidae). These furry mammals weigh from three ounces to one hundred pounds. Included are weasels, mink, martens, wolverines, otters, skunks, and badgers. Mustelids live worldwide, except for Antarctica, Australia, and the Sahara Desert. Their habitats range from the Arctic to tropical rain forests. Some live on the ground, others live in trees, and still others inhabit rivers or oceans.

Most weasel family members have long, slender bodies and all have short legs. Each mustelid paw has five toes, with sharp claws for grasping prey and burrowing. The smallest mustelids, least weasels, weigh three ounces. River otters, the largest mustelids, are four to six feet long (with tails) and weigh up to seventy pounds. Sea otters can weigh ninety to one hundred pounds. Beautiful, luxurious mustelid fur is sought for use in fur garments.

All mustelids have perineal glands which make unpleasant-smelling musks, causing their characteristic odors. Musks mark territory or are used for self-defense. The odor of mustelid musk is most offensive in skunks, which use it for self-defense.

Male and female mustelids often live alone, except when mating. Mating seasons vary with species and habitat. Pregnancy for mustelids ranges from one to nine months. Litters contain one to ten young, depending on species and food availability. Life spans of mustelids that reach old age are one to twenty-five years.

Weasels

Weasels, like most mustelids, are carnivores. They eat other animals, carrion, and insects. All weasels have keen eyesight, keen smell, and are excellent hunters. They are bloodthirsty, often killing for fun and leaving prey carcasses uneaten. Humans hunt weasels in response to their bloodthirsty natures and for their beautiful, soft fur. This is especially true of weasels that live in cold climates and grow white winter coats that collectively provide ermine.

Weasels are known for their unpleasant odor, from musk made in perineal glands. The long-tailed weasel is the best-known North American species. It has a white belly, a brown back, and a black tail tip. Males and females are 1.5 and 1 foot long, respectively. The least weasel, also North American, is the smallest known carnivore, only six to eight inches long.

Martens and Fishers

Martens are carnivorous mustelids, genus *Martes*. They occur in northern forests of the eastern and western hemispheres. Martens are long and graceful, with short legs and sharp-clawed toes. They live in hollow trees and take over woodpecker or squirrel holes. They eat rabbits, squirrels, birds, mice, eggs, and carrion. They mate in July and August. Nine months later, one to five young are born.

The most common North American marten is the pine marten (American sable). Pine martens are most plentiful in the northern Rockies, Canada, and Alaska. They also occur as far south as the Adirondacks, and west to Colorado. A pine marten is twenty-four to thirty inches long, including a six-inch tail. Its brown fur is thick and soft, with orange or white throat and chest patches. Pine martens are hunted for their valuable fur. Their enemies are lynx, owls, and humans. Similar animals, the baum and stone martens of Europe and Asia, have yellow and white throats, respectively.

Another important marten is the fisher, a pine marten subspecies and the largest of martens. Fishers are 4.5 feet long, including 16-inch tails, and weigh up to fifteen pounds. Their luxuriant fur is medium to dark brown, with gold or silver head and shoulder tops and creamy chest patches. Fisher fur is also quite valuable. The fisher habitat range is like that of pine martens, but not as far north. Fishers are the only animals that kill and eat porcupines without being hurt. Otherwise, their diets are like those of other martens.

The pine marten, or American sable, is the most numerous of North American mustelids. (Greg Vande Leest/Photo Agora)

Otters

Otters are mustelids that live mostly in water, either in rivers or oceans. River otters are found worldwide, except for Australia, New Zealand, and Antarctica. River otters of Europe, Asia, and North America have thick, four-foot-long, flexible bodies and tails about two feet long. Otter heads are broad and flat, with round ears and blunt snouts. Their nostrils are closed when diving. Beautiful otter fur is dark brown and desired for fur garments. As in other mustelids, otter legs are short. Otter feet are both webbed and powerfully clawed.

River otters live in waterside dens with underwater entrances. Their gestation period of two months produces two to three offspring, which stay with their mother for a year. These otters are rapid swimmers and expert divers. They eat fish, crayfish, snails, shellfish, frogs, insects, birds, and small water animals. River otters are friendly and playful.

Sea otters inhabit the Pacific coasts of Asia and North America. They live mostly in the water. Sea otters are 2.25

to 4 feet long and live on a diet of sea urchins, crabs, snails, and shellfish. They have one or two offspring per litter. Red-brown to black sea otter pelts may cost over two thousand dollars each. Sea otters have been so hunted that they are very cautious. They are protected species in U.S. waters.

Skunks

Skunks are found throughout the United States, in Mexico, and in southern Canada. Cat-sized, these stocky mustelids have long, pointy noses, arched backs, and short legs. Skunk fur is long, thick, shiny, and black, with white stripes down the back. Stripe patterns differ among species. Skunks have white forehead patches and long bushy tails, black on top and white underneath. Striped (common) skunks have long black fur and white stripes running from head to tail. They breed in late winter or early spring and have three to five offspring, after a 2.5-month gestation. Offspring suckle for two months and then their mother teaches them to hunt.

Skunk self-defense uses paired perineal glands on either side of its anus, near the tail. These glands are found in all other mustelids. In skunks they are relatively large and contain terrible-smelling liquid musk. Frightened skunks squirt musk distances up to twelve feet. The horrible odor keeps enemies away. Animals that are sprayed smell bad for days. Skunks inhabit hollow trees or burrows and eat rodents, birds, lizards, eggs, insects, honey, and bees.

Badgers

Badgers are two-foot-long mustelids with short, strong legs, squat, broad bodies, and forelegs having claws that are efficient burrowing tools. They are nocturnal, heavily furred, and very strong. Badgers live in deep burrows on and in prairies, woods, or hills. Like other mustelids, they have perineal glands which make fetid musk. Their pelts are valued garment furs. American badgers are found in western North America, east to Ohio and north to southern Canada. Their shaggy fur mixes gray, black, and brown hair. They eat field mice, squirrels, and gophers, digging animal prey out of underground homes.

European badgers are like American badgers in size and color, but have teeth designed for omnivorous diets. They live in forests in deep burrows, and their litter size is four or five young. These badgers eat fruits, nuts, eggs, birds, rodents, frogs, snails, worms, and insects, and love honey and wasp or bee larvae. Shaggy hair protects them from being stung. All badgers are quarrelsome. Caged together, they fight continually. They are also very brave and fight savagely, if cornered.

Wolverines

Wolverines are mustelids of the genus *Gulo*. They are among the most powerful animals of their size. Wolverines live in northern woods of North America, Europe, and Asia. They have long, ta-

Mustelid Facts

Classification:
Kingdom: Animalia
Subkingdom: Bilateria
Phylum: Chordata
Subphylum: Vertebrata
Class: Mammalia
Order: Carnivora
Family: Mustelidae
Subfamilies: Mustelinae (weasels and allies, ten genera, thirty-three species), Mephitinae (skunks, three genera, thirteen species), Lutrinae (otters, six genera, twelve species), Melinae (badgers, six genera, eight species), Mellivarinae (honey badger)

Geographical location: Worldwide, except Antarctica, Australia, and the Sahara

Habitat: From the Arctic to tropical rain forests; most live on the ground, some live in trees, and still others inhabit rivers or oceans

Gestational period: From one to nine months

Life span: From one to up to twenty-five years

Special anatomy: Paws with sharp digging and grasping claws, luxurious fur, webbed feet in otters

Mink

Mink, mustelid carnivores (genus *Mustela*), inhabit North America, from the Gulf of Mexico to the Arctic, and Northern Europe and Asia. They den near rivers and lakes, combining weasel land habitats and otter water habitats, and make foul-smelling musk. Mink, like otters, hunt in water for crayfish, frogs, and minnows. On the land, like weasels, they hunt rodents and snakes. Mink musk has a nauseating smell. Mink are nocturnal animals that live in burrows and are solitary except when mating (February to March). Mink gestation, 1.5 months, usually yields around six offspring.

Mink have short legs and slender, 2.5-foot bodies (including bushy, 6.0-inch tails). Males weigh two pounds, and females are half that weight. Bobcats, foxes, owls, and humans kill mink. Humans do this for their beautiful, soft, durable fur, highly valued for garments. The fur has two layers, an oily, water-repellant outer layer, and a thick, soft, warm inner layer. Wild mink are brown or blackish brown. Fur farms raise mink for genetically selected black, blue, silver gray, or white fur.

pering heads, heavy 2.5-foot-long bodies, bushy 8-inch tails, short legs, and large feet with sharp claws. Wolverine teeth are long and sharp. Their powerful jaws can crush the bones of prey. Wolverines look like small bears, and their dark, white-marked pelts make fine fur garments. This has made them rare, after ruthless hunting. Males and females can weigh fifty-five and thirty pounds, respectively.

Wolverines are solitary, except for mating in spring and summer. Nine months later, females give birth to around four offspring. Offspring nurse for ten weeks and stay with their mothers for a year. Wolverines do not stalk or chase. They pounce from trees or rocks and kill animals much larger than themselves, such as reindeer. Wolverines also eat smaller animals, birds, and carrion. Wolverine predators are bears, pumas, and humans. When attackers get near wolverines, they spray smelly musk. If this warning does not work, the wolverines fight attackers fiercely.

Mustelid Preservation

Several animal protection groups, such as the World Wildlife Fund and Friends of Animals, have long feared that many animal species will soon be extinct and that making and wearing fur garments is cruel. Thanks to their actions, the U.S. Congress passed an Endangered Species Conservation Act (1973) and its convention (1977). Therein, the United States and eighty other nations designed ways to control and monitor the import and export of fur of imperiled species. Endangered species were defined as in danger of extinction, while threatened species are likely to be endangered soon. Among the covered animals are otters and badgers. Other mustelids may be added soon. Under the act and convention, participant countries must stop fur movement in intercountry or interstate commerce unless they have proof that species involved are not threatened or endangered.

—*Sanford S. Singer*

See also: Endangered species; Fur and hair; Otters; Skunks; Urban and suburban wildlife.

Bibliography

Gilbert, Bill. *The Weasels: A Sensible Look at a Family of Predators.* New York: Pantheon Books, 1970. This text contains valuable information on mustelidae and weasels. It also has excellent illustrations.

Hirschi, Ron, and Galen Burell. *One Day on Pika's Peak.* New York: Dodd, Mead, 1986. A cute presentation of a day in the life of a family of weasels, showing their interactions with other animals.

King, Carolyn M. *The Natural History of Weasels and Stoats*. Ithaca, N.Y.: Comstock, 1989. An interesting book holding a lot of information on weasels and other mustelids. It contains an excellent bibliography and illustrations.

MacQuitty, Miranda. *Discovering Weasels*. New York: Bookwright Press, 1989. A brief introduction to weasel appearance, habits, and habitats, with nice illustrations.

Patent, Dorothy H. *Weasels, Otters, Skunks and Their Family*. Ithaca, N.Y.: Comstock, 1990. An interesting book full of weasel and Mustelid facts, with a fine bibliography and illustrations.

WHALE SHARKS

Type of animal science: Classification
Fields of study: Anatomy, behavior, ethology, marine biology, physiology, zoology

The whale shark, Rhincodon typus, *is the largest fish species in the world. Like all sharks and rays, its skeleton is composed of cartilage instead of bone. Unlike most sharks, however, the whale shark is a filter feeder, subsisting on plankton that it strains from the water column. Whale sharks are not commonly encountered, and much remains to be learned concerning their biology.*

Principal Terms

CARTILAGE: a tough and flexible tissue that constitutes the skeleton of all sharks and rays

CAUDAL FIN: the tail fin of fishes, which supplies the forward thrust in locomotion

CLASPER: a modification of the pelvic fins in male sharks and rays which acts as the male sexual organ in internal fertilization of the female

ELASMOBRANCHS: the suborder of the sharks and rays

GILL RAKERS: sievelike strainers found upon the gill arches of whale sharks, used to filter food from water taken in though the mouth

HETEROCERCAL: a type of caudal fin in which the spinal column extends into the upper lobe, producing an asymmetrical and distinctly sharklike tail

The whale shark, *Rhincodon typus*, is the largest extant species of fish in the world and may attain lengths in excess of twelve meters (forty feet). Despite its very large size, the whale shark is a generally placid and inoffensive creature that feeds upon plankton, swimming crustaceans, and small species of schooling fish in the water. The whale shark is a filter feeder, and it cruises through the water column with its mouth open. Any food in the water is caught by the large and extensive gill rakers, which are large, sievelike structures located upon the gill arches. The water then passes to the exterior of the animal through the gill arches, and the trapped food is swallowed. The whale shark occasionally exhibits more specialized feeding behavior, hanging almost vertically in the water and opening its mouth. The powerful suction caused by the opening of the mouth draws in water and any food in the vicinity.

Whale sharks are believed to be highly migratory, and are encountered either singly or in groups of up to several hundred individuals. Although comparatively little is known about the lives of these animals, they have been observed to congregate at various locations, such as Ningaloo Reef in Western Australia, at predictable seasonal times. These aggregations are believed to be associated with particularly favorable feeding conditions, such as the spawning cycles of corals. When they are not congregating at these sites, whale sharks are known to make long voyages. Satellite tracking of radio tags attached to whale sharks for periods as long as thirteen months have demonstrated that these fish may range far and wide, crossing the oceans.

Reproductive and Life Cycles

Although a few young whale sharks have been maintained in captivity for short intervals, much remains to be learned about their biology, including their reproductive cycles. As in all sharks and rays, fertilization is internal. The males possess pelvic fins that are modified to form claspers,

Whale Shark Facts

Classification:
Kingdom: Animalia
Subkingdom: Bilateria
Phylum: Chordata
Subphylum: Vertebrata
Superclass: Gnathostomata
Class: Chondrichthyes
Subclass: Elasmobranchii (sharks, skates, and rays)
Order: Orectolobiformes (carpet sharks)
Family: Rhincodontidae
Genus and species: Rhincodon typus
Geographical location: Worldwide, generally in warm temperate and tropical seas except for the Mediterranean, where it is conspicuously absent
Habitat: Coastal and oceanic
Gestational period: Unknown
Life span: Unknown
Special anatomy: The largest shark species, reaching total lengths of at least twelve meters (forty feet); broadly blunt and flattened snout with a very large mouth located almost at the very tip of the animal and containing many minute teeth; a small whisker, or barbel, located on the nostrils above either side of the mouth; eyes located just behind the mouth, comparatively very small; prominent longitudinal ridges extending down the length of the body trunk; large heterocercal tail that is asymmetric (as in all sharks), with the upper lobe larger than the lower lobe and a prominent lateral keel located where the tail joins the body trunk; body coloration characterized by a checkerboard pattern of light stripes and spots against a dark grayish or brownish background on the upper surfaces and pale white below

which are long, cylindrical structures. These are inserted into the female reproductive tract to release sperm, which fertilize the eggs of the female.

Whale sharks were once believed to be an oviparous (egg-laying) species, but more recent evidence has demonstrated that they are ovoviviparous. In this style of reproduction, the fertilized eggs are maintained within the female's uterus both prior to and after hatching. At about the time that the yolk of the embryonic egg has been completely absorbed, the embryos are released into the water to fend for themselves. A single pregnant female was found to have three hundred embryos contained within the uterus, ranging in size from forty-two to sixty-four centimeters (sixteen to twenty-six inches) in length. The length of the gestation period is unknown.

The life span of whale sharks is not known, and because of its immense size it is believed to have few natural enemies upon achieving adulthood. A recent report, however, has described a fatal attack on a whale shark by killer whales (orcas) in the Gulf of California, so these latter must be considered as whale shark predators. Humans may pose a threat to these gentle giants as well; small harpoon fisheries that fish for whale sharks exist in some Pacific cultures, including the Philippines.

—John G. New

See also: Fins and flippers; Fish; Marine animals; Migration; Sharks and rays; Reproduction; White sharks.

Bibliography

Last, P. R., and J. D. Stevens. *Sharks and Rays of Australia*. East Melbourne, Victoria, Australia: CSIRO, 1994. A comprehensive and beautifully illustrated book, limited to sharks and rays native to Australia, including the whale shark.

Parker, Steve, and Jane Parker. *The Encyclopedia of Sharks*. Willowdale, Ontario: Firefly Books, 1999. An introduction to shark biology, covering evolution, senses, internal organs, behavior, foods and feeding, and reproduction. Numerous photographs, drawing, maps, and diagrams.

The whale shark is the largest fish species in the world, yet it feeds on tiny plankton, crustaceans, and small fish. (Digital Stock)

Springer, V. G., and J. P. Gold. *Sharks in Question: The Smithsonian Answer Book.* Washington, D.C.: Smithsonian Institution Press, 1989. A good source of information concerning shark biology for the general public.

Tricas, T. C., K. Deacon, P. Last, J. E. McCosker, T. I. Walker, and L. Taylor. *Sharks and Rays.* Alexandria, Va.: Time-Life Books, 1997. A well-illustrated and useful reference book and introduction to the biology and behavior of sharks and rays.

WHALES, BALEEN

Type of animal science: Classification
Fields of study: Anatomy, marine biology, physiology, zoology

These huge marine mammals, in the order Cetacea, live in earth's oceans. All of them use baleen to capture a diet of small sea creatures, and they are the endangered prey of whalers.

Principal Terms

BALEEN: whalebone mouth plates that capture and strain small food animals eaten by baleen whales

ECHOLOCATION: batlike determination from sound echoes of positions of unseen objects; it is used by baleen whales

KRILL: tiny, shrimplike crustaceans eaten by baleen whales

UNGULATE: a hoofed mammal, such as a horse; includes animals deemed ancestors of whales

Baleen whales belong to the order Cetacea, the only mammals which spend their entire lives in the water of the earth's oceans. There are seventy-six cetacean species, and the baleen whales make up ten of them. Among baleen whales are blue (sulfur-bottomed) whales, the largest animals that have ever lived. They grow to lengths of over one hundred feet and can weigh 400,000 pounds (200 tons). This makes them more than fifty times the size and weight of a bull elephant and much larger than any dinosaur.

Like all whales, baleen whales are thought to be descendants of a land animal, believed to have been an early ungulate (hoofed mammal). The huge size of the baleen whale is possible because it lives in the water. This supports its mass and frees it from the limitations of land animals, which can only grow to the sizes and weights their legs will support or their wings can carry into the skies.

Why the ancestors of baleen whales entered the oceans is not understood. It is guessed that the return to the oceans was due to the need for a new food supply or to escape from predators. Most paleontologists believe that it happened sixty million years ago, twenty million years before the first whale fossils occur. No fossil that links baleen whales to their landbound ancestors has yet been found, although the search goes on.

Physical Characteristics of Baleen Whales

The most characteristic physical feature of baleen whales is that, rather than teeth, they have hundreds of baleen plates hung vertically from their upper jaws. The plates, with bristles on their inner edges, capture the krill the whales eat. A feeding baleen whale swims with its huge mouth open and engulfs several tons of seawater containing krill. Then the whale shuts its mouth, it presses its tongue against the back of the baleen bristles, and forces the water out of its mouth. This traps the krill on the baleen plates.

In all whales, evolution yielded streamlined, fishlike mammals. Furthermore, their front legs became paddle-shaped flippers whose bones resemble jointed limbs and digits. In contrast, the external hind limbs disappeared, although in many cases their vestiges are found internalized. The horizontal tail flukes that propel whales are not anatomically related to hind limbs. They are boneless and shaped by fibrous and elastic tissues. Nor are flippers related to fish tail fins, which differ in composition and are orientated vertically.

The whale body is surrounded by a thick blubber (fat) layer. This greatly enhances whale buoyancy, insulates, and is an excellent energy storehouse. The insulating ability of blubber may explain why whales living in warm waters have much thinner blubber layers than those living in cold waters. In addition, a whale's soft rubbery skin lacks pores, sweat glands, and sebaceous glands. It is hairless, except for small patches on areas near the chin and atop the head. These hair patches support the theory that baleen whales evolved from animals having hair or fur.

Whales, like all mammals, have lungs and breathe air through one or two nostrils, depending on species, located in a blowhole. Just prior to dives whales close their nostrils tightly. If kept submerged indefinitely, whales would drown. Whale nostrils are located on top of the head for functional convenience in diving. They connect directly to the lungs. Contrary to popular belief, a whale's visible "spout" is not water. It is exhaled, warmed water vapor and spent air from the lungs, plus any water present in the depression around the blowhole.

Four physiological adaptations allow whales to dive deeply and for prolonged time periods. First, they have more blood than land mammals and a huge capacity to store oxygen in blood and muscle. Second, each whale breath replaces 85 percent of the air in its lungs, compared with 15 percent in land mammals. Third, whales can resist the carbon dioxide buildup that triggers involuntary breathing in land mammals. Consequently, most baleen whales can hold their breath for an hour or more under water. Finally, they are able to restrict blood flow to various organs during deep dives, limiting oxygen flow to inessential sites. This protects the whale heart and brain from oxygen deprivation in long dives.

Baleen Whale Facts

Classification:
Kingdom: Animalia
Subkingdom: Bilateria
Phylum: Chordata
Subphylum: Vertebrata
Class: Mammalia
Order: Cetacea (whales)
Suborder: Mysticeti (baleen whales)
Families: Eschrichlidae (gray whale), Balaenopteridae (rorquals, two genera, six species), Balaenidae (right whales, two genera, three species)

Geographical location: Every ocean on Earth

Habitat: Oceans

Gestational period: Varies between ten and sixteen months, depending on species

Life span: Most estimates suggest eighty to one hundred years for animals that survive into old age

Special anatomy: Baleen in the mouth; lungs like land mammals; blow hole; flippers evolved from front legs; horizontal tail; nasal cavity that produces echolocation and vocalization

Special Senses and Baleen Whale Intelligence

Baleen whales have small eyes, lack external ears, and have brains much larger than those of humans. These physical characteristics have led to the belief that the whales use sound and hearing in the way vision and smell are used by land mammals. Two kinds of sounds are made by baleen whales: echolocation and vocalizations. Both are thought to arise from air moving between nostrils and the nasal passages that lead to the lungs

Echolocation sounds are thought to function like biosonar, as in terrestrial bats. That is, whale echolocation is perceived as the means by which they explore the world around them. On the other hand, vocalizations, such as the "songs of humpback whales," are perceived as the language by which members of the same whale species communicate. It is thought that the whale identifies the size, distance, and other characteristics of an object by the directing sounds made in a whale's head toward it and receiving sound waves that bounce back off it. Evidence for the role of echolocation is taken from observations made on whales in captivity. In addition, some beached whales have had parasites in their inner ears that lead scientists to believe that the whales ran aground due to losing the ability to echolocate shorelines and stay away from them.

The great ability of water to carry and amplify sound waves is deemed to be the reason why cetaceans have been able to discard the external ear that land mammals developed to gather airborne sounds. Operation of this system of sensing would have obvious use in navigating and in the capture of prey in dark, often murky ocean water. There it would provide the means to scan by sound for the information almost all land mammals get by seeing.

The unusual sensory capabilities of whales have given rise to considerable speculation as to their intelligence. This is stimulated by the observation that cetaceans are the only animals except elephants (with ten-pound brains) to have brains larger those of humans (three pounds). For example, an adult sperm whale has a twenty-pound brain. However, the relationship of brain size to intelligence, in this size range, is not clear.

Supporting the contention of brain size paralleling intelligence, whales and dolphins show considerable capacity for learning when studied in captivity. Their great playfulness, intraspecies communalism, and the affectionate care of offspring make many people strong advocates of cetacean "language" and great intelligence. These concepts remain unproven.

The Life Cycle of Baleen Whales

Baleen whales reproduce in a fashion similar to that in other mammals. Adults enter a period of courtship. This includes side-by-side swimming, body nuzzling, and body rubbing. Copulation soon follows. The pregnant female carries her unborn young for ten to sixteen months, depending on whale species. Next, a large, very well-developed calf (or occasionally, two) is born underwater. Healthy whale calves can swim well at birth and immediately find the ocean surface, with no help from the mother.

Baleen whales navigate by echolocation, but it is believed that the vocalizations of humpbacked whales are also a kind of language. (Digital Stock)

The calves nurse from two teats situated in a pocket located on either side of the mother whale's genital opening. Whale milk is white and very rich in minerals, protein, and fat; as a result, the calves grow quite fast. Usually a whale calf that weighs one to two tons at birth doubles its weight in the first week. It is reported that nursing occurs very rapidly. A calf approaches the mother whale, touches the nipple of the mammary gland, and milk immediately and copiously squirts into the calf's mouth. Mother whales are reportedly quite affectionate to and protective of their offspring.

Young whales are weaned between one and two years after birth, the age at which most of them leave their mothers. A whale is an adult, capable of reproduction, by six to ten years of age. The life spans of whales vary with maximum species size. The smaller species can live for thirty to forty years. The largest species can live for as long as eighty years. As with all other wild animals, not all whales—in fact, relatively few of them—survive to the ripe old ages noted.

Some Baleen Whale Species

BLUE WHALE (*Balaenoptera musculis*): up to 110 feet long; blue-gray with white spots on sides and belly; small dorsal fins and flippers eight feet long; baleen are short, coarse, and blue-black; lives in all oceans except the Arctic Ocean.

BOWHEAD WHALE (*Balaena mysticetus*): up to sixty-five feet long; black; arched, bowlike upper jaw; baleen are up to fourteen feet long; lives only in the Arctic Ocean.

CALIFORNIA GRAY WHALE (*Eschicthius robustus*): up to forty-five feet long, gray-black; flippers short and blunt; short, cream-colored baleen; lives in the North Pacific on both Asian and American sides.

FINBACK WHALE (*Balaenoptera physalus*): the most plentiful whale; up to ninety feet long; very streamlined body; dark gray-brown on back, white on belly; black dorsal fin; short flippers and baleen which are gray with black stripes; lives in all oceans except the Arctic Ocean.

HUMPBACK WHALE (*Megaptera novaengliae*): up to sixty-five feet long; black, marked with white splashes; many tubercles with hairlike bristles on jaws and atop head; flippers fifteen feet long; short, black baleen; lives in all oceans.

RIGHT WHALE (*Balaena glacialis*): up to fifty-five feet long; black; seven-foot-long black baleen; lives in the North and South Atlantic Oceans.

SEI WHALE (*Balaenoptera borealis*): up to forty-five feet long; bluish-black back, gray sides, white belly; long, very fine baleen; lives in all tropical oceans.

Baleen Whales as an Endangered Species

Past, uncontrolled whaling reduced the numbers of almost all baleen whale species so much that they are perceived as being endangered. At first, quotas on whales were set by the International Whaling Commission (IWC) for the purpose of managing whales so as to allow the continuation of the whaling industry. By the early 1990's, most major whaling nations belonging to the IWC had stopped whaling. It is hoped that these actions, in the absence of clandestine whaling, will allow baleen whales to make a comeback by natural increase. Only the future will tell whether this will happen.

—*Sanford S. Singer*

See also: Convergent and divergent evolution; Dolphins, porpoises, and toothed whales; Endangered species; Fins and flippers; Ingestion; Lactation; Mammals; Marine animals; Migration; Respiratory system; Ungulates.

Bibliography

Clapham, Phil. *Whales of the World*. Stillwater, Minn.: Voyageur Press, 1997. A brief, interesting book, with good illustrations and bibliographic references.

Corrigan, Patricia. *Where the Whales Are: Your Guide to Whale Watching Trips in North America*. Chester, Conn.: Globe Pequot Press, 1991. An interesting book, with maps, which helps to identify where one might go to see whales firsthand.

Cousteau, Jacques-Yves, and Yves Paccalet. *Jacques Cousteau: Whales*. Translated by I. Mark Perkins. New York: Abrams, 1988. A nice translation of Cousteau's book. It is handy and contains both illustrations and bibliographic references.

Darling, James. *Whales, Dolphins, and Porpoises*. Washington, D.C: National Geographic Society Press, 1995. A nice book, with interesting illustrations, and a good bibliography.

Evans, Peter G. H. *The Natural History of Whales and Dolphins*. New York: Facts on File, 1987. A solid book on whales, dolphins, and other cetaceans. Includes illustrations, and a solid bibliography.

WHITE SHARKS

Type of animal science: Classification
Fields of study: Anatomy, behavior, ethology, marine biology, physiology, zoology

The white shark, Carcharodon carcharias, *is the largest predatory fish and has become well known through its media-enhanced reputation as one of the most dangerous of sharks. This large and active shark is found worldwide, principally in temperate and subtropical coastal waters. Despite its fearsome reputation, much remains to be learned about the biology of these remarkable and formidable fish.*

Principal Terms

AMPHISTYLY: the mechanism by which the jaws of sharks are suspended from, but not directly fused to, the skull

CARTILAGE: a tough and flexible tissue that constitutes the skeleton of all sharks and rays

CAUDAL FIN: the tail fin of fishes, which supplies the forward thrust in locomotion

CLASPER: a modification of the pelvic fins in male sharks and rays which acts as the male sexual organ in internal fertilization of the female

ELASMOBRANCH: the classificatory subbranch of sharks and rays

HETEROCERCAL: a type of caudal fin in which the spinal column extends into the upper lobe, producing an asymmetrical and distinctly sharklike tail

OVIPHAGY: a mode of reproduction in which embryo sharks develop in the maternal uterus and feed on unfertilized eggs produced by the mother

The white shark (*Carcharodon carcharias*), also known as the great white shark, white pointer, or white death, is one of the best-known sharks principally due to interest surrounding the occasionally fatal interactions between humans and this species. Nevertheless, surprisingly little is known about their biology. White sharks are large, active fish that are only infrequently encountered by humans and which have not survived long in the few attempts that have been made to keep them in captivity. Attempts to study the biology of white sharks are therefore limited to brief observations of their biology and what can be inferred from the anatomy of captured dead sharks.

White Shark Anatomy and Physiology

Like all sharks, the white shark possesses a skeleton composed of cartilage, a tough elastic connective tissue found in all vertebrates. The cartilage in the skeleton of white sharks is strengthened by deposits of calcium carbonate, but there is no true bone as in most other fishes. Like most sharks, the white shark has five pairs of gill slits located just in front of the broad pectoral fin and a heterocercal tail, in which the spinal column extends into the upper lobe. The tail, or caudal fin, of the white shark, as in all members of the family Lamnidae, is almost symmetrical and possesses a pronounced lateral keel at the point where the tail is attached to the body trunk. Such a lateral keel may improve the efficiency of the tail in swimming and turning and is a common element of many large, fast-swimming fishes. The upper jaw of white sharks is suspended from the skull by ligaments rather than fused directly to the skull, as in most terrestrial vertebrates. This amphistylic suspension of the jaws allows them to be pushed forward and out, increasing the efficiency of the biting mechanism in a mouth that is located below and behind the broadly conical snout. White sharks, like all sharks, lack a gas-filled swim bladder to provide buoyancy in the water and rely on the lifting action of their fins and body shape to prevent them from sinking. The streamlined, spindle-shaped body, broad fins, and powerful, symmetrical tail of the white shark suit it well to a fast and powerful swimming style, and white sharks, like others in the family Lamnidae, are among the most active sharks.

White sharks possess special modifications of the circulatory system in the body wall and central nervous system that are known as rete mirabile. These modifications permit the shark to return the heat generated by metabolism to the body core, rather than losing it to the surrounding water as in most fish. The rete mirabile allows white sharks to maintain a core body temperature that is higher than the surrounding water; this may provide an advantage to the shark by raising its metabolic rate and permitting greater activity in the cooler waters which they typically inhabit.

White Shark Reproduction and Feeding Habits

Little is known about the reproductive biology of white sharks. Like all sharks and rays, fertilization is internal, and the males possess special modifications of the pelvic fins called claspers, which are inserted into the female oviduct. White sharks are believed to be viviparous ("born alive"), with juvenile sharks developing within the uterus of the mother until they are ready to be born. During early embryonic development the developing white sharks are nourished by the yolk sac of the egg. When this is consumed, the developing shark consumes unfertilized eggs produced by the mother, a reproductive strategy known as intrauterine oöphagy. There have been relatively few captures of pregnant females or free-swimming juveniles, and the duration of gestation and the size of the juveniles at birth are unknown. The best estimate of white shark size at birth is between 1.2 and 1.4 meters.

White sharks are known to reach a length of six meters (twenty-one feet) and may possibly grow larger. Their life span is unknown. As the white shark grows, the nature of its diet changes; juvenile white sharks appear to feed principally upon fish, whereas marine mammals play an important role in the diet of mature sharks. The high energy yield of the blubber in seals and sea lions may make them attractive prey for these very large predators. White sharks hunt seals and sea lions by swimming deep below them and searching for their silhouettes against the bright surface. Once prey is spotted, the shark attacks with a very swift lunge from below and inflicts an incapacitating or mortal wound with its bite.

White Shark Facts

Classification:
Kingdom: Animalia
Subkingdom: Bilateria
Phylum: Chordata
Subphylum: Vertebrata
Superclass: Gnathostomata
Class: Chondrichthyes
Subclass: Elasmobranchii (sharks, skates, and rays)
Order: Lamniformes
Family: Lamnidae (mackerel sharks, also porbeagle salmon and mako sharks)
Genus and species: Carcharodon carcharias

Geographical location: Worldwide, generally in temperate and subtropical seas, occasionally in tropical seas

Habitat: Continental shelf and occasionally inshore

Gestational period: Unknown

Life span: Unknown

Special anatomy: Large (up to six meters long) and moderately stout-bodied, with a blunt, conical snout and an almost symmetrical caudal fin; pronounced lateral ridge or keel where the tail is attached to the body; large teeth, broadly triangular and serrated; gray-blue or gray-brown (often bronze) on the upper body surfaces, with a white belly; a black spot often present at the axil (armpit) on the bottom surface of the pectoral fin

White sharks have been responsible for fatal attacks on humans, but humans are not a normal part of their diet. It is more likely that the shark mistakes a human for a marine mammal or may merely be curious. White sharks themselves are in considerably more danger from humans, and several nations and states have passed laws protecting these sharks from the effects of human depredations.

—*John G. New*

See also: Fins and flippers; Fish; Marine animals; Migration; Sharks and rays; Reproduction; Whale sharks.

Bibliography

Ellis, R., J. E. McCosker, and A. Giddings. *Great White Shark*. New York: HarperCollins, 1991. An engrossing and beautifully illustrated book about white sharks

Klimley, A. P., and D. G. Ainley, eds. *Great White Sharks: The Biology of "Carcharodon carcharias."* San Diego, Calif.: Academic Press, 1996. A valuable and interesting reference to current research on white sharks. Intended principally for scientists but accessible to the lay reader.

Springer, V. G., and J. P. Gold. *Sharks in Question: The Smithsonian Answer Book*. Washington, D.C.: Smithsonian Institution Press, 1989. A good source of information concerning shark biology for the general public.

Tricas, T. C., K. Deacon, P. Last, J. E. McCosker, T. I. Walker, and L. Taylor. *Sharks and Rays*. Alexandria, Va.: Time-Life Books, 1997. A well-illustrated and useful reference book and introduction to the biology and behavior of sharks and rays.

White sharks have jaws that are specially adapted to increase their biting power. (Digital Stock)

WILDLIFE MANAGEMENT

Type of animal science: Ecology
Fields of study: Ecology, environmental science, ethology, wildlife ecology

Wildlife management strives to allow the use of ecological communities for human benefit while preserving their ecological components unharmed. It also seeks to restore biological communities by managing habitats and controlling the taking of organisms for sport or economic gain.

Principal Terms

CARRYING CAPACITY: the number of individual animals that a habitat can support

COMMUNITY: all the living organisms existing in an area at a particular time

FURBEARERS: mammals that are harvested for their fur, such as muskrat, mink, and beaver

GAME: economically important animals, usually birds or mammals; it includes those taken for recreation or products and those that damage human property

HABITAT: a specific type of environment or physical place where an animal lives; it usually emphasizes the vegetation of an area

HOME RANGE: the physical area that an animal uses in its daily activities to get all its needs, such as food and water

SUCCESSION: change in a plant or animal community over time, with one kind of organism or plant being replaced by others in a more or less predictable pattern

SUSTAINED ANNUAL YIELD: the harvest of no more animals than are produced, so that the total population remains the same

WILDLIFE: traditionally, the term included only mammals and birds that were hunted or considered economically important; today, it includes all living organisms

Wildlife management, also known as game management, is often compared with farming or forestry, because one of its goals is to ensure annual "crops" of wild animals. Aldo Leopold, in 1933, defined "game management" as the art of making land produce sustained annual crops of wild game for recreational use. At that time, animals considered to be game included deer and animals such as coyotes that do damage to domestic animals or crops. Now, however, the term "wildlife" has replaced game, and virtually all living organisms, including invertebrates and plants, are included in management considerations.

Approaches to Wildlife Management

The process of wildlife management has moved through a sequence of six approaches: the restriction of harvest (by law); predator control; the establishment of refuges, reserves, and parks; the artificial stocking of native species and introduction of exotic ones; environmental controls, or management of habitat; and education of the general public. All six are used in modern wildlife management programs, but most emphasis is placed on habitat management and control of harvest.

Primitive man practiced a form of wildlife management simply by setting fires. These fires stimulated the growth of new grasses that lured grazing animals to the areas near tribal camps. It was then easier to kill the animals for food. Tribal taboos often regulated the use of animals, but the first written wildlife law is probably contained in the biblical Mosaic law (in Deuteronomy 22:6). The Egyptians hunted for sport, and the Romans

Preserves

Putting aside land for the benefit of its animals and plants was a novel concept in the early 1800's. Prior to this period, there was too much undeveloped wilderness, and no one gave much thought to saving it in its natural state. The attitude was to develop the wilderness so that it could be put to productive use. In addition, the wilderness was a menacing place where strange and unknown forces lurked. However, as urban and rural areas overtook the wilderness, attitudes began to change.

Yellowstone was the first national park, established in 1872. Establishment of national parks in Australia, Canada, New Zealand, and other countries soon followed. The term "national park" has had a variety of meanings. One version of the national park concept is the wildlife preserve (or reserve). These are parks having game conservation or management as their primary objective. The first national game preserves may be those established in South Africa during the 1890's. While many national game preserves are now found throughout Africa, regional, state, or local game preserves exist in many other countries.

In the United States, game preserves were established in the early 1900's for placing captive-bred bison back into their natural habitat after they had become extinct in the wild. The first of these preserves was the Wichita Forest Reserve in Oklahoma, established in 1905. State wildlife agencies have also set aside lands as game management areas. Some of these areas were initially for the use of hunters, but after the 1960's these areas began to serve nongame species and conservation as well. Others are still maintained for sport hunting and are managed in an effort to control game species populations. At the same time these areas are used for hunting some species, they also preserve habitat for all of the area's species. The hunting preserve has a long history going back to ancient societies, such as those in Mesopotamia, Egypt, and China, starting about 3000 B.C.E.

Although preserves are often set aside for specific wildlife species, they save habitat for all the species found there. Thus, wildlife preserves protect all forms of animals and plants, as do similar areas with different names (such as national park, national seashore, forest reserve, scenic river, wilderness area, and so on). Since these areas change over time, they must be managed. This management is based on ecological and wildlife management principles, but it is difficult to manage for the needs of all the species simultaneously. It is also important to link preserves with corridors so they are incorporated into a self-sustaining environmental system. Maintaining undeveloped areas as preserves is becoming increasingly important, not only for the health and well-being of the animals and plants, but for humans as well.

—*Vernon N. Kisling, Jr.*

and Greeks had game laws. The first comprehensive wildlife management program was in the Mongol Empire: Marco Polo reported in the thirteenth century that the Great Khan protected animals from hunters between March and October and provided food for animals during the winter.

During feudal times in Europe, wildlife belonged to the royal family; today it legally belongs to the landowner. In most other countries of the world, wildlife belongs to the state, province, or federal government. In the United States, wildlife belongs to the state, as originally granted legally by the Magna Carta, in 1215, to the people of England—the right was transferred to the state when independence was won by the colonies from England.

Only in the last century has the philosophy of wildlife management been not only to preserve but also to increase wildlife abundance. All fifty states of the United States have departments responsible for wildlife conservation. An appointed board of directors or commission oversees the actions of the departments. Groups for wildlife law enforcement, research, management, and information and education make recommendations to the board of directors regarding wildlife management actions. The federal government of the United States also has many agencies that manage

This cow elephant has been tranquilized to see whether a contraception program to limit population growth is succeeding. Many wildlife management programs attempt to maintain populations at the optimum size for their environment. (AP/Wide World Photos)

wildlife on public lands. The U.S. Fish and Wildlife Service is involved with animals that cross state lines, including migratory birds such as waterfowl, marine mammals, and any plants and animals listed as rare or endangered by the National Environmental Protection Act. Other agencies, such as the U.S. Forest Service, Bureau of Land Management, Soil Conservation Service, and the U.S. National Park Service, do extensive wildlife work. Many private organizations, such as the National Wildlife Federation, the Audubon Society, and the Sierra Club actively promote wildlife conservation.

Wildlife management decisions involve the entire range of biological, sociological, political, and economic considerations of human society. Today, the wildlife resource in the United States is managed primarily either for consumptive use (such as sport hunting) or for nonconsumptive use (such as bird watching). Virtually all wildlife management problems are related to the large human population of the earth. Some specific problems are habitat losses (for example, the destruction of tropical rain forests), pollution, diseases introduced by domestic animals into wildlife populations, and the illegal killing of animals for their parts, such as the poaching of elephants for their ivory.

The Process of Management

A wildlife manager must first determine the physical and biological conditions of the organism or organisms being managed. Issues include what the best habitat for the animal is and how many animals this habitat can support. The stage of ecological succession determines the presence or absence of particular animals in an area. All animals need food, water, and protection from weather and predators. Special needs, such as a hollow tree in which to raise young, for example, must be fulfilled within the animal's home range. Wildlife managers attempt to remove or provide items that are most limiting to a population of animals. In many respects, solving wildlife management problems is an art; it is similar to medicine in that it often must deal with symptoms (birds dying, for example) and imprecise information.

The stage of ecological succession may be maintained by plowing lands, spraying unwanted plants with a chemical to kill them, or using fire, under controlled conditions, to burn an area to improve the habitat for a certain wildlife group. Refuges and preserves may be set aside to assure that some of the needed habitat is available; nest boxes and water supplies may even be provided.

Periodic surveys of the number of animals in a population provide guidelines for their protection. If animals are more abundant than the lowest carrying capacity, a controlled harvest may be allowed. Sustained annual yield assures that no more than the population surplus is taken. Wildlife laws protect the animals, provide for public safety, often set ethical guides for sporting harvest, and attempt to provide all hunters with an equitable chance of obtaining certain animals (for example, by setting bag limits). If proper wildlife management procedures are followed, no animal need become rare or endangered by sport hunting. Market hunting, the taking of animals for the sale of their products, such as meat or hides, has been stopped in the United States since the 1920's and is also illegal in most other areas of the world. There are almost no societies left that are true subsistence hunters—that is, living exclusively on the materials produced by the wildlife resource.

The Need for Wildlife Management

The proper management of wildlife resources, based on sound ecological principles, is essential to the well-being of humans. All domestic plants and animals came from wild stock, and this genetic reservoir must be maintained. Maintaining the web of life that includes these organisms is necessary for man's survival. Wildlife resources are used by at least 60 percent of the citizens of the United States each year, and about 6 percent are sport hunters. Wildlife provides considerable commercial value from products, such as meat; it also offers aesthetic values of immeasurable worth. Seeking and observing wildlife provides needed relief from the everyday tensions of human life. Moreover, by observing wildlife reactions to environmental quality, investigators can monitor the status of the biological system within which man lives. Wildlife populations serve as a crucial index of environmental quality.

Wildlife management is a dynamic force that, to be effective, must reflect an understanding of and respect for the natural world. It cannot be practiced in a vacuum but must encompass the realm of complex human interactions that often have conflicting goals and values. Aldo Leopold once defined conservation as man living in harmony with the land; successful wildlife management will help assure that this occurs.

—David L. Chesemore

See also: Biodiversity; Biogeography; Cloning of extinct or endangered species; Competition; Demographics; Ecological niches; Ecology; Ecosystems; Endangered species; Extinction; Fauna: Africa; Fauna: Antarctica; Fauna: Arctic; Fauna: Asia; Fauna: Australia; Fauna: Caribbean; Fauna: Central America; Fauna: Europe; Fauna: Galápagos Islands; Fauna: Madagascar; Fauna: North America; Fauna: Pacific Islands; Fauna: South America; Habitats and biomes; Mark, release, and recapture methods; Population analysis; Population fluctuations; Population genetics; Predation; Zoology; Zoos.

Bibliography

Anderson, S. H. *Managing Our Wildlife Resources*. 3d ed. Upper Saddle River, N.J.: Prentice Hall, 1999. A very good introductory text on wildlife management. Provides an excellent overview of the field and discusses management strategies.

Bailey, James A. *Principles of Wildlife Management*. New York: John Wiley & Sons, 1984. An introductory text of wildlife principles. A complex but readable treatment of wildlife work. Has excellent examples of wildlife problems as they relate to ecological theory.

Bissonette, John A., ed. *Wildlife and Landscape Ecology: Effects of Pattern and Scale*. New York: Springer-Verlag, 1997. Looks at the effects of landscape-level patterns and demographic processes on wildlife management planning.

Cooperrider, Allen Y., R. J. Boyd, H. R. Stuart, and Shirley L. McCulloch. *Inventory and Monitoring of Wildlife Habitat*. Washington, D.C.: U.S. Government Printing Office, 1986. Excellent discussion of how wildlife management is done. Details how specific habitats, such as rangelands, are managed. Also includes chapters on the management of animal groups.

Dasmann, R. F. *Wildlife Biology*. New York: John Wiley & Sons, 1981. Provides a good overview of wildlife management principles and basic ecological ideas that govern what wildlife managers do.

Di Silvestro, Roger, ed. *Audubon Wildlife Report, 1986*. New York: National Audubon Society, 1986. An excellent summary, in five volumes, on worldwide wildlife conservation. Provides in-depth information on major animals, groups, and problems of wildlife conservation. An important reference for anyone interested in wildlife conservation.

Giles, R. H., Jr. *Wildlife Management*. San Francisco: W. H. Freeman, 1978. A more advanced treatment of wildlife principles and ecological areas, including population ecology. It has specific chapters on muskrat, waterfowl, forest, and fisheries management.

Leopold, Aldo. *Game Management*. New York: Charles Scribner's Sons, 1939. Reprint. Madison: University of Wisconsin Press, 1986. This is the historical book that officially launched wildlife management as a discipline. The natural history information in it is somewhat out of date, but the philosophy of the book still provides the base for wildlife management today.

Matthiessen, Peter. *Wildlife in America*. New York: Viking Press, 1987. A review of wildlife management successes and failures in North America. Provides a good historical summary and discusses major wildlife management problems.

Robinson, W. L., and E. G. Bolen. *Wildlife Ecology and Management*. 4th ed. Upper Saddle River, N.J.: Prentice Hall, 1999. One of the best introductory wildlife management books available. Extensive bibliography and glossary.

WINGS

Types of animal science: Anatomy, evolution
Fields of study: Anatomy, biophysics, developmental biology, evolutionary science, ornithology, paleontology, systematics (taxonomy), zoology

Insects, reptiles, dinosaurs, birds, and mammals have developed different types of wings to take them into the air. The anatomy of each of these wings is quite different. Although pterosaurs, dinosaurs, birds, and bats used their forelimbs to support their wings, their forelimbs evolved very differently.

Principal Terms

CAMBER: the degree to which wings are convex on their top and concave on their underside

LIFT: the upward force that is developed by moving wings, which opposes the pull of gravity

SLOTTING: the separation of primary feathers at the tip of the wings

THRUST: the forward force that is developed by engines, rotors, or moving wings, which pushes planes, helicopters, or flying animals and which opposes drag

It is believed that insect wings evolved from bilateral, dorsal flaps called gill plates on thoracic and abdominal segments. These gill plates are seen on some fossilized early insects. During the evolution of gill plates into winglike appendages, the appendages grew larger and became restricted to the second and third thoracic segments. Some early insects used these winglike appendages to sail on the surface of ponds or glide in gentle updrafts and air currents.

Fly Wings

In the fruit fly *Drosophila*, each wing develops from a packet of epithelial cells called the imaginal disk. The imaginal disk grows into a wide, baglike structure that flattens out to become a transparent wing only a few cells thick. Hemolymph-filled veins along the anterior wing margin and within the wing support the fragile wing and supply the wing cells with nutrients and water.

Drosophila powers its single pair of wings quite differently than the more primitive dragonflies and grasshoppers that have two pairs of wings. Flies have internal dorsoventral (vertical) muscles on each side of the body that pull the top of the thorax down when they contract. This moves each of the wings up because the wing bases are inserted into the top of the thorax. Contracting longitudinal muscles (running along the length of the thorax) force the top of the thorax back to its original position. This causes each of the wings to move down. Wing movement in most advanced flying insects is associated with the extremely rapid deformation of the thorax. Houseflies can flap their wings up to two hundred times per second, whereas gnats can flap up to one thousand times per second.

The wings of true flies not only flap up and down but they also alter their angle of attack while moving forward during the down stroke and moving backward during the upstroke. These complex movements provide both lift and thrust and require a number of highly evolved muscle groups.

Lift

For many insects, simple wing flapping cannot generate sufficient lift for flight or for hovering. This is especially true of tiny insects or for insects that have small wing size to body-weight ratios, such as bumblebees. Some flying insects are less than one millimeter long and have extremely small wings that have difficulty moving through air. The viscosity of the air prevents the development of sufficiently small leading-edge vortices. So how do insects produce sufficient lift for flight and hovering?

Beginning with wings above the body, the downstroke moves the wings down and forward through the air at a high attack angle that produces lift. Leading and lagging edge vortices that lower the air pressure above the wing cause lift. As the wing moves down and forward, it continues to rotate slightly, creating larger vortices above the wing that increase the lift. The leading edge vortex begins near the base of the wing and develops quickly toward the tip. Air entering the vortex moves from the base of the wing to the tip. The upstroke provides insects with additional lift. In the upstroke, the wing rotates so that the leading edge is still ahead of the trailing edge. The wing moves up and backward through the air. During the upstroke, lift is created by air reentering the wake created by the down stroke. This moving air pushes against the wing and provides lift. Lift is the sum of "delayed stall" during the downstroke and "wake capture" on the upstroke. In most insects, wing curvature also plays a small role in the development of lift. Most insect wings do not remain flat or rigid during flight or hovering.

Unlike airplanes, which use wings only for lift and have propellers or jet engines for thrust, insects use their wings to generate both lift and thrust. Although lift is generated during both the downstroke and upstroke, thrust is generated only on the upstroke. In fact, during the downstroke there is a force on the insect that slows it or causes it to hover.

Flying Fish

Some fish have greatly elongated their pectoral fins into winglike appendages that can be extended and used for gliding through the air. In water, the pectoral fins are kept folded against the body. Powerful tails propel these fish up to ten miles an hour in the water. After swimming out of the water, they extend their pectoral fins and glide for forty to two hundred yards, a flight that lasts two to fifteen seconds. They sometime reach air speeds of thirty miles per hour if they encounter a tail wind. Some flying fish of the southern Atlantic Ocean have large, oval wings supported by an array of long, bonelike rods that radiate from the base of the pectoral fin. Their pectoral fins are nearly half the length of the fish. Some flying fish have not only elongated their pectoral fins but also their pelvic fins into wings. The pelvic fins are about a quarter of the length of the fish. It is claimed that some flying fish of the Amazon basin actually fly short distances by buzzing their pectoral fins.

Pterosaur Wings

The first pterosaurs appeared in the fossil record about 215 million years ago. They were a very successful group of reptiles, surviving until the mass extinction that marks the Cretaceous-Tertiary boundary 65 million years ago. These flying animals evolved from a population of tetrapod, carnivorous reptiles. Their wings consisted of leathery skin supported by forelimbs, body trunk, and hind legs. The leathery skin of the wings was reinforced by stiff fibers that ran from the front of the wing to its trailing margin.

The forelimbs of these flying reptiles consisted of an upper arm bone (humerus), two forearm bones (radius and ulna), wrist bones (carpals), palm bones (metacarpals), bones of three claw-tipped fingers (phalanges), and the bones of a fourth, very long, clawless finger (phalanges). The first, second, and third finger bones are the thumb, the index finger, and the middle finger. The greatly elongated palm and fourth finger bones along the distal front edge of the wing was longer than the upper and lower arm combined and supported about three-quarters of the wing.

The wings of all pterosaurs were long and sickle-shaped, yet they could fold up like a fan along the sides of the body. Pterosaurs and their wings varied in size. The smallest pterosaurs were about the size of a starling, with a wing span of

The downstroke of wings from above the body provides the lift that allows birds to fly. (PhotoDisc)

about one foot, whereas the largest animal, belonging to the genus *Quetzalcoatlus*, was about twenty-six feet in length from bill to trailing feet, with a wing span of nearly forty feet. The elongated palm and fourth finger bones were over fifteen feet. *Quetzalcoatlus* was the largest flying animal ever, weighing up to two hundred pounds. The long, thin, swept-back, sickle-shaped wings suggest that the pterosaurs were rapid flyers and consummate gliders.

Bird Wings

Approximately 155 million years ago, birds began to evolve from a population of feathered, bipedal, meat-eating dinosaurs. Their wings consisted of feathers supported by the forelimbs. The forelimbs of these early, birdlike animals consisted of an upper arm bone (humerus), two forearm bones (radius and ulna), wrist bones (carpals), palm bones (metacarpals), and bones of the three claw-tipped fingers (phalanges).

In modern birds, the wing bones of the palm and fingers have changed significantly. The palm and first finger bones have been fused and greatly shortened so that a single, small thumb bone protrudes near the wrist. The feather or feathers originating in the skin at the end of the thumb is known as the alula. The palm bones associated with the second and third fingers have partially fused at their ends to become a bone that roughly resembles a fused miniature radius and ulna. The second finger that supports the feathers forming the tip of the wing has become shorter. The third finger has been reduced to a single, tiny bone. The long feathers protruding from the skin covering the palm and second finger are called primaries, whereas the long feathers attached to forearm are known as secondaries. Feathers attached to the upper arm are called tertiary feathers.

Wing flapping is powered by muscles that stretch from the base of the humerus to the large, keel-like sternum. Muscles attached to the top of the humerus base and to the sternum pull the wings up, whereas muscles attached to the bottom of the humerus base and the sternum pull the wings down.

Wing Shapes for Special Tasks

A bird's lifestyle and habitat have selected for wings that may be divided into a number of major categories. Birds that live in forested or densely wooded habitats or birds that prey on flying insects and other fast-moving animals have short, broad wings, sometimes described as elliptical wings. Elliptical wings allow birds to carry out rapid, intricate maneuvers. These wings have a high degree of camber and extensive slotting. Highly cambered wings provide greater lift than more flattened wings. The separation of primary feathers at the wing tip provides most of the needed thrust by acting as miniwings. In general, the primary feathers that constitute the wing tip generate most of the thrust and even some lift.

Gliding Reptiles

The oldest known flying reptile, *Coelurosauravus*, is from the Upper Permian. It had membranous wings supported by twenty-four to twenty-eight long, bonelike rods originating in the skin along the side of the animal. The rods were hollow, with thin walls, and probably developed from ossification of dermal (skin) tissue. Skin connected the rods.

The most anterior seven rods began as a bundle along the side of the animal. The first bundle of rods was followed by four more bundles containing five, four, three, and two rods respectively. The most posterior six or seven rods were not bundled. The rods were not extensions of the ribs, since these lizards have only thirteen thoracic vertebrae. The rods and wings folded against the body when not in use; however, the rods became radially arranged when the wing was extended. The front edge of the wing formed a fifty to sixty degree angle with the body and was not attached to the forelimbs. The trailing edge of the wing was supported by the distal ends of the posterior rods and was not attached to the hindlimbs. There were no large muscles that could power flapping motion of the wings. Thus, these reptiles were incapable of powered flight, only gliding.

The wings of the extant "flying" gecko, *Ptychozoon*, appear to be constructed in much the same way as those of the extinct *Coelurosauravus*, except that there are fewer rods supporting the wings. A gliding lizard in the genus *Draco* that lives in tropical forests of Southeast Asia has winglike folds of skin that are supported by five to seven rods that extend from the body.

The first finger (thumb) has one or more feathers, parallel to the anterior wing margin, which can be raised above the front of the wing to eliminate air turbulence above the wing. While turbulence significantly decreases lift, the raising of the alula decreases turbulence and increases lift.

Sparrows, finches, cuckoos, barn owls, warblers, grouse, and similar birds have elliptical wings that develop a high degree of camber and slotting. Because of these wing characteristics, birds with elliptical wings are capable of intricate maneuvers. Pigeons have broad wings for maneuverability but also elongated, pointed tips to increase their speed. To muffle the noise generated when barn owls swoop down on prey, their flight feathers have developed fringed edges and their coverts have become soft and downy.

Large birds of prey, such as vultures, buzzards, eagles, condors, swans, and storks, have high-lift wings. Their wings are wide and highly cambered with extensive slotting at the tips. The slightest updraft of air provides lift for these heavy birds. The inner flight feathers provide most of the lift within a thermal, whereas the long primary feathers that resemble fingers are used for maneuvering and creating thrust as well as additional lift. The larger the bird, the fewer wing-beats per second. Large vultures flap their wings about once each second.

Many marine birds that live along the shore and fly long distances to find food, as well as many migrating birds, have nearly flat, narrow, long, pointed wings. These wings are associated with rapid flying and, in some cases, with hovering. Peregrine falcons, gulls, geese, ducks, swifts, and swallows are examples of fast-flying birds. Long, narrow, swept-back wings are found in some birds that develop high speeds, such as swifts, swallows, and falcons. The common swift flies up to five hundred miles each day, gathering insects for its chicks. The Arctic tern is known for its long migration from the North American Arctic via Europe and Africa to Antarctica, approximately eleven thousand miles. The peregrine falcon is an extremely rapid flyer, sometimes reaching speeds of 175 miles per hour when it dives in pursuit of other birds.

Kingfishers, hummingbirds, and kestrels are both fast and capable of hovering. Hummingbirds beat their wings forty to eighty times each second, depending on their size. In addition, they can hover and fly backward for short distances. To achieve these feats, the wing bones evolved quite differently than in other birds. Hummingbird wing bones are mostly hand bones, as the forearm

Features of Bird Wings

Wing Type	*Wing Length to Width Ratio*	*Camber*	*Lifestyle and Habitat*	*Slotting*	*Birds*
Elliptical wings for maneuvering	Low (1.0-2.5) short, wide wings	High	Prey on flying insects and other rapidly moving animals and/or live in densely wooded habitats	Moderate to extensive	Wallcreepers, warblers, rollers, sparrows, cuckoos, thrushes, owls, blackbirds, woodpeckers, flycatchers, finches, sparrows, grouse, pheasant, kingfishers, doves, pigeons
High-lift wings for soaring	Moderate (2.5-2.8) very large, long wide wings	High	Heavy birds of prey or carrion eaters	Extensive	Storks, hawks, secretary birds, buzzards, swans, vultures, condors, eagles, pelicans
Wings for speed, gliding, or hovering	High (2.8-4.8) long, narrow, pointed wings, often swept-back	Low	Very fast moving birds that obtain their food on-the-move and/or migrate long distances	Very little to none	Sandpipers, kestrels, cormorants, loons, cranes, geese, gulls, herons, skimmers, hummingbirds, swallows, curlews, swifts, pintails, peregrine falcons, arctic terns, ducks
Wings for long-distance gliding	Very high (4.8-7.0)	Low, nearly flat	Mostly marine birds that fly long distances to feed or migrate long distances	None	Frigate birds, albatrosses
Wings for swimming	High (3.5-5.5)	Flat	Flightless swimmers that use wings as flippers	None	Penguins
Wings for display or warmth			Flightless birds that use wings to keep warm and/or to display their beauty		Rheas, ostriches, emus, kiwis

and upper arm became extremely short. The wrist joint and elbow became rigid, so that the wing only rotates at the shoulder. Their wings are thin, flat, pointed and can become highly slotted. When these birds hover or fly backward, they obtain lift on both the up- and downstrokes of the wing, somewhat like insects. Larger birds usually only generate lift on the downstrokes.

Accomplished gliders, such as the albatrosses and frigate birds, have nearly flat, extremely long, narrow wings. These wings efficiently create lift from updrafts and surface air movements over water and land. These marine birds, which fly long distances when feeding or during migrations, have very long, slender, pointed wings. Each of these birds catches updrafts and is able to glide long distances without using much energy flapping their wings.

Bat Wings

Bats evolved from tetrapod, insectivorous mammals some time after the Cretaceous-Tertiary mass extinction 65 million years ago. Their wings consist of elastic, leathery skin supported by forelimbs, body trunk, and hind legs. The upper arm and forearm support the proximal half of the extended wing, while the elongated hand and digit bones support the distal half of the wing like the struts of an umbrella. Greatly elongated palm bones and digits two, three, four, and five provide the outer wing struts. The thumb, midway along the anterior margin of the wing, is a mere stump. Depending upon the bat, sometimes the thumb has a claw the bat uses to hold onto its perch.

Bats are extremely agile fliers because they can alter the camber and shape of their wings. The skin between the body and the fifth finger and the tail membrane generate most of the lift. The skin between the second and fifth fingers produces forward thrust. Highly maneuverable bats have relatively short, broad wings, whereas migrating bats have exceptionally long, narrow, pointed wings. These long wings increase lift and allow for extended gliding, but bats with long wings are not as agile as those with broader wings. Tail membranes are continuations of the wings and are used for sudden turns and changes of direction. The tail membrane is controlled by the back legs. The tail membrane is also used during landings to brake and help flip the bat upside-down so it can attach to its roost. The bat uses both its wings and its tail membrane to stall as it lands.

Some species of bats are strong enough to take off from the ground, but most initiate their flight from an elevated roost. Although bats do not have a keeled sternum like birds, some are able to take off from the ground and most are very agile fliers.

—*Jaime Stanley Colomé*

See also: *Archaeopteryx*; Bats; Birds; Feathers; Flight; Insects; Locomotion; Pterosaurs; Respiration in birds.

Bibliography

Dalton, Stephen. *Borne on the Wind: The Extraordinary World of Insects in Flight*. New York: Reader's Digest Press, 1975. This book, filled with beautiful colored photographs of flying insects, clearly distinguishes different wing types and how they are powered. Many sequential pictures taken fractions of a second apart illustrate the intricate wing movements and complex maneuvers.

Dickinson, Michael H., Fritz-Olaf Lehmann, and Sanjay P. Sane. "Wing Rotation and the Aerodynamic Basis of Insect Flight." *Science* 284, no. 5422 (June 18, 1999): 1954-1960. A somewhat technical discussion of how insects with small wings are able to generate enough lift to fly.

Dowswell, Paul, John Malam, Paul Mason, and Steve Parker. *The Ultimate Book of Dinosaurs*. Bath, England: Parragon, 2000. This book, targeted at children, contains a large number of excellent drawings of reptiles, pterosaurs, and dinosaurs (including *Archaeopteryx*). Short descriptions and comparisons are very useful.

Frey, Eberhard, Hans-Dieter Sues, and Wolfgang Munk. "Gliding Mechanism in the Late Permian Reptile *Coelurosauravus*." *Science* 275, no. 5305 (March 7, 1997): 1450-1452. A good description and interpretive figure of an extinct flying reptile.

Graham, Gary L. *Bats of the World*. New York: St. Martin's Press, 1994. A small book with good drawings and descriptions.

Padian, Kevin, and Luis M. Chiappe "The Origin of Birds and Their Flight." *Scientific American* 278, no. 2 (1998): 38-47. A good outline of how birds evolved from dinosaurs.

Sereno, Paul C. "The Evolution of Dinosaurs." *Science* 284, no. 5423 (June 25, 1999): 2137-2146. A very nicely illustrated paper on the evolution of the dinosaurs and birds.

Welty, Joel Carl, and Luis Baptista. *The Life of Birds*. 4th ed. New York: Saunders College Publishing, 1988. A textbook for college students, with everything they need to know about birds.

WOODPECKERS

Types of animal science: Anatomy, behavior, classification, ecology
Fields of study: Anatomy, ornithology, wildlife ecology

Woodpeckers are scansorial (tree-creeping) piciform birds which are found in areas with trees all over the world, except in Madagascar, Australia, New Zealand, and Oceania. They are members of the family Picidae and consist of 215 species, of which 25 are found in North America.

Principal Terms

ALTRICIAL: type of development in which hatchlings are born naked, blind, nearly immobile, and helpless

BRISTLES: specialized feathers that protect the nostrils

DISPLAY: visual signal used by a bird to trigger behavior in courtship or attacks

DRUMMING: type of nonvocal communication produced by banging bill on a hollow tree trunk or other noise-producing object

RECTRICES: tail feathers

REMNANTS: nesting material not deliberately placed in nest by birds

ZYGODACTYL: type of foot with two toes directed forward and two toes directed backward

Woodpeckers are found worldwide in areas with trees and a temperate climate. They range in size from 3.5 inches to 2 feet. The largest North America species is the pileated woodpecker, which is 16.5 inches long with a wingspan of 29 inches. Most species have a dominant plumage pattern of black or brown and white, with additional splashes of red or yellow. Woodpeckers are scansorial, much like the ancient bird *Archaeopteryx*. To facilitate tree climbing, they have sharply curved claws and strong zygodactyl feet. Most birds have four toes, with one toe (hallux) directed backward and the others forward (anisodactyl foot). This type of foot is ideal for perching, but in woodpeckers the fourth toe is directed backward along with the hallux to produce a foot with two toes forward and two toes rearward, more suited to tree climbing and clinging. A few woodpecker species have lost one of the rear toes and are referred to as three-toed woodpeckers. Rectrices act as props for the woodpecker when climbing, resulting in a hitching motion as the bird moves. These stiff tail feathers are critical for climbing activities, and during molting, the center two feathers are preserved until the rest of the tail feathers are replaced, whereas in other birds the center feathers are the first to be lost.

Stout, chisel-like bills are used to peel bark and excavate wood in search of insects or larvae to eat. Bills are also employed to carve out roosting and nesting cavities in trees and for drumming. Both the skull and bill are designed to absorb the shock of repeated pounding on wood. The sturdy feet and claws together with the rectrices form a triangular brace for the hard pounding by the bill. Woodpecker tongues are very long and wrap around the skull to anchor at the base of the bill. Extension of the tongue to retrieve food is accomplished by a complex system featuring long hyoid (tongue-base) bones. Some woodpecker tongues have barbs to help extract insects and larvae from chiseled holes, while sapsucker tongues are shorter, with fine, hairlike processes to aid in capturing sap and associated insects. Nostrils of woodpeckers are protected from the "sawdust" that excavating bills create by bristles or by being reduced to narrow slits.

Wings are tapered, with a low aspect ratio (ratio of length to width). This wing configuration is designed for rapid takeoff and swift evasive flight maneuvers to capture prey, such as flying insects, or to escape predators. The eyes are positioned on the side of the head, giving the bird a wide field of vision to help spot predators. This ocular arrangement produces a predominantly monocular vision in which the environment is seen by only one eye. Binocular vision, providing depth perception important for flying and landing, is present in a relatively narrow field of vision straight ahead.

Woodpecker Behavior

Some species of woodpeckers live up to fifteen years in the wild. Woodpeckers are monogamous (one male mates with one female). Some of the earliest sounds emanating from the woods in the spring are the drumming of woodpeckers. Woodpeckers vocalize with calls rather than songs. Drumming functions instead of song to proclaim territorial boundaries and attract mates. Both males and females drum. Dry branches, hollow tree trunks or logs, or any other object capable of producing a loud noise may be selected. Most woodpeckers drum in a burst with a rate characteristic of the species. For example, hairy woodpeckers drum at a rate of twenty-six drums per second, but the smaller downy woodpecker drums at a rate of fifteen drums per second. Sapsuckers drum more rhythmically, with a varying rate slowing at the end. Territoriality and courtship are also announced by displays. Bowing, bobbing, and side-to-side head motions, along with partial spreading of wings and tail, are performed during these displays. Some of the most lively and spectacular displays are given by flickers in the spring.

Woodpeckers use their stiff tail feathers for support while they peck. (Adobe)

Nearly all woodpeckers nest in unlined tree cavities, which they excavate. Typically, both male and female birds excavate a new site each year, and old sites are abandoned, to be used by other species of cavity-nesting birds that lack the carving talents of woodpeckers. Cavity nesting protects the eggs and young birds from predators and bad weather. Holes often face the sun to help with warming. Eggs are laid on wood chips and "sawdust" that are deposited at the bottom of the nesting cavity as a consequence of excavation. These remnants are not supplemented by other nesting materials. An average of four (a range of three to nine) eggs are laid. The eggs are white since the protection afforded by the nest cavity obviates the need for camouflage coloration. Incubation takes about two weeks and is shared by both male and female birds during the day, but is done solely by the male at night. Incubation begins prior to the last egg being laid, resulting in asynchronous hatching. Hatchlings are born immobile, downless, and with eyes closed, and are fed by the parents (altricial development). Feeding is by regurgitation and is performed by both parents until the birds are completely developed, in about four weeks. First-born birds have an advantage and are more likely to survive.

Woodpeckers eat insects lodged in the trunks or limbs of trees, and some species also feed on the ground or catch insects in the air. Nuts, fruit, seeds, and tree sap are also consumed, depending upon the particular species. Ants are the insect preferred by many woodpeckers. It should be noted that the effects of woodpeckers on trees are rarely harmful and are usually beneficial as a result of pest control. Nest holes are drilled in diseased or dead trees.

Bird watchers can easily identify woodpeckers at a distant by their characteristic posture and flight. They are usually seen clinging to branches or tree trunks using their rigid tail feathers as props. Woodpeckers fly in a pattern of moderate rises and falls. This bounding flight pattern is produced when a short burst of wing beats (rising) alternates with a closed-wing glide (falling). Woodpeckers are not long-distance flyers, and most species are permanent residents or migrate modest distances, which may vary with the abundance of food. Most species are solitary or paired, but the acorn woodpecker lives in communal groups of up to sixteen birds. This clown-faced bird has a complex social structure and is known for its acorn hoarding. Birds harvest acorns in the fall and ram them into holes drilled in one or two "granary" trees, usually at the center of their territory. As many as fifty thousand acorns are each carefully matched to a hole which will result in a tight fit. As the acorns shrink in size with age they are relocated into slightly smaller holes to produce another snug fit. The secure fit makes it difficult for robbers to steal acorns from the tree without alerting the vigilant acorn woodpecker family.

Woodpecker Facts

Classification:
Kingdom: Animalia
Phylum: Chordata
Class: Aves
Order: Piciformes
Family: Picidae (woodpeckers and allies)
Subfamily: Picinae (true woodpeckers, thirty-three genera, seventy-two species)
Geographical location: Every continent except Australia and Antarctica
Habitat: Wooded areas with temperate climates
Gestational period: One to seven days
Life span: Up to fifteen years
Special anatomy: Feathers, including stiff retrices and bristles; two wings, low-aspect ratio; zygodactyl feet; stout bill; long tongue

Environmental Ecology

The largest (with a length of twenty-one inches) and most spectacular North American woodpecker was the ivory-billed woodpecker, now thought to be extinct. When European settlers first came to the United States, this species was found throughout the southeast, as far north as North Carolina and up the Mississippi Valley to southern Illinois and Ohio. These birds were killed for their bills, which were prized for head ornaments by tribal leaders and warriors among Canadian Indians. Hunters could get two or three buckskins in trade for one bill. An even greater impact resulted from logging. These huge birds required large trees for nest cavities which are up to nine inches across and fourteen inches deep. In the twentieth century, ivory-bills had been exterminated from all but a few isolated areas in Louisiana, Florida, and South Carolina. Their last stronghold was in the Singer Tract of large cyprus trees in Louisiana. Unfortunately, the Singer Tract was heavily logged with the help of German prisoners of war in 1941, destroying much of the habitat for the few remaining birds. The last authenticated sighting of an ivory-billed woodpecker was in April, 1944, in northwest Louisiana. Since that time there have continued to be unconfirmed sightings, and a glimmer of hope remains that some of these magnificent birds may yet live in the fifty thousand remaining acres of the Singer Tract, or in other forests in the southeastern United States or Cuba.

—H. Bradford Hawley

See also: Beaks and bills; Birds; Feathers; Flight; Molting and shedding; Nesting; Respiration in birds; Wildlife management; Wings.

Bibliography

Bent, Arthur Cleveland. *Life Histories of North American Woodpeckers*. 1939. Reprint. Bloomington: Indiana University Press, 1992. Fascinating observations of woodpecker behavior are detailed in this classic book.

Ehrlich, Paul R., David S. Dobkin, and Daryl Wheye. *The Birder's Handbook*. New York: Simon & Schuster, 1988. This book is a treasury of information on birds in an easy-to-read format.

Gill, Frank B. *Ornithology*. 2d ed. New York: W. H. Freeman, 1995. This textbook of basic ornithology is the gold standard.

Short, Lester L. *Woodpeckers of the World*. Riverton, N.J.: Weidner and Sons, 1983. Definitive coverage of world woodpeckers, with high quality color plates of every known species.

WORMS, SEGMENTED

Type of animal science: Classification
Fields of study: Anatomy, invertebrate biology, physiology

Segmented worms are a class (Annelida) whose body parts are arranged in a longitudinal series of rings, or segments. Many are hermaphroditic.

Principal Terms

HERMAPHRODITIC: a condition where both male and female reproductive organs are located in the same animal

NEPHRIDIA: excretory tubules specialized for excretion and osmoregulation, with an external opening and with or without an internal opening

PARAPODIA: paired lateral projections on each side of most segments, variously modified for locomotion, respiration, or feeding

SEPTA: a thin, muscular tissue used to separate segments into a series of ringlike cavities

SETAE: a needlelike, chitinous structure of the integument, used for locomotion

The Annelida are segmented worms in which the body wall, coelom (body cavity), epidermis, circular muscle, longitudinal muscle, and peritoneum are arranged into a longitudinal series of rings or segments. True segmented animals exhibit metamerism, a repetition of a structure or organ from segment to segment. Each segment has the same fundamental structures as all the others. With the exception of the digestive system, the major organ systems of the annelids are metameric in structure. Young annelids generally have few segments, but as they grow, new segments are formed by the division of the terminal segment. Annelids represent the most highly organized animals capable of complete regeneration.

General Characteristics of the Annelids

The mouth lies between the first and second segments and forms one segment called the prostomium. In leeches, the mouth contains suckers for attaching to the body of a host. The brain originates in the prostomium and develops a pair of circumpharyngeal nerve rings that reach around the pharynx to form the ventral nerve cord, which appears as a chain of ganglia, one pair in each segment. In the Polychaetes, a pair of swimming or crawling parapodia are located on most of the segments. Both the Polychaetes and Oligochaetes contain external setae to assist in locomotion.

The annelid body is covered with a thin cuticle. Each segment has a ring of circular and longitudinal muscles that contract to either elongate or shorten the segment. A spacious coelom, divided by septa, lies between the body wall and an internal digestive tract. The coelom is filled with fluid and serves as a hydrostatic skeleton in all annelids except the leeches. The coelom also contains the circulatory and excretory systems. A system of large vessels (hearts) pump blood through a ventral vessel into capillary beds that invade all of the tissues. The blood is returned to the hearts via the dorsal vessel. Each segment, except the first and last, contains a pair of nephridia, which collect wastes and deliver them to the outside.

Reproduction in the Annelids

In many species of polychaete worms, fertilization is external and takes place in the open sea water. The palolo worm provides a good example of polychaete reproduction. Through most of the year, the worms exist as sexually immature ani-

Each section of a segmented worm, such as this marine fireworm, contains the same fundamental structures as all the other sections. (OAR/National Undersea Research Program)

mals called atokes, but during the breeding season, the posterior segments develop gonads, and the coelom becomes filled with gametes. On the night of breeding, individuals back out of their holes, and the posterior portion, called an epitoke, breaks free. The epitokes swim to the surface for a few minutes and burst, shedding eggs or sperm and leaving a rapidly disintegrating body. Reproduction in the palolo worms is tied to an annual cycle designating the month, a lunar rhythm designating the day, and a diurnal cycle designating the hour of reproduction. Over 90 percent of the population breeds within a single two-hour period of the entire year.

Oligochaete worms are hermaphroditic, with a pair of testes in the tenth and eleventh segments and a pair of ovaries in the thirteenth segment. During copulation, two worms exchange sperm. Once the eggs are fertilized, the clitellum, a swollen glandular region of the epidermis, secretes a membrane that slips forward along the body so that the eggs are laid directly into it as it passes. Finally, the cocoon slips off the head, and the eggs develop into tiny worms, which later emerge from the cocoon. The reproductive system of leeches is very similar to that of the Oligochaetes.

—D. R. Gossett

See also: Fertilization; Flatworms; Hermaphrodites; Hydrostatic skeletons; Regeneration; Roundworms.

Segmented Worm Facts

Classification:
Kingdom: Animalia
Subkingdom: Bilateria
Phylum: Annelida
Classes: Polychaeta (with parapodia and numerous setae), Archiannelida (small marine annelides with simple body), Oligochaeta (parapodia absent and few setae), Hirudinea (leeches, parapodia and setae absent)
Orders: Polychaeta—Errantia (palolo worms), Sedentaria (lugworms); Oligochaeta—Lumbricus (earthworms); Hirudinea—Rhynchobdellida (no jaws, pharynx eversible, colorless blood), Gnathobdellida (three jaws, red blood)

Geographical location: Found all over the world

Habitat: Polychaeta—mostly marine, found near the shore or on the bottom of shallow areas, with a few species in living in brackish water or freshwater; Archiannelida—marine; Oligochaeta—freshwater and terrestrial forms found burrowing in soil or leaf mold; Hirudinea—freshwater and terrestrial environments often found attached to the body of a host

Gestational period: Varies among species, but most species lay eggs within a few days after fertilization; eggs usually hatch within a few days to a few weeks after being deposited

Life span: Varies among species; can be as short as a year in some polychaetes and up to several years in some earthworms

Special anatomy: Elongated, metameric (segmented) bilateral invertebrates with appendages (parapodia) and chitinous setae in many species but lacking in Hirudinea; possess a true coelom (body cavity lined with epithelial tissue) divided by septa with a closed circulatory system, complete digestive system and an excretory system (nephridia) in each segment; leeches have specially adapted mouth parts for attaching to the body of the host

Bibliography

Conniff, R. "The Little Suckers Have Made a Comeback." *Discover* 8 (1987): 84-94. An interesting article that describes the medical uses of leeches in microsurgery.

Dales, R. P. *Annelids*. 2d ed. London: Hutchinson, 1967. An older but excellent and concise description of the major groups of Annelids.

Hickman, C. P., L. S. Roberts, and A. Larson. *Integrated Principles of Zoology*. Boston: McGraw-Hill, 2001. An introductory zoology text that gives a very good review of the characteristics of flatworms.

Kingman, J., and P. Kingman. "The Dance of the Luminescent Threadworms." *Underwater Naturalist* 22 (1993): 36. An excellent description of the reproductive swarming of these Polychaete worms off the coast of Belize during the full moon of July and August.

Lent, C. M., and M. H. Dickinson. "The Neurobiology of Feeding in Leeches." *Scientific American* 258 (1988): 98-103. A description of how feeding behavior in leeches is controlled by the nervous secretion of serotonin.

ZOOLOGY

Type of animal science: Fields of study
Fields of study: Zoology

Zoology is the branch of biology devoted to the study of the animal kingdom. The study of zoology encompasses the analysis of and classification of animals.

Principal Terms

COMPARATIVE ANATOMY: the branch of natural science dealing with the structural organization of living things

ECOLOGY: the study of the interactions between animals and their environment

EMBRYOLOGY: the study of the development of individual animals

EVOLUTIONARY ZOOLOGY: the study of the mechanisms of evolutionary change and the evolutionary history of animal groups

MORPHOLOGY: the study of structure; includes gross morphology, which examines entire structures or systems, such as muscles or bones; histology, which examines body tissues; and cytology, which focuses on cells and their components

PHYLOGENETICS: the study of the developmental history of groups of animals

PHYSIOLOGY: the study of the functions, activities, and processes of living organisms

SYSTEMATICS: the delineation and description of animal species and their arrangement into a classification

TAXONOMY: the classification of organisms in an ordered system that indicates natural relationships

ZOOGEOGRAPHY: the study of the distribution of animals over the earth

Attempts at classification are known from documents in the collection of the Greek physician, Hippocrates, as early as 400 B.C.E. However, the Greek philosopher Aristotle (384-322 B.C.E.) was the first to devise a system of classifying animals that recognized commonalities among diverse organisms. Aristotle arranged groups of animals according to mode of reproduction and habitat. After observing the development of selected animal groups, he noted that general structures appear before specialized ones, and he also distinguished between sexual and asexual reproduction. Aristotle was also interested in form and structure, and concluded that different animals can have similar embryological origins and different structures can have similar functions.

In Roman times, Pliny the Elder (23-79 C.E.) compiled four volumes on zoology widely read during the Middle Ages. Some scholars have deemed those volumes little more than a collection of folklore, myth, and superstition. One of the more influential figures in the history of physiology, the Greek physician Galen (c. 130-c. 201 C.E.), dissected farm animals, monkeys, and other mammals and described many features accurately, although scholars have noted that some of these features were then wrongly applied to the human body. His misconceptions, especially with regard to the movement of blood, remained virtually unchanged for hundreds of years. In the seventeenth century, the English physician William Harvey established the true mechanism of blood circulation.

The Foundations of Zoology

Until the Middle Ages, zoology was little more than a collection of folklore and superstition. However, during the twelfth century, zoology began to emerge as a science. The thirteenth century German scholar and naturalist St. Albertus Magnus refuted many of the superstitions associated with biology and reintroduced the work of Aristotle. The anatomical studies of Leonardo da Vinci in the fifteenth century have been noted as being far ahead of their time. His dissections and comparisons of the structure of humans and other animals led him to several important conclusions. For example, Leonardo noted that the arrangement of joints and bones in the leg are similar in both horses and humans, thus embracing the concept of homology, or the similarity of corresponding parts in different kinds of animals, suggesting a common grouping. A Flemish physician of the sixteenth century, Andreas Vesalius, is considered the father of anatomy for establishing the principles of comparative anatomy.

Throughout most of the seventeenth and eighteenth centuries, classification dominated zoology. The Swedish botanist Carolus Linnaeus developed a system of nomenclature still in use today, referred to as the binomial system of genus and species. Linnaeus also established taxonomy as a discipline. His work was built on that of the English naturalist John Ray and relied upon the form of teeth and toes to differentiate mammals and upon beak shape to classify birds. Another leading figure in systematic development of this era was the French biologist Comte Georges-Louis Leclerc de Buffon. The study of comparative anatomy was further developed by men such as Georges Cuvier, who devised a systematic organization of animals based on specimens sent to him from all over the world.

A cell is the smallest structural unit of an organism capable of independent functioning. Although the word "cell" was introduced in the seventeenth century by the English scientist Robert Hooke, it was not until 1839 that two Germans, Matthias Schleiden and Theodor Schwann, proved that the cell is the common structural unit of living things. The concept of the cell provided impetus for progress in embryology and animal physiology, including the concept of homeostasis, referring to the stability of the body's internal environment.

The formation of scientific expeditions in the eighteenth and nineteenth centuries gave scientists the opportunity to study plant and animal life throughout the world. The most famous scientific expedition was the voyage of the HMS *Beagle* in the early 1830's. During this voyage, Charles Darwin observed the plant and animal life of South America and Australia and developed his theory of evolution by natural selection. Although Darwin recognized the importance of heredity in understanding the evolutionary process, he was unaware of the work of a contemporary, the Austrian monk Gregor Mendel, who first formulated the concept of particulate hereditary factors, later called genes. Mendel's work was not widely disseminated until 1900.

Electronic Zoology Resources

CONSERVATION INTERNATIONAL: www.conservation.org. This Web site provides information about biodiversity conservation in the world's endangered ecosystems, including a map of global biodiversity hotspots, profiles of hotspots, and many other resources.

INTERNATIONAL COMMISSION ON ZOOLOGICAL NOMENCLATURE (ICZN): www.iczn.org. This Web site describes the official body responsible for providing and regulating the system for ensuring that every animal has a unique and universally accepted scientific name.

INTERNATIONAL SPECIES INFORMATION SYSTEM (ISIS): www.worldzoo.org. This Web site offers information about this organization that helps zoological institutes manage their living collections by providing software for records keeping and collection management, and then pools this information.

WILDLIFE CONSERVATION SOCIETY: www.wcs.org. This Web site offers information about the organization's conservation activities.

Modern Zoology

During the twentieth century, zoology has become more diversified and less confined to such traditional issues as classification and anatomy. Zoology, broadening its span to include such areas of study as genetics, ecology, and biochemistry, has become an interdisciplinary field applying a wide variety of techniques to obtain knowledge about animal kingdom. The current study of zoology has two main focuses, taxonomic groups, and the structures and processes common to these groups. Studies of taxonomy concentrate on the different divisions of animal life. Invertebrate zoology deals with multicellular animals without backbones; its subdivisions include entomology (the study of insects) and malacology (the study of mollusks). Vertebrate zoology, the study of animals with backbones, is divided into ichthyology (the study of fish), herpetology (amphibians and reptiles), ornithology (birds), and mammalogy (mammals). Taxonomic groups also subdivide paleontology, the study of fossils. In each of these fields, researchers investigate the classification, distribution, life cycle, and evolutionary history of the particular animal or group of animals under study. Most zoologists are also specialists in one or more of the related disciplines of morphology, physiology, embryology, and ecology.

Animal behavioral studies have developed along two lines. The first of these, animal psychology, is primarily concerned with physiological psychology and has traditionally concentrated on laboratory techniques such as conditioning. The second, ethology, had its origins in observations of animals under natural conditions, concentrating on courtship, flocking, and other social contacts. One of the important recent developments in the field is the focus on sociobiology, which is concerned with the behavior, ecology, and evolution of social animals such as bees, ants, schooling fish, flocking birds, and humans.

—*Mary E. Carey*

See also: Anatomy; Animal kingdom; Biology; Demographics; Ecology; Embryology; Ethology; Evolution: Animal life; Evolution: Historical perspective; Genetics; Marine biology; Paleoecology; Paleontology; Physiology; Systematics; Veterinary medicine; Wildlife management.

Bibliography

Allaby, Michael, ed. *The Concise Oxford Dictionary of Zoology*. 2d ed. New York: Oxford University Press, 1999. This book contains some six thousand entries and covers subjects such as animal behavior, physiology, genetics, cytology, evolution, earth history, and zoogeography, and reflects the current emphasis on ecology in the study of animals. Included are biographical notes on important figures in the history of zoology.

_______, ed. *The Dictionary of Zoology*. 2d ed. New York: Oxford University Press, 1999. A comprehensive review of field of zoology.

Anderson, Donald Thomas. *Invertebrate Zoology*. New York: Oxford University Press, 1999. An up-to-date textbook for undergraduate students studying the biology and evolution of invertebrate animals. Emphasizes function, physiology, and reproductive biology, rather than the more traditional comparative anatomy. Recent advances in the cladistic analysis of invertebrate taxonomy are incorporated into the classifications used in the text.

Griffin, Donald R. *Animal Minds*. Chicago: University of Chicago Press, 1992. The author presents the view that the significance of animal consciousness falls into three categories—philosophical, ethical, and scientific—and discusses examples of each type.

Proctor, Noble S., and Patrick J. Lynch. *Manual of Ornithology: Avian Structure and Function*. New Haven, Conn.: Yale University Press, 1993. A visual guide to the structure and anatomy of birds. The text is informative and written at a level appropriate to undergraduate students and to bird lovers in general. Covers systematics and evolution, topography, feathers and flight, the skeleton and musculature, and the digestive, circulatory, respiratory, excretory-reproductive, sensory, and nervous systems of birds, as well as field techniques for watching and studying birds. References, comprehensive bibliography, over two hundred excellent illustrations.

Taylor, Barbara. *Animal Encyclopedia*. New York: Dorling Kindersley, 2000. A visual encyclopedia for a juvenile audience, covering more than two thousand animal species, complete with stunning close-up color photos of the animals in their natural habitats. Includes a fact section about animal life, including behavior, camouflage, adaptation, migration, breeding, and conservation.

Wernert, Susan J., ed. *Reader's Digest North American Wildlife: An Illustrated Guide to Two Thousand Plants and Animals*. Pleasantville, N.Y.: Reader's Digest Association, 1998. Covers more than two thousand plants and animals of all types. Spanning the land from Florida to the Northwest Territories, topics range from field, forest, pond, prairie to all the natural communities that make North American flora and fauna so diverse. It serves as both an at-home reference and a guide to the most common and conspicuous wild plants and animals of the North American continent.

Whitaker, John O. *National Audubon Society Field Guide to North American Mammals*. 2d ed. New York: Alfred A. Knopf, 1996. This classic guide contains nearly 375 photographs and up-to-date information on the characteristics, behavior, and habitats of 390 species that breed in the United States and Canada. Easy to use with thumb-tab keys, 135 drawings of tracks and anatomical features, and three hundred range maps.

Wilson, Edward O. *Sociobiology: The New Synthesis*. Cambridge, Mass.: The Belknap Press of Harvard University Press, 2000. In the introduction to this twenty-fifth anniversary edition, Wilson shows how research in human genetics and neuroscience has strengthened the case for a biological understanding of human nature. Human sociobiology, now often called evolutionary psychology, has in the last quarter of a century emerged as its own field of study, drawing on theory and data from both biology and the social sciences

ZOOPLANKTON

Type of animal science: Classification
Fields of study: Ecology, invertebrate biology, marine biology

Zooplankton are a free-floating collection of aquatic organisms that are carried by water currents. In contrast to phytoplankton, zooplankton are heterotrophic. Zooplankton is a term that includes organisms from many animal and protist phyla.

Principal Terms

AUTOTROPH: an organism that has the ability to make its own food from inorganic substances
BENTHIC: organisms that live at or near the bottom of the ocean
HETEROTROPH: an organism that requires the ingestion of food for survival
NEKTON: an aquatic organism that has the ability to swim
PHYTOPLANKTON: small, aquatic free-floating plants

Zooplankton is found in almost all major fresh and marine water bodies. Many species, however, are restricted to specific territories depending on conditions of light, temperature, salinity, chemical composition, and turbulence. Their range also varies geographically and vertically within the water column.

Zooplankton Types

The diversity of zooplankton makes classification difficult. Holoplanktonic zooplankton are free-floating their entire lives and are mostly invertebrates. Flagellated, ciliated, and amoeboid protozoa as well as several species from the animal phyla Cnidaria, Ctenophora, Chaetognatha, Mollusca, Annelida, Urochordata, Echinodermata, and Arthropoda make up the majority of collected organisms. Meroplanktonic zooplankton are the free-floating forms of organisms having swimming nekton or attached benthic stages as part of their life cycle. They consist mostly of egg and larval stages of marine benthic invertebrates that include worms, snails, clams, barnacles, crabs, starfish, and sea urchins, as well as many marine fish, such as tuna. Body forms, life cycles, diet, and location vary among species

Zooplankton are often divided according to size. The most common divisions and sizes for zooplankton are nanoplankton (2 to 20 micrometers), microplankton (20 to 200 micrometers), mesoplankton (0.2 to 20 millimeters), macroplankton (2 to 20 centimeters), and megaplankton (20 to 200 centimeters). Although generally small, even jellyfish several meters in size are considered zooplankton because they are unable to swim effectively against the current.

Copepods are one of the most studied forms and are classified as holoplanktonic zooplankton. Copepods have a segmented body with three distinct sections. Several pairs of legs and large antennae are used to create feeding currents that capture phytoplankton, especially diatoms, and other small zooplankton. Copepods are classified as mesoplankton and are nearly five millimeters in size. They are the most important herbivore in the ocean. Advanced collection techniques, however, have uncovered the presence of smaller zooplankton classifications that may be an important ecological first link.

Ecological Importance

Zooplankton diet varies greatly among species. There are carnivorous and herbivorous zooplank-

Zooplankton Facts

Classification:
Kingdom: Protista or Animalia
Phyla: Cnidaria, Ctenophora, Annelida, Nemertea, Arthropoda, Chaetognatha, Mollusca, Phoronida, Bryozoa, Echinodermata, Hemichordata, Chordata, Urochordata
Geographical location: Large bodies of freshwater and salt water
Habitat: Mainly in surface waters
Gestational period: Variable
Life span: Variable
Special anatomy: Highly variable between different phyla; generally small with a flattened body and many projections; zooplankton may also store oils used for buoyancy

ton, as well as those that feed on nonliving organic material. Herbivorous forms feed on phytoplankton through a process called grazing. This process prevents damaging blooms of phytoplankton from increasing in numbers. The link between phytoplankton and zooplankton is also important in passing the energy fixed by autotrophs to higher trophic levels of the food chain. Herbivorous zooplankton are eaten by carnivorous zooplankton, other invertebrates, or fish. The food chain may terminate in the top carnivores that include large fish, birds, and mammals, including humans.

Adaptations

Because of their small size, zooplankton are very sensitive to water conditions. Maintaining position in the water column is an important adaptation for many zooplankton species because they lack the ability to swim strongly enough to counteract the water current. Their density is often slightly greater than that of sea water. In addition, high salinity and low temperatures increase the viscosity of surrounding water, making it harder for zooplankton to move. A larger surface area and flattened body form covered by many projections or spines counteracts these obstacles and prevents the zooplankton from sinking. Several zooplankton also increase their buoyancy by storing low-density materials, such as certain oils. Small body movements, including the action of flagellae and cilia, also counteract the sinking motion.

Vertical Migration

The depth at which zooplankton exist is variable. While some float at the surface of the water, others may inhabit depths greater than four thousand meters. The concentration, or biomass, of zooplankton decreases with increasing depth. Many zooplankton exhibit a daily rhythmic vertical movement called diel vertical migration. They may migrate downward from the water surface during the day and upward at night. These movements may avoid predation and may conserve energy by slowing metabolism in colder deeper waters. Seasonal migrations from deeper waters in the winter months to surface waters in the spring are common. This migration may decrease metabolism and conserve energy throughout the winter when there is a lack of food.

—Paul J. Frisch

See also: Arthropods; Echinoderms; Marine animals; Marine biology; Mollusks; Protozoa.

Bibliography

Cerullo, Mary M. *Sea Soup: Zooplankton*. Gardiner, Maine: Tilbury House, 2001. This book is a wonderful reference for children. It answers many questions about zooplankton and their role in the environment.

Lalli, Carol M., and Timothy R. Parsons. *Biological Oceanography: An Introduction*. 2d ed. New York: Pergamon Press, 1997. Zooplankton ecology is studied is great detail in this book. The many phyla and class divisions of zooplankton are explained in good detail, complete with pictures.

Nybakken, James W. *Marine Biology: An Ecological Approach*. 5th ed. San Francisco: Benjamin/Cummings, 2001. This is a standard text for those studying marine biology. It specifically covers zooplankton and its role in marine ecology.

Sanders, Robert W. "Zooplankton." In *McGraw-Hill Encyclopedia of Science and Technology*. 8th ed. New York: McGraw-Hill, 1997. This is a broad science reference. The section on zooplankton covers several biological phenomena of zooplankton.

Todd, C. D., M. S. Laverack, and G. A. Boxshall. *Coastal Marine Zooplankton: A Practical Manual for Students*. 2d ed. New York: Cambridge University Press, 1996. This is a comprehensive field manual providing numerous photographs and taxonomy of many zooplankton species.

ZOOS

Type of animal science: Ecology
Fields of study: Conservation biology, environmental science, ethology, population biology, wildlife ecology, zoology

Keeping wild animals has evolved, over the past five thousand years, from animal collections maintained by ancient societies to modern zoological gardens and aquariums with significant programs in wildlife appreciation, education, science, and conservation.

Principal Terms

AQUARIUM: self-contained and self-supporting aquatic environmental exhibits, maintained either independently or in association with a zoo

EX SITU: conservation and research out of the animal's natural environment (at a zoo)

FROZEN ZOO: frozen tissue bank maintained at a zoo that contains wild animal tissue and reproductive samples for use in future breeding programs

IN SITU: conservation and research in the animal's natural habitat

MENAGERIE: a French word for zoos first used in the early 1700's to describe the keeping (and later exhibition) of animals

VIVARIA: a Latin word for a structure housing living animals, first used by the Romans to describe the places holding elephants and other animals for their shows

ZOO: an English abbreviation of the term "zoological garden"; first used in the early 1800's to describe the early British zoos, it tended to replace the word menagerie in the 1900's

Ancient rulers maintained wild animal collections beginning about 3000 B.C.E. in Mesopotamia (in the area that is now Iraq), Egypt, China, and possibly the Indus Valley (now Pakistan). From around 3000 B.C.E. in Mesopotamia, Sumerian, Babylonian, and Assyrian kings formed collections of native and exotic wildlife in royal parks. These parks were a combination protomenagerie, hunting reserve, and garden park. The parks were used for falconry, hunting wild beasts, entertaining guests, and personal pleasure. Animals kept in these parks often included elephants, wild bulls, lions, apes, ostrich, deer, gazelle, and ibex. Kings and wealthy individuals also had fishponds, flight cages for birds, and pets. Keepers and veterinarians were employed to care for these animals.

From around 2700 B.C.E., Egyptian pharaohs and wealthy Egyptians had animal collections as well. Native species were caught locally, and exotic species were obtained through commerce and tribute. These animals included lions, leopards, hyena, gazelle, ibex, baboons, giraffe, bears, and elephants. There were also fishponds and bird flight cages. Exotic animals came from the so-called Divine Land (Palestine, Syria, and Mesopotamia), Nubia (on Egypt's southern border), and Punt (the Ethiopia area).

In China, each ruling dynasty created animal reserves, beginning with the Shang dynasty, the first to unify the region, from about 1520 B.C.E. These royal parks were large, walled-in natural areas, where wild animals roamed freely and were maintained by park administrators, keepers, and veterinarians. Animals in these parks included fish, turtles, alligators, birds, camels, horses, yaks, deer, elephants, rhinoceroses, and possibly giant pandas.

Ancient and Medieval Animal Collections

The Greco-Roman societies, from 1100 B.C.E. to 476 C.E., were the next to evolve urban centers, overtaking the Mesopotamian and Egyptian civilizations and their animal collections, while the Chinese society and collections continued in isolation from the Western world. Greek curiosity, travel, and trade were favorable for the development of animal collections; however, their ruling city-states did not have the wealth or influence to develop large collections. There were pets, temple collections with animals used in processionals, showmen and animal trainers with animals used in itinerant animals acts, and small collections featuring native animals, including the large ones, such as bears and lions, that could still be found in the area. The Roman Republic (509-27 B.C.E.) provided its rulers and wealthy citizens with the opportunity to maintain small collections of native species. The first exotic animals seen in the Republic were Indian elephants taken in battle (280 B.C.E.). Hunts and processionals, which began during the Republic period, evolved into increasingly elaborate spectacles during the Empire period (27 B.C.E.-476 C.E.). Much has been made of these shows because of the large number of animals that were displayed and killed, but it is their appearances in these spectacles that reveal the introduction of exotic species to Rome. Unfortunately, little is known about the collections in which these animals were kept. Large collections kept in vivaria by emperors, municipalities, and military units supplied animals for the spectacles. Public entertainment also included itinerant performing animal acts. Private collections of native and exotic animals were kept in villa gardens, fresh and salt water ponds, bird enclosures, cages, parks, and hunting reserves.

Persian and Arabic societies between about 546 B.C.E. and 1492 C.E. also had gardens containing wild animals. During this period, Persian and Arabic collections extended throughout the Middle East, India, Asia, northern Africa, and Spain. Meanwhile, medieval Europe (476-1453 C.E.) saw the fall of the Roman infrastructure and the rise of monarchies, monasteries, and municipalities. These centers of administrative, religious, and social life were also centers for medieval animal collections. Kings and wealthy barons kept elephants, lions, bears, camels, monkeys, and birds, especially falcons. Monasteries maintained modest collections for economic and aesthetic reasons. Towns had animals in moats (when no longer needed for defensive purposes), pits, and cages; they often kept animals used in their coat of arms, such as deer, lions, bears, and eagles.

Collections even existed in the Americas, although it is not known when they began. Both the Aztecs of Central America and the Incas of South America had extensive animal collections. Montezuma's estates, in what is now Mexico, included a large bird building, which included freshwater and saltwater ponds, as well as a staff of three hundred keepers. A separate collection combined birds of prey, reptiles, and mammals, along with its own staff of three hundred keepers. The Incas also had animal collections, along with domesticated herds of guanacos and vicuñas.

Early Modern Menageries

Europe emerged into its Renaissance period as the sixteenth century dawned. The accompanying age of exploration brought to light many new species from distant lands. The proliferation of sixteenth century European animal collections was greatly influenced by these discoveries, increasing the collections in both size and number. Renaissance collections first developed in Italy, where the importation of animals from Asia and Africa was already well established, and then spread throughout Europe as the continental nation-states developed their own trade routes. Living animal collections, an essential part of royal courts, became commonplace among wealthy collectors. However, obtaining, shipping, and caring for wild, living animals was expensive and difficult, especially since little was known about the needs of the newly discovered species. Nevertheless, these new animals were status symbols and were of immense interest, and so the collections grew.

During the seventeenth and eighteenth centuries these collections, known as menageries, developed throughout Europe and the European colonies. The earliest colonial menageries began as acclimatization farms and animal holding areas at colonial botanical stations located at the more important colonial posts. A tremendous increase in these stations occurred during the nineteenth century, as the exchange of plants and animals became commonplace.

As the nineteenth century advanced, menageries could be found everywhere, due to the intense interest in the many new species and the exotic places from which they came. Knowledge about these animals increased and their transport was greatly improved. As a result, menageries evolved into zoological gardens during the early 1800's, and aquariums developed in the mid 1800's.

Early Zoos and Aquariums

The establishment of the Zoological Society of London's zoological garden at Regent's Park in 1828 was a significant event in the history of animal collecting and may be considered a transition in the evolution of zoos. This collection was intended from its inception to surpass any then in existence, with an emphasis on education and research. Private European menageries evolved into public zoological gardens, going from collections for the few (royalty and wealthy collectors) to zoological gardens for all citizens. In the United States, exotic animals were first introduced in 1716, when a lion arrived at Boston. Other species were gradually introduced over the remainder of the eighteenth century, and menageries containing many species appeared toward the end of this century. Traveling and circus menageries became popular in the early 1800's.

A few small urban menageries were already established when the Zoological Society of Philadelphia (chartered in 1859) opened its zoological garden in 1874. Most of these urban menageries eventually closed, but a few, such as those at Lincoln Park in Chicago and Central Park in New York City, continued and eventually became modern zoological gardens. Throughout the rest of the world, menageries were developing into zoological gardens, although often hindered by local economies and politics.

Robert Warington and Philip H. Gosse first developed the modern aquarium during the early 1850's in England. Gosse also worked with the Zoological Society of London to establish the first public aquarium at the London Zoo in 1853. Other public aquariums, along with the home aquarium craze, swept Europe and the United States soon after.

The turn of the century brought about a tremendous increase in the number of zoos and, to a lesser extent, aquariums that lasted up through World War I. It was a period during which zoos and aquariums improved their programs in conservation, animal husbandry, research, and education. Beginning in the 1890's, conservation of wildlife became an important concern in the United States and the European colonies. Europe had been dealing with conservation issues for many centuries, but these other regions saw their seemingly limitless resources quickly disappear. The New York Zoological Park (the Bronx Zoo), along with other United States zoos, played a significant role in conserving the American bison, which had become extinct in the wild. European zoos did likewise, saving the European bison (wisent), Père David's deer, and Przewalski's horse from extinction.

Animal husbandry research improved when the Penrose Research Laboratory opened at the Philadelphia Zoo about 1901-1905. In situ field research began at the New York Zoological Park with the inception of its Department of Tropical Research in 1916. Exhibition techniques received further attention when Hagenbeck's Tierpark (in Stellingen, Germany) opened in 1907. This was an important event in the trend toward moated, open exhibits (rather than buildings), ecological exhibits (rather than systematic arrangements), and mixed species exhibits (rather than single species exhibits). Hagenbeck's Tierpark featured panorama exhibits based on zoogeographic themes, with a series of back-to-back moated displays designed in such a way that, from the visitor's perspective, the animals appeared to be together in one space.

World War I affected zoos because of a loss of employees to the war effort, a loss of revenues to operate the facilities, a loss of paying visitors, difficulty in finding food for the animals, and a loss of some animals. After recovering from the social and economic impacts of this war there was another surge in the number of zoos and aquariums up through World War II. As zoos and aquariums increased in numbers, the professional management of these institutions improved and became more organized.

During the 1930's, a studbook for the European wisent was established, the first of many species-specific wild animal studbooks. As of 1997, there were 150 mammal, bird, and reptile international studbooks and world registers. In 1887, the Verband Deutscher ZooDirektoren (Association of German Zoo Directors) formed in Germany, the first of several early professional associations. In the United States, the American Association of Zoological Parks and Aquariums (now the American Zoo and Aquarium Association) formed in 1924. In 1999 there were forty-six national and regional associations.

World War II repeated the problems faced during the previous war but to a much greater extent, because in addition to the previous problems, there was more physical destruction of European zoos. The physical, social, and economic damage from this war took longer to correct than did the first war's. Improvements began in earnest as soon as the war ended, however, and the postwar period became a time of scientific advances, improved technology, better animal husbandry, new exhibit designs, improved education, more intense conservation programs, and professionalism.

Modern Zoos and Aquariums

Many new zoos and aquariums were built in the decades after World War II, and many older zoos and aquariums were renovated. In addition, many advances were made in areas important to animal husbandry, such as veterinary medicine, chemical immobilization and transportation, animal nutrition, reproductive biology, conservation techniques, biotechnology, materials and exhibit design, and water management technology for aquariums.

Zoo staff training became more formal and systematic beginning in the 1950's. The work had become a profession, and the required knowledge had increased, making formal training a necessity. In 1959, a zoo school was established at the Wrocław Zoo, Poland, as well as at several German zoos in the 1960's and 1970's. The American Zoo and Aquarium Association published a zookeeper training manual in 1968 and began a series of training classes in 1975. In 1972, the first of sev-

Wild Animals as Diplomatic Gifts

Animals have always been used as diplomatic gifts and tribute. Because wild animals are difficult to capture, ship, and maintain, they have been coveted luxury items. Rarely seen exotic animals, especially the large mammals, were impressive prizes in royal collections. King Manuel I of Portugal sent his rare Indian rhinoceros as a gift to Pope Leo X in 1515, hoping to influence the Pope's decisions on the placement of the demarcation line dividing Portugal's and Spain's colonial possessions. Muhammad Ali of Egypt provided giraffes to Paris, London, and Vienna in 1826 in order to secure their governments' favor when he waged war against their ally, Greece. In less dramatic ways, kings stayed on the good side of fellow rulers with gifts, including animals. Subordinates to rulers also provided wild animal gifts to their rulers as favors. If a ruler was very interested in an animal collection, he often used diplomatic envoys and other government officials to obtain particular species. Just about every country has a national collection, to which wild animal gifts are sent. In the United States, the National Zoological Park in Washington, D.C., has received living animals presented to the president since it was established in 1889. Even today, China still uses giant pandas as diplomatic gifts, although such gifts are now accompanied with a mandate to contribute to panda conservation programs.

Zoo Biology

Until recently, there has been little training needed or given at zoos. Most zoo employees learned their work through word-of-mouth from experienced workers. During this time, very little was understood about wild animals or any of the fields of study now associated with wildlife husbandry. Knowledge about wild animal care has evolved and grown along with the knowledge in those sciences relevant to this care. As more was learned about species' needs in the wild, animal behavior, animal health, nutrition, population biology, other zoological studies, and the scientific and professional aspects of zookeeping emerged. Heini Hediger, former director of the Bern, Basel and Zürich zoos, reviewed and explained this growing base of knowledge in a series of books written from the 1940's to the 1960's. Hediger's concept was known as zoo biology, a practical and scientific approach to wildlife husbandry that laid the foundation for much of today's studies and animal management. There is now a journal with the same name *Zoo Biology*, which encourages the study and publication of information related to this field. There are also numerous training programs for zookeepers and other zoo employees, both within zoos and independently at community colleges. While taking care of wild animals is fun, it is also a professional and a scientifically managed business.

eral academic programs in zookeeper training began at the Santa Fe Community College in Gainesville, Florida.

Professional associations increased as well. The American Association of Zoo Veterinarians formed in 1946, the American Association of Zoo Keepers formed in 1967, and the Association of British Wild Animal Keepers was founded in 1972. Since then, several other segments of the zoo and aquarium work force have formed professional associations.

Conservation efforts intensified as the seriousness of the endangered species problem increased. More attention was paid to endangered species and environmental problems in the 1960's and 1970's, particularly after activities were held at many zoos and aquariums for the first Earth Day on April 22, 1970. Government laws and regulations concerning endangered species and wildlife importation increased significantly during these decades as well. Conservation efforts at zoos and aquariums included the establishment of international and regional studbooks, regional species survival programs, taxon advisory groups, conservation assessment and management plans, International Union for Conservation of Nature and Natural Resources (IUCN) species survival commission action plans, fauna interest groups, breeding consortia, scientific advisory groups, species survival plans, and, as of 1997, over 1,200 conservation research studies at American zoos and aquariums.

Recent trends include a popular increase in the number of new aquariums, the development of more naturalistic exhibits, participation in in situ conservation research, development of frozen zoos, and the use of biotechnology to bring back extinct species. Zoos and aquariums have evolved over the past five thousand years into important cultural and conservation institutions. This evolutionary process will continue into the future.

—*Vernon N. Kisling, Jr.*

See also: Breeding programs; Cloning of extinct or endangered species; Domestication; Ecology; Endangered species; Genetics; Population genetics; Reproductive strategies; Urban and suburban wildlife; Wildlife management.

Bibliography

Bell, Catharine, ed. *Encyclopedia of the World's Zoos*. Chicago: Fitzroy Dearborn, 2001. Numerous authoritative articles on institutions, personalities, animal collections, and topics related to zoos and aquariums.

Croke, Vicki. *The Modern Ark: The Story of Zoos, Past, Present, and Future.* New York: Charles Scribner's Sons, 1997. A critical look at zoos and the work they do, including a brief history, but mostly a modern perspective.

Hoage, R. J. and William A. Deiss, eds. *New Worlds, New Animals: From Menagerie to Zoological Park in the Nineteenth Century.* Baltimore: The Johns Hopkins University Press, 1996. Papers from a conference on the history of zoos at an important point in time, when they were emerging as modern zoological parks.

Kisling, Vernon N., Jr., ed. *Zoo and Aquarium History: Ancient Animal Collections to Zoological Gardens.* Boca Raton, Fla.: CRC Press, 2001. Comprehensive and authoritative history of the world's zoos, from 3000 B.C.E. to the present.

Norton, Bryan G., Michael Hutchins, Elizabeth F. Stevens, and Terry L. Maple, eds. *Ethics on the Ark: Zoos, Animal Welfare, and Wildlife Conservation.* Washington, D.C.: Smithsonian Institution Press, 1995. Papers dealing with animal-related ethics by zoo professionals, humane society professionals, and other concerned individuals.

Wemmer, Christen M., ed. *The Ark Evolving: Zoos and Aquariums in Transition.* Front Royal, Va.: Smithsonian Institution Conservation and Research Center, 1995. Conference papers on the ever-changing role of zoos and aquariums.

GLOSSARY

Abdomen: The part of the body between the thorax and the pelvis, containing the viscera; the posterior part of an arthropod (such as an insect or crustacean).

Abiotic: The physical part of an ecosystem or biome, consisting of climate, soil, water, oxygen, and carbon dioxide availability, and other physical components.

Ablation: The technique of removing a gland to determine its function and observe what effects its removal will precipitate.

Absorption: The movement of nutrients out of the lumen of the gut into the body.

Abundance: In ecology, the number of organisms living in a particular environment.

Acceleration: The appearance of an organ earlier in a descendant's development than in the ancestor, as a result of an acceleration of development in ontogeny.

Acclimatization: A process by which animals are adapted to new environmental conditions.

Accommodation: Changing the shape of the eye's lens in order to keep objects at different distances focused on the retina.

Acetylcholine: A neurotransmitter produced by a nerve cell that enables a nerve impulse to cross a synapse and reach another nerve or muscle cell.

Acid rain: An excessive release of nitrogen oxides or sulfur dioxide as the result of the burning of primarily fossil fuels; these combine with water to form nitric or sulfuric acids that dissolve into rainfall.

Acidosis: A body fluid pH of less than 7.4 at 37 degrees Celsius.

Actin: One of the two major types of contractile proteins; it forms the thin myofibrils of the sarcomere.

Active or wide foraging: Moving about in search of prey.

Aculeus: The sting, either a single or double hollow barb that delivers the venom to prey.

Adaptation: In evolutionary biology, a heritable structure, physiological process, or behavioral pattern that gives an organism a better chance of surviving and reproducing; in physiology, the decrease in the size of the response of a sense organ following continuous application of a constant stimulus.

Adaptive radiation: The rapid evolution of new species following invasion of a new geographic region or ecological niche, or exploitation of a new ecological opportunity.

Adenosine triphosphate (ATP): The primary energy storage molecule in cells of all organisms; links energy-producing reactions with energy-requiring reactions.

AER. *See* **Apical ectodermal ridge (AER).**

Aerobic: Requiring free oxygen; any biological process that can occur in the presence of free oxygen.

Aerobic metabolism: A set of chemical reactions that require oxygen to make ATP, as opposed to anaerobic metabolism, which does not require oxygen.

Aggression: A physical act or threat of action by one individual that reduces the freedom or genetic fitness of another.

Aging: A process common to all living organisms, eventually resulting in death or conclusion of the life cycle.

Agnatha: A class of vertebrates that includes all forms in which jaws are not developed; the group to which the earliest vertebrates belong.

Airfoil: The wing of a flying animal or airplane that the provides lift and/or thrust needed for flight.

Alates: Recently molted winged adult termites.

Alkalosis: A body fluid pH greater than 7.4 at 37

degrees Celsius, the opposite of acidosis.

Allele: Alternative version of a single gene, located at the same position on a chromosome, that can be expressed as a new physical trait in organisms.

Allele frequency: The relative abundance of an allele in a population.

Allelochemic: A general term for a chemical used as a messenger between members of different species; allomones and kairomones are allelochemics, but hormones and pheromones are not.

Allogrooming: Mutual grooming or grooming between two individuals.

Allomone: A chemical messenger that passes information between members of different species, resulting in an advantage to the sender.

Alloparenting: Performance of parenting duties by an individual not the parent of the offspring (though usually a relative).

Allopatric: Populations of organisms living in different places and separated by a barrier that prevents interbreeding.

Allozyme: One of two or more forms of an enzyme determined by different alleles of the same gene; usually analyzed by gel electrophoresis.

Altricial: Animals born in a helpless state and completely dependent on the parent(s).

Altruism: A behavior that increases the fitness of the recipient individual while decreasing the fitness of the performing individual.

Alveoli: The milk-producing areas within the mammary glands.

Alveolus: The thin-walled, saclike lung structure where gas exchange takes place.

American Sign Language (ASL): An American system of communication for the deaf that employs hand gestures and facial expressions.

Amino acid: An organic molecule with an attached nitrogen group that is the building block of polypeptides.

Amniote: Animals with eggs in which embryos develop within fluid-filled membranes (the amnion), allowing eggs to be laid on land; amniotes include reptiles, birds, and mammals.

Amphistyly: The mechanism by which the jaws of sharks are suspended from, but not directly fused to, the skull.

Amplexus: A form of pseudocopulation seen in amphibians, where the male mounts and grasps the female so that their cloacae are aligned, and eggs and sperm are released into the water in close proximity and at the same time.

Amygdala: Subcortical brain structure related to emotional expression.

Anabolism: A series of chemical reactions that builds complex molecules from simpler molecules using energy from ATP.

Anaerobic metabolism: Metabolism in the absence of oxygen that leads to the production of lactic acid, a strong acid.

Analogue: An individual structure shared by two or more species that is of only superficial similarity; thus, it is not indicative of a common ancestor.

Anatomy: The science of the structure of organisms and their parts.

Androgens: The general term for a variety of male sex hormones made by the testes, such as testosterone and dihydrotestosterone, responsible for male secondary (anatomical) sex characteristics and masculine behavior.

Angle of attack: The angle at which an airfoil meets the air passing it.

Animal husbandry: Care and welfare of domestic animals.

Anisogamy: Reproduction using gametes unequal in size or motility.

Anosmia: The clinical term for the inability to detect odors.

Antagonism: Any type of interactive, interdependent relationship between two or more organisms that is destructive to one of the participants.

Antennae: A pair of segmented sensory appendages located above the mouth parts.

Anterior pituitary gland: The front portion of

the pituitary gland, which is attached to the base of the brain; the source of luteinizing hormone (LH) and follicle-stimulating hormone (FSH).

Anthropogenic disturbance: A change (usually a reduction) in population size caused by human activities.

Anthropomorphism: Attributing human characteristics or states of mind to animals.

Antibody: Protein produced by lymphocytes, with specificity for a particular antigen.

Antidiuretic hormone (ADH): A hormone produced in the hypothalamus that controls reabsorption of water in the loop of Henle.

Antigen: A chemical that stimulates the immune system to respond in a very specific manner.

Antipredator benefits: Benefits that come from actions that protect individuals from being killed.

Antlers: Branched, temporary horns made of solid bone, shed and regrown yearly.

Aorta: The major arterial trunk, into which the left ventricle of the heart pumps its blood for transport to the body.

Apes: Large, tailless, semierect anthropoid primates, including chimpanzees, gorillas, gibbons, orangutans, and their direct ancestors—but excluding humans and their direct ancestors.

Apical ectodermal ridge (AER): A thickened ridge of ectodermal (outer tissue) cells that appears along the distal edge of the limb as it first begins its development as a limb bud.

Aposematic coloration: Bright warning coloration that toxic species use to advertise their distastefulness to would-be predators.

Appendicular skeleton: One of two main divisions of vertebrate skeletal systems, composed of the bones of the pelvic girdle, the shoulders, and the limbs.

Apterous: Insects without wings, such as fleas.

Aquaculture: The artificial growth of animals or plants that live in the water; the culture of something living in water.

Arachnid: A class of arthropods with jointed legs and hard external skeleton that includes mites, scorpions, spiders, and ticks.

Arboreal: Living in trees.

Archenteron: The primitive gut cavity formed by the invagination of the blastula; the cavity of the gastrula.

Arteriole: The finest branch of an artery.

Artery: A blood channel with thick muscular walls which transports blood from the heart to various parts of the body.

Arthropods: Invertebrate animals having jointed legs, a chitinous exoskeleton, and a ventral nerve cord; includes crustaceans, insects, and arachnids.

Artiodactyls: Hoofed mammals with even numbers of toes.

Asexual reproduction: Reproduction without the union of male and female sex cells.

Assortative mating: A type of nonrandom mating that occurs when individuals of certain phenotypes are more likely to mate with individuals of certain other phenotypes than would be expected by chance.

Asthenosphere: The region below the lithosphere where rock is less rigid than that above and below it.

Astragalus: A pulley-shaped bone between the legs and ankles of antelopes.

Atoll: A remnant horseshoe or ring-shaped barrier reef surrounding a sunken island.

ATP. *See* **Adenosine triphosphate (ATP)**.

Atria: The two chambers of the heart, which receive venous blood from the body (via the right atrium) or oxygenated blood from the lungs (left atrium).

Auditory nerve: The cranial nerve that conducts sensory impulses from the inner ear to the brain.

Australopithecines: Nonhuman hominids, commonly regarded as ancestral to humans.

Autotomy: The self-induced release of a body part.

Autotrophs: Organisms that have the ability to make their own food from inorganic substances; also known as primary producers.

Axial skeleton: A main division of vertebrate

skeletal systems, made up of the bones of the skull, the vertebral column, the ribs, and the sternum.

Axon: That part of a nerve cell through which impulses travel away from the cell body.

Baleen: Whalebone mouth plates that capture and strain small food animals eaten by baleen whales.

Banding: Technique for studying the movement, survival, and behavior of birds by means of identification tags.

Barrier reef: A reef that is separated from land by a lagoon.

Basal metabolic rate (BMR): The rate of metabolism measured when the animal is resting and has had no meals for twelve hours; used to compare different species.

Basilar membrane: The flexible partition connected to the cochlea along which are attached neural hair cells.

Batesian mimicry: An evolutionary trend in which an edible species mimics the form of a distasteful species to avoid predation.

Behavior: An animal's movements, choices, and interactions with other animals and its environment.

Behavioral ecology: The systematic study of the strategies animals use to overcome environmental problems and the adaptive value of those strategies.

Behavioral thermoregulation: Maintaining relatively constant body temperature by shuttling between warm and cool microhabitats.

Behaviorism: The school of psychology that focuses on the investigation of overt behaviors and rejects allusion to inner processes as a means of explaining behavior.

Benthos: The area of the ocean floor; organisms associated with the sea bottom.

Bet-hedging: A reproductive strategy in which an organism reproduces on several occasions rather than focusing efforts on a single or few reproductive events.

Bifurcation: The division of a Y-shaped and connected mesenchymal structure into a single proximal chondrogenic focus and two distal chondrogenic foci; this can lead to the formation of separate chondrification centers in a developing digit.

Big game: Large animals, usually mammals such as deer or predators such as mountain lions; some wildlife departments may list other animals as big game.

Bilateral symmetry: An arrangement of body parts of an organism down a central axis which, when divided down the midline, produce right and left mirror images.

Bile salts: Organic compounds derived from cholesterol that are secreted by the liver into the gut lumen and that emulsify fats.

Binocular vision: The ability to utilize image information from both eyes to form a single image with depth information.

Binomial nomenclature: The two-word system used for naming every individual species.

Bioclimatic zone: A zone of transition between differing yet adjacent ecological systems.

Biodiversity: Variety of life found in a community or ecosystem; includes both species richness and the relative number of individuals of each species.

Biogenetic law: Ernst Haeckel's term for his generalization that the ontogeny of an organism recapitulates the adult stages of its ancestors (recapitulation).

Biological clock: A timekeeping mechanism that is endogenous (a part of the animal) and capable of running independently of exogenous timers such as day-night cycles or seasons, although the clock is normally set by them.

Biological magnification: The increasing accumulation of a toxic substance in progressively higher feeding levels.

Biological rhythm: A cyclical variation in a biological process or behavior, often with a duration that is approximately daily, tidal, monthly, or yearly.

Biology: The science of life and living organisms.

Bioluminescence: Production of visible light by living organisms.

Biomass: The weight of organic matter in an environment or ecosystem, often expressed in terms of grams per square meter per year.

Biome: A terrestrial ecosystem that occupies an extensive geographical area and is characterized by a specific type of plant community, such as deserts.

Biosphere: The sum of all the occupiable habitats for life on earth.

Biostratigraphy: The dating of rocks using fossils.

Biotechnology: Methods used to manipulate biological processes (such as reproduction).

Biotic: The living part of an ecosystem or biome, consisting of all organisms.

Bipedal: Walking on only two feet, as humans do.

Biserial dermal armor: Bony plates running along side of the vertebral column.

Bivalve: A mollusk having two shell halves.

Blastema: A region of surviving, proliferating cells at the edge of a damaged tissue.

Blastula: An early stage of an embryo which is shaped like a hollow ball in some animals and a small, flattened disc in others; contains a cavity called the blastocoel.

Blood: The fluid connective tissue within blood vessels that carries raw materials to cells and carries products and wastes from them.

Blood vessels: Membranous tubes through which blood flows; arteries carry blood from the heart, veins carry blood to the heart, and capillaries are tiny vessels in which exchange takes place.

Blubber: Thick layer of fat under the skin.

BMR. *See* **Basal metabolic rate (BMR)**.

Body mass: The average weight of females of a species, expressed in kilograms.

Boltrun: Mole burrow tunnel used as an emergency exit.

Bond: The tie or relationship between opposite-sex partners in a pair bond.

Bonding behaviors: Behavior patterns that establish, maintain, or strengthen the pair bond.

Bone: The dense, semirigid, calcified connective tissue which is the main component of the skeletons of all adult vertebrates.

Book lungs: A system of blood-filled diverticula that are surrounded by air pockets located in a chamber called the atrium.

Brachiation: A form of locomotion in which the body is held suspended by the arms from above; also called arm-swinging.

Bridge: The portion of the shell that connects the carapace to the plastron.

Bristles: Specialized feathers that protect the nostrils.

Bronchus (*pl.* bronchi): An individual tube that is part of a lung and leads to one of the smaller lung parts.

Brood: All the immature insects within an insect colony, including eggs, larvae, and, in the Hymenoptera, the pupal stage; also, to cover young with the wings.

Brood-parasite. *See* **Nest-parasite**.

Brood pouch: A temporary external pouch created by folding the skin of the abdomen together; used to carry young as they continue to develop.

Browser: An animal that feeds on leaves and twigs from trees.

Brucellosis: Cattle disease caused by brucellae bacteria.

Budding: A form of asexual reproduction that begins as an outpocketing of the parental body, resulting in either separation from or continued connection with the parent, forming a colony.

Bulla: Hollow bony area.

Burrowing insectivore: An insect-eating animal that usually lives in nests formed by digging holes or tunnels in the ground.

Calcareous: A material composed primarily of calcium compounds.

Calcification: Calcium deposition, mostly as calcium carbonate, into the cartilage and other bone-forming tissue, which facilitates its conversion into bone.

Calorie: The traditional unit of heat; one calorie

is the amount of heat required to raise the temperature of one gram of water 1 degree Celsius.

Camouflage: Patterns, colors, and/or shapes that make it difficult to differentiate an organism from its surroundings.

Cancellous bone: Spongy bone that is composed of an open, interlacing framework of bony tissue oriented to provide maximum strength in response to normal strains and stresses.

Canine teeth: Four elongated, pointed teeth that grasp and kill prey.

Capillaries: The very fine vessels in various tissues, which connect arterioles with venules; it is here that respiratory gases, nutrients, and wastes are exchanged between blood and tissues.

Carangiform swimming: A method of swimming where the tail is moved while the body is held rigid.

Carapace: Hard case covering the back of an animal, such as the chitinous outer covering of a crustacean shell or insect exoskeleton, or the portion of the shell that covers a turtle's back.

Carbohydrate: An organic molecule containing only carbon, hydrogen, and oxygen in a 1:2:1 ratio; often defined as a simple sugar or any substance yielding a simple sugar upon hydrolysis.

Carbonic anhydrase: An enzyme used in the mineralization process to convert carbon dioxide to bicarbonate.

Cardiac output: The amount of blood ejected by the left ventricle into the aorta per minute.

Carnassial teeth: Pairs of large, cross-shearing teeth designed to sheer flesh; the characteristic that unites all members of the order Carnivora.

Carnivore: A member of the meat-eating order Carnivora, which includes dogs, cats, weasels, bears, and their relatives.

Carrion: Corpses of dead animals.

Carrying capacity: The maximum number of animals that a given area can support indefinitely.

Cartilage: A soft, pliable typically deep-lying tissue that constitutes the endoskeletons of primitive vertebrates, such as sharks, as well as the embryonic skeletons and jointing structures of adult higher vertebrates.

Caste: One of the recognizable types of individuals within an insect colony, such as queens, workers, soldiers, and males or drones; usually these individuals are physically and behaviorally adapted to perform specific tasks.

Castoreum: A secretion from the castor glands of beavers, used in scent marking.

Catabolism: A series of chemical reactions that break down complex molecules into simple components, usually yielding energy.

Catarrhini: A primate group including Old World monkeys, apes, and humans, with reduced tails and only two pairs of premolar teeth.

Catastrophism: A scientific theory which postulates that the geological features of the earth and life thereon have been drastically affected by natural disasters of huge proportions in past ages.

Caudal fin: The tail fin of fishes, which supplies the forward thrust in locomotion.

Cecum: Structure in the digestive tract that aids in digestion and water retention.

Cell-mediated immunity: Production of lymphocytes that specifically kill cells with foreign antigens on their surfaces.

Cellular respiration: The release of energy in organisms at the cell level, primarily by the use of oxygen.

Cellulose: Fibrous polysaccharide that chiefly constitutes the cell walls of plants.

Celsius: A scale for measuring temperature in which freezing is zero degrees and boiling is one hundred degrees, abbreviated C.

Centrum: The spool-shaped body of a vertebra.

Cephalothorax: Combined head and thorax of an arthropod.

Cercal organs: Tufts of hair supplied with nerves, located on insect's abdomens, which respond to aerial sounds.

Cetaceans: Plant-eating marine mammals, such as whales, dolphins, and porpoises.

CFCs. *See* **Chlorofluorocarbons (CFCs).**

Chelate: Pincerlike.

Chelicerae: Appendages with pincers.

Chemical pollutants: Harmful chemicals manufactured and released to the environment; normally referring to those that contaminate ecosystems.

Chemoreceptor: Specialized nervous tissue that senses changes in pH (hydrogen ions) and oxygen.

Chemotaxis: An oriented response toward or away from chemicals.

Chitin: A transparent, horny substance of invertebrate exoskeletons.

Chlorofluorocarbons (CFCs): A group of very stable compounds used widely since their development in 1928 for refrigeration, coolants, aerosol spray propellants, and other uses; once risen in stratosphere, they cause ozone depletion.

Chlorophyll: One of several forms of photoactive green pigments in plant cells that is necessary for photosynthesis to occur.

Cholera: Gastrointestinal diseases.

Chondrification: The process by which undifferentiated connective cells transform into chondrocytes (cells that make cartilage) and begin forming extracellular matrix.

Chordates: A phylum of organisms characterized by the presence of a notochord, a dorsal nerve cord, and gill slits.

Chorion: The outer cellular layer of the embryonic sac of reptiles, birds, and mammals; the term was coined by Aristotle.

Chromatophores: Pigment-producing cells.

Chromosome: A molecule of DNA that contains a string of genes, which consist of coded information essential for all cell functions, including the creation of new life; chromosomes are passed from one generation to the next by the gametes.

Ciliated: Bearing short, hairlike organelles on the surface of cells, used for motility.

Circadian rhythm: A physiological or behavioral cycle that occurs in a twenty-four-hour pattern.

Circular muscle: Muscle fibers that run in a circular pattern around the body perpendicular to the long axis of the body.

Clade: A group of animals and their common ancestor.

Cladistics: A method of analyzing biological relationships in which advanced characters of organisms are used to indicate closeness of origin.

Clasper: A modification of the pelvic fins in male sharks and rays which acts as the male sexual organ in internal fertilization of the female.

Class: The taxonomic category composed of related genera; closely related classes form a phylum or division.

Classification: The arranging of organisms into related groups based on specific relationships.

Clavicle: The collarbone, connecting the top of the breastbone to the shoulder.

Cleavage: The process by which the fertilized egg undergoes a series of rapid cell divisions which result in the formation of a blastula.

Cleidoic egg: A shelled egg equipped with internal membranes that make terrestrial reproduction possible.

Cline: A gradual, continuous variation from one population of a species to the next that is related to differences in geography.

Cloaca: A bodily opening at the end of the gut into which both the waste disposal and reproductive systems open; the common opening through which the products of the urinary, intestinal, and reproductive systems are expelled from the body.

Clock sense: An inherent awareness of time or time intervals used, for example, to compensate for celestial movements in navigation.

Clone: An organism that is genetically identical to the original organism from which it was derived.

Closed circulation: A circulation system made

of arteries, capillaries, and veins that returns blood flow back to the heart.

Clutch: Number of eggs in the nest.

Cnidocyte: Specialized cells on the body or tentacles of jellyfish that contain nematocysts.

Coalitions: Short-term alliances designed to gain access to a contested resource, often by fighting.

Cochlea: The vertebrate neural organ which transduces sound waves into nerve impulses.

Coelenteron: The fluid-filled gastrovascular cavity of Cnidarians.

Coelom: The body cavity of higher invertebrates and vertebrates, where mesodermal tissues enclose a fluid-filled space.

Coevolution: Joint evolutionary change caused by the close interaction of two or more species; each species serves as the natural selection agent for the other species.

Cognition: Transformation and elaboration of sensory input.

Cognitive ethology: Scientific study of animal intelligence.

Cohort: A group of organisms of the same species, and usually of the same population, that are born at about the same time.

Cold-blooded: Referring to animals whose body temperatures equal the temperature of their surroundings.

Collagen: A fibrous protein very plentiful in bone, cartilage, and other connective tissue.

Collembola: Primitive, wingless insects.

Colony: A cluster of genetically identical individuals formed asexually from a single individual.

Colostrum: The precursor to milk that is formed in the mammary gland during pregnancy and immediately after birth of the young.

Commensalism: A type of coevolved symbiotic relationship between different species that live intimately with one another without injury to any participant.

Commitment: The "decision" by an embryonic cell to develop in a certain way, which may be reversed if the cell is removed from its normal surroundings.

Community: A population of plants and animals that live together and interact with one another through the processes of competition, predation, parasitism, and mutualism, making up the biotic part of an ecosystem.

Compact bone: A dense type of bone, often termed lamellar bone, formed of a calcified bone matrix having a concentric ring organization.

Comparative anatomy: The branch of natural science dealing with the structural organization of living things.

Comparative psychology: The branch of psychology that uses comparative studies of animals as a means of investigating phenomena such as learning and development.

Competition: Interactions among individuals that attempt to utilize the same limited resource.

Compound eyes: Eyes that are made up of multiple lenses or light detectors.

Conditioning: The behavioral association that results from the reinforcement of a response with a stimulus.

Connective tissue: Any fibrous tissue that connects or supports body organs.

Consort pair: A temporarily bonded pair within a polygamous group; also called consortship.

Conspecific: A member of the same species.

Constriction: A method of killing prey using increasingly tight coils around the body to trigger stress-induced cardiac arrest.

Consumer: An organism that eats other organisms.

Consumer food chain: A simplified description of the grazing and predator/prey relationships within an ecosystem.

Continental drift: Theory that the continents have moved slowly apart from an early landmass, explaining why many species appear to be closely related while separated by wide expanses of ocean.

Continuous growth: Growth in a population in which reproduction takes place at any time during the year rather than during specific time intervals.

Contractile vacuole: The excretory organ of several one-celled organisms.

Controlling site: A sequence of nucleotides generally fifteen to sixty nucleotides long, to which a transcriptional activator or repressor binds.

Convection: A transfer of heat from one substance to another with which it is in contact.

Convergent evolution: The process by which unrelated animals tend to resemble one another as a result of adaptations to similar environments.

Copulation: Mating; the insertion of the male's penis into the female's vagina to fertilize her ova.

Core temperature: Internal body temperature around the heart, brain, and spinal cord.

Corona radiata: The layers of follicle cells that still surround the mammalian egg after ovulation.

Corpora allata: A gland in insects that synthesizes and secretes juvenile hormone (JH).

Corpus luteum (pl. corpora lutea): The structure on the ovary that is formed from the follicle after the egg has been released; it secretes progesterone.

Cortex: The main part of a hair, made of pigment-containing cells, surrounding a central medulla.

Costal grooves: Parallel grooves or folds on the side of a salamander's body.

Countercurrent exchanger: The process where a medium (air or water) flowing in one direction over a tissue surface encounters blood flowing through the tissue in the opposite direction; this improves the gas diffusion by maintaining a concentration gradient.

Countercurrent mechanism: A heat exchange system in which heat is passed from fluid moving in.

Countershading: A form of crypsis involving dark coloration on top and light coloration on the underside.

Coverts: Feathers covering the bases of the large feathers of the wings and tail.

Crepuscular: Active after sunset and in early morning.

Crest: Tuft of feathers on the head.

Critical period: A very brief period of time in the development of an animal during which certain experiences must be undergone; the effects of such experiences are permanent.

Crop: A specialized part of a bird's digestive system that holds and softens food.

Crossbridge: A structure seen in electron micrographs of contracted muscle; the point of attachment of a myosin "head" to actin.

Cross-pollination: The transfer of pollen grains and their enclosed sperm cells from the male portion of a flower to a female portion of another flower within the same species.

Crown: External tooth surface above gums.

Crustaceans: Lobsters, shrimps, crabs, and barnacles.

Cryptic coloration: Any color pattern that blends into the background.

Ctenoid scales: Scales with comblike serrations on rear edge, found on many bony fishes.

Cuckold: A partnered male who is helping his mate to raise offspring which are not genetically his own.

Cud: Food regurgitated and chewed a second time after its initial ingestion.

Cuspid: A tearing tooth found in the mouth of a carnivorous animal.

Cuticle: The outer arthropod exoskeleton consisting of several layers of secreted organic matter, primarily nonliving chitin; or the outermost layer of a hair, made of scales.

Cuttlefish: A squidlike marine mollusk, eaten by Odontoceti.

Cycloid scales: Thin, flat bony scales with a smooth surface; rounded in shape, found on herrings, minnows, trout, and other primitive teleosts.

Cyclostomes: The modern agnathans, comprising lampreys and hagfish.

Cyst: A secreted covering that protects small invertebrates from environmental stress.

Cytoplasm: The living portion of the cell that is contained within the cell membrane.

Darwinism: Branching evolution brought about by natural selection.

Death: The cessation of all body and brain functions.

Decapods: Animals with ten appendages (from Greek *deca*, "ten," plus *poda*, "foot or leg").

Decomposers: Microbes such as fungi and bacteria that digest food outside their bodies by secreting digestive enzymes into the environment.

Deep sea: Water depths below six hundred feet, also below penetration of light.

Definitive host: The host in which a symbiont (the organism living within the host) matures and reproduces.

Delayed implantation: An extended period after fertilization when an embryo stops developing, before it attaches to the uterine wall and resumes embryonic development.

Deme: A local population of closely related living organisms.

Dendroclimatology: The study of tree-ring growth as an indicator of past climates.

Denning: The period of winter sleep during which a bear does not eat, drink, urinate or defecate.

Density: The number of animals present per unit of area being sampled; for example, ten mice per hectare or five moose per square kilometer.

Density-dependent growth: Growth in a population in which the per capita rates of birth and death are scaled by the total number of individuals in the population.

Density-dependent population regulation: The regulation of population size by factors or interactions intrinsic to the population; the strength of regulation increases as population size increases.

Dental battery: Unit of teeth in the upper and lower jaws consisting of the cutting teeth and the rows of replacements below them.

Dental comb: Forward-projecting lower incisors and canines that are used for grooming and feeding.

Dental formula: The types of teeth in one quarter of the mouth, expressed as the ratio of incisors to canines to premolars to molars.

Dentine: The ivory portion of a tooth or scale; dentine or dentinelike substances such as cosmine are found in the scales of most fishes.

Dentition: Referring to the teeth.

Deoxyribonucleic acid (DNA): The genetic material of cells, having the molecular form of a twisted double helix that is linked by purine and pyrimidine base pairs; carries the inherited traits and controls for cell activities.

Dermis: Layer beneath the epidermis, primarily connective tissue but also containing nerves and blood vessels.

Determination: An event in organismal development during which a particular cell becomes committed to a specific developmental pathway.

Detritus feeders: An array of small and often unnoticed animals and protists that live off the refuse of other living beings, such as molted shells and skeletons, fallen leaves, wastes, and dead bodies.

Deuterostomes: Echinoderms, hemichordates, and chordates, a group linked by features of cell development including retention of the blastopore as anus.

Dewlap: Loose fold of skin hanging from the throat of some cattle.

Diapause: A resting phase in which metabolic activity is low and adverse conditions can be tolerated; also, an interruption in embryonic development.

Diapsids: A group of reptiles in which the temporal region of the skull is characterized by two openings, a supratemporal opening and an infratemporal opening.

Diastole: Period of heart muscle relaxation and declining pressure.

Differentiation: The process during

development by which cells obtain their unique structure and function.

Diffusion: The net movement of molecules from an area of high concentration to one of lower concentration as a result of random molecular movements; the passive movement of a gas across a membrane from a region of high pressure to one of low pressure: the process by which larger organic nutrients are broken down to smaller molecules in the lumen of the gut.

Digit: A finger, toe, or related bony animal structure.

Digitigrade: Walking on the toes, with the heel raised.

Dilution effects: The reduction in per capita probability of death from a predator due to the presence of other group members.

Dimorphism: Existence of two distinct forms within a species.

Dingo: The wild dog brought to Australia by the aborigines.

Dioecious: Having two separate sexes, namely male and female.

Diploid: A cell containing two sets of chromosomes, usually one derived from the father and one derived from the mother; the normal condition of all cells except reproductive ones.

Diplosomite: Millipede trunk segment, formed by the fusion of two body segments.

Discrete growth: Growth in a population that undergoes reproduction at specific time intervals.

Discrete signals: Signals that are always given in the same way and indicate only the presence or absence of a particular condition or state.

Disharmonic: Ecologically unbalanced.

Disjunct: Pertaining to the geographic distribution pattern in which two closely related groups are widely separated by areas that are devoid of either group.

Dispersal: The movement of organisms from one geographic area to another; movements may be the result of an animal's own efforts or the consequence of being passively transported by natural or human-mediated means; dispersal is limited by barriers.

Display: A social signal, particularly a visual signal, exchanged between animals.

Disruptive coloration: Use of stripes, spots, or blotches to break up the body outline and blend into a complex background.

Distal: Occurring near the outer end of a limb.

Distemper: Viral disease affecting respiration.

Diurnal: Awake and functional during the daylight hours.

Divergence: The evolution of increasing morphological differences between an ancestral species and offshoot species caused by differing adaptive pressures.

Diversity: The number of taxa (classification groups) associated with a particular place and time.

Domestication: A process by which animals are adapted biologically and behaviorally to a domestic (human) environment in order to tame and manipulate them for the benefit of humans.

Dominance: The physical control of some members of a group by other members, initiated and sustained by hostile behavior of a direct, subtle, or indirect nature.

Dominance hierarchy: A social system, usually determined by aggressive interactions, in which individuals can be ranked in terms of their access to resources or mates.

Dominance social behavior: Organization around a dominant leader, whom the rest of the group follows.

Dominant: Requiring only one copy of a gene for expression of the trait.

Dominant species: A species in a community that acts to control the abundance of its competitors because of its large size, extended life span, or ability to acquire and hold resources.

Dopamine: Neurotransmitter involved in movement and reward systems.

Dorsal: At the hind (posterior) end of a living organism.

Drag: A force that acts in the opposite direction of the movement of a body through a fluid medium; sources of drag vary but include friction and pressure suction.

Drone: A fertile male social insect.

Drumming: Type of nonvocal communication produced by banging bill on a hollow tree trunk or other noise-producing object.

Dryopithecines: Extinct Miocene-Pliocene apes (sometimes including *Proconsul*, from Africa) found in Europe and Asia; their evolutionary significance is unclear.

Dung showering: Behavior by bulls to show dominance over other males.

Duodenum: The first part of the small intestine, where it joins the stomach.

Eccrine gland: A type of sweat gland that helps maintain body temperature; the glands are located on the palms and soles, forehead, neck, and back.

Ecdysis: Molting; the process of removing (escaping from) the old exoskeleton.

Ecdysone: A hormone that triggers both molting and metamorphosis in insects as well as in many other species of animals.

Echolocation: The ability of animals to locate objects at a distance by emitting sound waves which bounce off an object and then return to the animal for analysis.

Eclipse plumage: The drab plumage of male birds following the postbreeding molt, in which their bright courtship feathers are replaced by dull earthy feathers that provide inconspicuous coloring.

Ecological niche: The sum of environmental conditions necessary for the survival of a population of any species, including food, shelter, habitat, and all other essential resources.

Ecology: The study of the interactions between animals and their environment.

Ecomorph: Species of different phyletic origins (at most distantly related) with similar structural and behavioral adaptations to similar niches.

Ecosystem: A biological community and the physical environment contained in it.

Ectoderm: The outermost of the three fundamental tissue types that appear during the development of most animals; it will form the skin and all of its derivatives, nervous tissue, and many other tissues.

Ectoparasite: A parasitic organism that attaches to the host on the exterior of the body.

Ectotherm: An animal that depends on environmental heat sources, usually solar radiation, to maintain body temperature.

Ediacarian fauna: A diverse assemblage of fossils of soft-bodied animals that represents the oldest record of multicellular animal life on earth; also called Ediacaran fauna.

Effector: That part of a nerve which transmits an impulse to an organ of response.

Egg tooth: A hard, calcified structure on the tip of the bill of a bird embryo that is used to help the bird break its shell during hatching.

Ejaculation: The process of expelling semen from the male body.

Elapids: A snake classification which includes cobras and rattlesnakes that have short, fixed front fangs.

Elasmobranch: The classificatory subbranch of sharks and rays.

Electroencephalogram (EEG): A chart of brain wave activity as measured by electrodes glued to the surface of the skull.

Electron transport chain: A series of electron carrier molecules found in the membrane of mitochondria; oxygen is used and ATP is made at this site.

Electrophoresis: A technique for separating molecules when they are placed in an electrical field; the separation is usually based on their charge and weight.

Electroreceptors: Sensors in the bill of the platypus that detect the weak electric field given off by animals.

Elfin forest: A stunted forest growing at high elevations in warm, moist climates.

Elytra: Rigid front wings of beetles that cover the second, functional pair of wings.

Embryo: A fertilized egg as it undergoes divisions from one cell to several thousand cells, but before the individual is completely differentiated into a fetus.

Embryo polarity genes: Genes whose expression in maternal cells results in products being stored in the egg that establish polarity, such as the anterior-posterior axis, after fertilization.

Embryology: The study of the development of individual animals.

Embryonic induction: The point at which one embryonic tissue signals another embryonic tissue to develop in a certain way.

Emergent vegetation: Aquatic vegetation that grows tall enough to be visible above the water.

Emigration: The movement of animals out of an area; one-way movement from a habitat type.

Encounter effects: The reduction in the probability of death from a predator due to a single group of N members being more difficult to locate than an equal number of solitary individuals.

Endangered species: A species of animal or plant that is threatened with extinction.

Endemic: Belonging to or native to a particular place.

Endocannibalism: A form of human cannibalism in which members of a related group eat their own dead.

Endocrine glands: Glands that produce hormones and secrete them into the blood; the hypothalamus of the brain, the pituitary gland, and the testes are all endocrine glands.

Endocrine system: An array of ductless glands scattered throughout the mammalian body that produce and secrete hormones directly into the bloodstream.

Endocuticle: Usually the thickest layer of the cuticle, found just outside the living epidermal cell layer and made of untanned proteins and chitin.

Endogenous: Refers to rhythms that are expressions of only internal processes within the cell or organism.

Endometrium: An inner, thin layer of cells overlying the muscle layer of the uterus.

Endoparasitic: A parasitic organism that attaches to an interior portion of the host's body.

Endoskeleton: The type of skeleton that is found beneath the musculature of an animal's body; it provides mechanical support for the body as a whole, some protection for vulnerable internal organs and tissues, and sites for muscle attachment.

Endotherm: An animal that, by its own metabolism, maintains a constant body temperature (warm-blooded); birds and mammals are endotherms.

Energy pyramid: A graphical representation of the energy contained in succeeding trophic levels, with maximum energy at the base (producers) and steadily diminishing amounts at higher levels.

Enterocytes: The cells that line the lumen of the small intestine.

Entrainment: The synchronization of one biological rhythm to another rhythm, such as the twenty-four-hour rhythm of a light-dark cycle.

Environment: All the external conditions that affect an organism or other specified system during its lifetime.

Environmental constraints: The physical demands placed upon any species by its surroundings that ultimately determine the success or failure of its adaptations and consequently its success as a species; also called pressures.

Enzyme: A protein that acts as a catalyst under appropriate physiological conditions to break down bonds of a large protein, fat, or carbohydrate.

Epicuticle: The outermost and thinnest layer of the arthropod cuticle, composed mainly of the hardened protein cuticulin.

Epidermis: The outer, protective layer of epithelial cells in the skins of vertebrates.

Epifauna: Animals that live on the sea floor.

Epihyal bone: A hyoid bone whose presence or

absence determines whether a cat generally purrs or roars.

Epiphragm: Covering or sealing membrane.

Epithelium: The tissue that covers and lines all exposed surfaces of an organism, including internal body cavities such as the viscera and blood vessels.

Equids: Members of the horse family (Equidae) including horses, asses (including the donkey), zebras, and their crosses (including the mule and the hinny).

Erection: The process of enlargement and stiffening of the penis because of increased blood volume within it.

Esophagus: The part of the oral cavity (pharynx) that transfers morsels to the stomach; it is usually a long, muscular tube with no digestive function other than transport.

Essentialism: The Platonic-Aristotelian belief that each species is characterized by an unchanging "essence" incapable of evolutionary change.

Estivation: Similar to hibernation; period of reduced activity or dormancy triggered by dry and/or hot environmental conditions.

Estrogen: A hormone secreted by the ovary and placenta for development of the uterus.

Estrus cycle: Hormonally controlled changes that make up the female reproductive cycle in most mammals; ovulation occurs during the estrus (heat) period.

Ethology: The study of an animal's behavior in its natural habitat.

Eukaryote: A higher organism, whose cells have their genetic material in a membrane-bound nucleus and possess other membrane-bound organelles.

Euryhaline: The ability of an organism to tolerate wide ranges of salinity.

Eusocial: A social system with a single breeding female; other members of the colony are organized into specialized classes (exemplified by bees, ants, and termites).

Evolution: A process, guided by natural selection, that changes a population's genetic composition and results in adaptations.

Evolutionarily stable strategy: A behavioral strategy that will persist in a population because alternative strategies, in the context of that population, will be less successful.

Evolutionary zoology: The study of the mechanisms of evolutionary change and the evolutionary history of animal groups.

Ewe: Female sheep.

Ex situ: Conservation and research out of the animal's natural environment (at a zoo).

Exocannibalism: A form of human cannibalism in which unrelated humans are eaten.

Exocuticle: A thick middle layer in the cuticle made up of both chitin and rigid, tanned proteins termed sclerotin.

Exogenous: Rhythms that originate outside the organism in the environment.

Exoskeleton: A jointed and segmented, relatively thin, hard covering made up of chitin and protein that surrounds and protects the entire inner body in most arthropods.

Exotic species: Organisms that are not naturally found in a place but have been artificially introduced, whether by accident or intentionally.

Exponential growth: A pattern of population growth in which the rate of increase becomes progressively larger over time.

External fertilization: The union of eggs and sperm in the environment, rather than in the female's body.

External genitals: The external reproductive parts of the female.

Extinct: No longer found anywhere on earth.

Extirpated: Not found in an immediate local area but found elsewhere on earth.

Extracellular fluid: The fluid outside cell membranes, including fluid in spaces between cells (interstitial), in blood vessels (plasma), in lymph vessels (lymph), and in the central nervous system (cerebrospinal).

Facial disk: The distinctive concentric circles of feathers that encircle the eyes of owls, helping direct sound toward the ears.

Fahrenheit: A scale for measuring temperature

in which freezing is 32 degrees and boiling is 212 degrees, abbreviated F.

Fangs: Enlarged teeth that are hollow like a hypodermic needle or grooved to facilitate the injection of venom.

Fast muscle: Muscle cells that respond quickly to nervous impulses; in invertebrates, these muscle fibers have short sarcomeres and a low ratio of thin to thick myofibrils.

Fate map: A map of determined, but undifferentiated, tissue by which specific cell regions can be identified as giving rise to specific adult structures.

Fecundity: The number of offspring produced by an individual.

Feedback: In endocrinology, this usually refers to one hormone controlling the secretion of another that stimulates the first, usually in the form of negative feedback, in which the second hormone inhibits the first.

Female: An organism that produces eggs, the larger of two different types of gametes.

Fenestration: A latticework of openings on the sides of the skull.

Feral: Referring to domestic animals that have reverted to a wild or semiwild condition, such as cats, dogs, or caged birds that have been released or escaped and now survive in the wild.

Fertilization: The process by which the egg and sperm unite to form the zygote.

Fetus: A differentiated but undeveloped individual with organ systems usually identifiable as a member of a species.

Fibula: The smaller of two bones between the knee and ankle.

Field observations: Observing behavior in naturalistic settings.

Filtration: The process of diffusion of plasma from the blood to the glomerulus and nephron.

Fin-fold theory: Theory that fins initially evolved as long folds of tissue extending around the body.

Fire ecology: An ecosystem that depends on periodic fires to clear underbrush; the seeds of many plants in such an ecosystem require fire in order to germinate.

Fission: The division of an organism into two or more essentially identical organisms; an asexual process.

Fitness: The ability of an organism to produce offspring that, in turn, can reproduce successfully; the fitness of organisms increases as a result of natural selection.

Fixed action pattern: A complex motor act involving a specific, temporal sequence of component acts.

Flagellate: A protozoan that uses long, whiplike structures called flagella for locomotion.

Flagellum: A long cell extension used in locomotion.

Fledgling period: Period after hatching, during which a nestling grows flight feathers and learns to fly.

Flehmen: Behavior involving curling and wrinkling of lips and nostrils, with the activation of the Jacobson's organ.

Flipper: Finlike structures of marine mammals that have evolved from the forelimbs of their terrestrial ancestors.

Fluid: A substance, either liquid or gas, that flows or conforms to the outline of its container.

Flyway: An established route that migratory birds take between their wintering and nesting grounds.

Follicle: The saclike organ from which a hair grows; its blood vessels nourish the hair.

Food chain: An abstract chain representing the links between organisms, each of which eats and is eaten by another.

Food pyramid: Diagram representing organisms of a particular type that can be supported at each trophic level from a given input of solar energy in food chains and food webs.

Food web: A network of interconnecting food chains representing the food relationships in a community.

Fossil: A remnant, impression, or trace of an animal or plant of a past geological age that has been preserved in the earth's crust.

Four-fin system: The combined activity of paired fins in some bony fishes that makes them highly maneuverable.

Fovea: Specific area with an exceptionally dense concentration of light-sensitive cells in eyes of animals.

Fratricide: Deliberate killing of one sibling by another sibling.

Free-living: An organism that does not have to spend a portion of its life cycle attached to another organism.

Free-running: Denotes a rhythm that is not entrained to an environmental signal such as a light-dark cycle.

Frenulum: A spinelike device that connects the front and hind wings in moths.

Frequency: The number of repetitions of a rhythm per unit time, such as a heart rate of seventy beats per minute.

Frequency-dependent predation: Predation on whichever species is most common in a community; a frequency-dependent predator will switch prey if necessary.

Frill: An elaborate crest at the back of the skull used for visual display but not for protection.

Frontal bone: The bone which, vertically, makes up the forehead and is important, horizontally, to formation of the top (roof) of the orbital and nasal cavities.

Frozen zoo: Frozen tissue bank maintained at a zoo that contains wild animal tissue and reproductive samples for use in future breeding programs.

Functional response: The rate at which an individual predator consumes prey, dependent upon the abundance of that prey in a habitat.

Funnel: An opening in a cephalopod mantle, providing oxygen and propulsion.

Furcula: Fused clavicles, the wishbone of birds.

Fusion-fission community: A society whose members are of both sexes and all ages, which can form and dissolve subgroupings.

Game: Economically important animals, usually birds or mammals; it includes those taken for recreation or products and those that damage human property.

Gamete: A haploid sex cell that contains one allele for each gene; sperm and egg cells are gametes that fuse to form a diploid zygote.

Ganoid scales: Heavy, dense scales containing ganoine found in primitive bony fishes.

Gap rule genes: Expressed in the zygote, these genes divide the anterior-posterior axis of fruit flies into several regions.

Gastroliths: Stones found in the gut region that aided in the digestion of coarse plant food.

Gastronomic: Pertaining to the art of fine dining.

Gastrula: The stage of development during which the endoderm (gut precursor) and the mesoderm (muscle and connective tissue precursor) are internalized.

Gastrulation: The transformation of a blastula into a three-layered embryo, the gastrula; initiated by invagination.

Gelding: Castrated male, usually a horse.

Gemmule: An asexual reproductive structure that becomes a new sponge.

Gene: A sequence of approximately one thousand to ten thousand DNA nucleotide pairs on a chromosome that encodes a messenger ribonucleic acid (mRNA) for eventual protein production.

Gene flow: The movement of genes from one population to another through migration and hybridization between individuals belonging to adjacent populations.

Gene pool: The array of alleles for a gene available in a population; it is usually described in terms of allele or genotype frequencies.

Generalized: Not specifically adapted to any given environment.

Genetic code: The three-nucleotide base sequences (codons) that specify each of the twenty types of amino acids; there can be more than one codon for a particular amino acid.

Genetic diversity: The total number and distribution of alleles and genotypes in a population; a population with a very high genetic diversity would have many alleles and genotypes, all evenly distributed or with approximately equal frequency.

Genetic drift: Change in gene frequencies in a population owing to chance.

Genital tubercle: A small swelling or protuberance toward the front of an embryo's genital area; it is destined to become the penis tip or clitoris.

Genitalia: The external sex structures.

Genome: All the genes of one organism or species.

Genotype: The complete genetic makeup of an organism, regardless of whether these genes are expressed.

Genotype frequency: The relative abundance of a genotype in a population; to calculate, count the number of individuals with a given genotype in the population and divide by the total number of individuals in the population.

Genus (pl. genera): A group of closely related species; for example, *Felis* is the genus of cats, and it includes the species *Felis catus* (the domestic cat) and *Felis couguar* (the cougar or mountain lion).

Geoffroyism: An early theory of evolution in which heritable change was thought to be directly induced by the environment.

Geographically isolated: Living in different habitats.

Geological periods: The twelve divisions in successive layers of sedimentary and volcanic rocks, which are differentiated by the distinctive fossils present within each division; because of many recently discovered fossils and careful dating, the beginning and ending dates for these periods tend to vary somewhat among different references.

Germ layers: The embryonic layers of cells which develop in the gastrula—ectoderm, mesoderm, and endoderm.

Gestation: Duration of pregnancy.

Gigantothermy: A form of metabolism in which internal temperature is maintained by the large mass of the animal.

Gill: An evaginated organ structure where the membrane wall turns out and forms an elevated, protruding structure; typically used for water respiration.

Gill rakers: Sievelike strainers found upon the gill arches of whale sharks, used to filter food from water taken in though the mouth.

Gilt: A young female pig that has not produced a litter.

Gizzard: A part of a bird's stomach that uses ingested pebbles (gastroliths) to grind up food.

Gland: A tissue composed of similar cells that produce a hormone.

Global extinction: The loss of all members of a species; that is, extinction whereby all populations of a species disappear or are eliminated.

Glomerulus (pl. glomeruli): A capsule fitting around capillary blood vessels that receives the filtrate from the blood and passes it into the tubule.

Gnathostomata: All vertebrates in which jaws are developed.

Gonad: The organ responsible for production of gametes—the testis in the male, the ovary in the female.

Gonadotropin: A hormone that stimulates the gonads to produce gametes and to secrete other hormones.

Gonochorism: Sexual reproduction in which each individual is either male or female, but never both; the opposite of hermaphroditism.

Gorget: Patch of feathers between the bird's throat and breast.

Gracile: Slender and light-framed, as opposed to robust.

Gradualism: The idea that transformation from ancestor to descendant species is a slow, gradual process spanning millions of years.

Granular glands: One of many kinds of glands in the skin of amphibians; granular glands secrete toxins for defense from predators.

Gray matter: The part of the central nervous

system primarily containing neuron cell bodies and unmyelinated axons.

Grazer: An organism that feeds primarily on grasses.

Greenhouse effect: The process by which certain gases, such as carbon dioxide and methane, trap sunlight energy in atmosphere as heat, resulting in global warming as more gases are released to atmosphere by human activities.

Gregarious: Forming groups temporarily or permanently.

Growth: The increased body mass of an organism that results primarily from an increase in the number of body cells and secondarily from the increase in the size of individual cells.

Guanine: A chemical deposited in the skin of freshwater eels as they return to the sea.

Gulper: A fish that captures and ingests its prey in one swallow.

Gymnosperm: A plant whose seeds are borne on seed scales arranged in cones.

Habitat: The physical environment, usually that of soil and vegetation as well as space, in which an animal lives.

Habitat selection: Process of choosing a home range, territory, nesting site, or feeding site on the basis of specific features of the habitat that the raptor is best adapted to exploit.

Hand: Ten-centimeter (four-inch) measuring unit.

Haplodiploidy: Sex determination found in the Hymenoptera, where males arise from unfertilized eggs and females from fertilized eggs.

Haploid: Having one chromosome of each type; gametes (eggs or sperm) are usually haploid.

Harem: A group of breeding females controlled by a single male.

Haversian systems: Narrow tubes surrounded by rings of bone, called lamellae, that are found within compact bones of animals having endoskeletons; the tubes contain blood vessels and bone.

Heart: A discrete, localized pumping structure within the circulatory system.

Heat: That part of the estral cycle when the female is receptive to male copulatory behavior.

Heliotherm: An animal that uses heat from the sun to regulate its body temperature.

Hemimetabolous: Incomplete metamorphosis in which juveniles resemble wingless adults.

Hemoglobin: A protein in vertebrate red blood cells that carries oxygen and carbon dioxide.

Hemolymph: The transport fluid of organisms with open circulation systems in which there is no clear distinction between blood and intercellular tissue fluid.

Hemotoxin: A substance that causes blood vessel damage and hemorrhage.

Herbivores: Animals that eat plants and show specializations of teeth and digestive tracts to do so.

Heritability: The extent to which variation in some trait among individuals in a population is a result of genetic differences.

Hermaphrodite: A single organism that produces both eggs and sperm.

Hermatypic coral: Reef-building coral species, belonging to the Cnidarian order of Scleractinia.

Heterocercal tail: A tail in which the spine extends into the upper lobe, giving a distinctly sharklike impression.

Heterochrony: Any phenomenon in which there is a difference between the ancestral and descendant rate or timing of development.

Heterodont: Having two or more types of teeth, such as molars and incisors.

Heterotroph: An organism that requires the ingestion of food for survival.

Heterozygote: A diploid organism that has two different alleles for a particular trait; a person with blood type A could be heterozygous by having an A allele and an O allele.

Hexapod: Six-footed; a general term for an insect.

Hibernacula: The winter habitats of brown bats.

Hibernation: A sustained period of torpor (lack

of activity) where an animal reduces its metabolic rate.

Hierarchy: A social structure in which animals are dominated by those higher on the linear ladder.

Higher insects: Generally larger insects with increasing levels of morphological complexity.

Home range: Geographic area used by an individual, pair, or group for their daily, seasonal, and sometimes their yearly activities; the defended portion of the home range is called a territory.

Homeobox: One of 180 nucleotide pairs that code for a protein called the homeodomain, found in such diverse organisms as insects, frogs, and humans; they are known to influence body plan formation in fruit flies.

Homeosis: A process that results in the formation of structures in the wrong place in an organism, such as a leg developing in place of a fly's antenna.

Homeostasis: The dynamic balance between body functions, needs, and environmental factors which results in internal constancy.

Homeotherm: An animal that maintains a constant, steady body temperature.

Homeotic selector genes: Genes that determine the identity and developmental fate of segments established in fruit flies by a hierarchy of genes.

Hominid: An anthropoid primate of the family Hominidae, including the genera *Homo* and *Australopithecus*.

Homocercal tail: A type of tail at which the spine ends at the base of the tail, which consists of two equal lobes.

Homodont: Having teeth all of the same type.

Homologue: An individual structure shared by two or more different species that is indicative of a common ancestor.

Homozygote: A diploid organism that has two identical alleles for a particular trait; a person with blood type A would be homozygous if he had two A alleles.

Hormone: A blood-borne chemical signal, either protein or steroid, from one area of the body to another.

Horn: Hard, smooth, keratinous material forming an external covering.

Hornet: A social wasp.

Host: By convention, the larger of two species involved in a symbiotic association.

Humans: Hominids of the genus *Homo*, whether *Homo sapiens sapiens* (to which all varieties of modern humans belong), earlier forms of *Homo sapiens*, or such presumably related types as *Homo erectus* and the still earlier (and more problematic) *Homo habilis*.

Humidity: The amount of water vapor in the air, often considered with temperature (as in relative humidity).

Humoral immunity: Production of antibodies specifically reactive against foreign antigens, carried in body fluids (humors).

Hybrid: An organism resulting from the crossing of two species.

Hybrid vigor: The tendency of hybrids to be larger and more durable than their parent species; also called heterosis.

Hydrologic cycle: Earth's cycle of evaporation and condensation of water, which produces rain and maintains oceans, rivers, and lakes.

Hydrostatic skeleton: A system in which fluid serves as the support by which muscles interact.

Hygroscopic: Able to retain moisture.

Hyoid bones: Series of connected bones at the base of the tongue.

Hyperdactyly: A condition whereby the number of digits is increased above the normal five to create a wing.

Hypermorphosis: A phenomenon in which the rate and initiation of growth in the descendant are the same as in the ancestor but in which the cessation of development takes place later.

Hyperosmotic: A solution with a higher osmotic pressure and more osmotically active particles than the solution with which it is being compared.

Hyperphalangy: A condition whereby the

number of phalanges is increased in each digit.

Hyperventilation: An increase in the flow of air or water past the site of gas exchange (lung, gill, or skin).

Hypoosmotic: A solution with a lower osmotic pressure, fewer osmotically active particles relative to the same volume, than the solution with which it is being compared.

Hypophysis: The pituitary gland, or master gland, which produces and secretes at least eight protein hormones influencing growth, metabolism, and sexual development.

Hypothalamus: A brain region just below the cerebrum that interconnects the nervous and endocrine systems of mammals, thereby controlling most hormone production and many body functions.

Hypoxia: From two Latin words, *hypo* and *oxia*, meaning "low oxygen."

Ichnology: The study of trace fossils.

Imaginal disk: Flat sheets of cells within an insect larva; these cells will change shape during metamorphosis and form the external structures of the adult.

Immigration: The movement of animals into an area; a one-way movement into a habitat type.

Imprinting: A specialized form of learning characterized by a sensitive period in which an association with an object is formed.

Impulse: A message traveling within a nerve cell to another nerve cell or to a muscle cell.

In situ: Conservation and research in the animal's natural habitat.

In situ hybridization: A technique used to visualize the location of specific DNA or RNA sequences; typically, a radioactively tagged sequence of nucleic acids is paired (hybridized) to a complementary sequence of nucleic acids.

Inbreeding: Mating between relatives, an extreme form of positive assortative mating.

Incisors: Teeth that are located in the front of the mouth and whose function is to tear, hold, and cut the prey.

Indicator species: A species monitored by biologists as a means of ascertaining the health of the ecosystem in which it lives.

Industrial melanism: The rapid rise in frequency of the melanic form in many moth species downwind of manufacturing sites, associated with the advent of industrial pollution.

Inertia: The property of an object with kinetic energy to move in a straight line unless acted upon by an outside force.

Infauna: Animals that live in the sea floor.

Innate: Any inborn characteristic or behavior that is determined and controlled largely by the genes.

Insectivore: A member of the order Insectivora of small, nocturnal mammals, including shrews, moles, and hedgehogs; also, any animal that feeds on insects.

Insight learning: Using past experiences to adapt and to solve new problems.

Instinct: Any behavior that is completely functional the first time it is performed.

Integumentary processes: Surface outgrowths from the cuticle, primarily rigid nonarticulated processes or movable articulated processes.

Interbreeding: The mating of closely related individuals which tends to increase the appearance of recessive genes.

Interference: The act of impeding others from using some limited resource.

Interfertile: Able to breed and produce fertile offspring.

Intermediate host: An animal species in which nonsexual developmental stages of some commensals and parasites occur.

Interneuron: A central nervous system neuron that does not extend into the peripheral nervous system and is interposed between the sensory and motor neurons.

Interstitial fluid: The fluid found in between cells.

Intestine: The part of the digestive system involved in completing the process of digestion and absorption of nutrients; usually divided into the small intestine and the large intestine, which opens to the exterior by way of the anus.

Intracellular fluid: The fluid compartment within the cell membrane.

Intrinsic rate of increase: The growth rate of a population under ideal conditions, expressed on a per individual basis.

Introgression: The assimilation of the genes of one species into the gene pool of another by successful hybridization.

Invagination: The turning of an external layer into the interior of the same structure; formation of archenteron.

Invertebrates: Animals lacking an internal skeleton.

Involuntary: Functioning automatically; not under conscious control.

Iridescent: Showing the colors of the rainbow depending on light reflection.

Irruption: A sudden increase in the size of a population, usually attributed to a particularly favorable set of environmental conditions.

Isogamy: Reproduction in which all gametes are equal in size and motility.

Isosmotic: A solution having the same osmotic pressure, the same number of osmotically active particles relative to the same volume, as the solution with which it is being compared.

Ivory: A white or honey-colored, bony substance.

Jacobson's organ: A sense organ in the mouth, which detects reproductive chemical signals.

Juvenile hormone (JH): A species-specific hormone which controls whether a molt will produce a larger larva or initiate metamorphosis.

K strategy: A reproductive strategy typified by low reproductive output; common in species living in areas having limited critical resources.

Kairomone: A chemical messenger that passes information between members of different species, resulting in an advantage to the receiver.

Keratin: A tough, fibrous major component of hair, feathers, nails, hooves, horns, and scales.

Keystone species: A species that determines the structure of a community, usually by predation on the dominant competitor in the community.

Kin selection: A phenomenon by which acts of altruism can help pass on genes for altruism by improving the survival of kin and their offspring.

Kinetic skull: A highly moveable arrangement of bones that allows independent action of the snout and jaws on both sides.

Kingdom: The broadest category of organisms; the system currently used recognizes five kingdoms—Monera, Protista, Fungi, Animalia, and Plantae.

Knuckle-walking: Terrestrial locomotion, in which the animal walks on the knuckles of the forelimbs and soles of the hind feet.

Krill: Small, shrimplike sea creatures.

Labial folds: The paired ridges of tissue on either side of the embryo's genital area, which become penis and scrotum in males and labia in females.

Labium: The sheath that contains the slender, styletlike mouthparts of the mosquito, including the mandibles, maxillae, and hypopharynx.

Lactation: The process of producing and delivering milk to the young; also, the time period during which milk is produced.

Lactogenesis: The synthesis of milk within the mammary gland.

Lacunae: Small spaces among tissue cells through which hemolymph flows in open circulatory systems.

Lamarckism: An early evolutionary theory in which voluntary use or disuse of organs was thought to be capable of producing heritable changes.

Lamellae: Any one of several structures in the context of gas-exchange organs, usually found in gills; or, toothlike structures in the beak, forming a strainer that permits birds to retain food particles while still enabling water to flow from the closed mouth.

Laminate structure: Having a layered shell, as in the exoskeletons of crustaceans and the valves of clams.

Larva: The reproductively immature feeding stage in the development of many species of animals, including those insects which undergo complete metamorphosis.

Larynx: The vocal mechanism of mammals, consisting of a structure of cartilage at the upper end of the trachea containing the vocal folds.

Lazarus taxa: Groups that apparently disappear during a mass extinction only to appear again later.

Lecithotrophy: Nutrition of developing offspring from yolk reserves within the egg.

Lek: A territory used by a certain animal for mating.

Lepidotrichia: Modified scales that form the supporting rays of the fins of bony fishes.

Leptocephalus: The larval form of most eel species, bearing little resemblance to the adult eel environment.

Leptospirosis: Bacterial infection of spirochetes spread by urine.

Lexigrams: Symbols associated with objects or places in keyboard communication experiments with primates.

Lichens: Organisms formed by algae and fungi that are a source of food for tundra animals.

Life cycle: The sequence of development beginning with a certain event in an organism's life (such as the fertilization of a gamete), and ending with the same event in the next generation.

Life expectancy: The probable length of life remaining to an organism based upon the average life span of the population to which it belongs.

Life span: The maximum time between birth and death for the members of a species as a whole.

Life table: A chart that summarizes the survivorship and reproduction of a cohort throughout its life span.

Lift: An aerodynamic force created through differential flow above and below a structure.

Limb bud: Thickened epithelial cells along the lateral body fold that are underlain by mesoderm, creating a paddle-shaped extension from the trunk.

Limbic system: Brain structures related to the regulation of emotions.

Lipid: An organic molecule, such as a fat or oil, composed of carbon, hydrogen, oxygen, and sometimes phosphorus, that is nonpolar and insoluble in water.

Lipophilic: Fat soluble or water insoluble.

Litter: The offspring produced in a single birth; also referred to as a clutch.

Liver: An organ derived from the gut that secretes bile; it is connected to the gut by a duct through which its secretions enter the gut.

Local extinction: The loss of one or more populations of a species, but with at least one population of the species remaining.

Locomotion: The ability of an organism to move from one place to another as needed.

Logistic growth: A pattern of population growth that involves a rapid increase in numbers when the density is low but slows as the density approaches the carrying capacity.

Long-term pair bond: Pair-bonding that continues beyond a single reproductive period.

Longitudinal muscle: Muscle fibers that run along the longitudinal or anterior-posterior axis of the body.

Loop of Henle: A slender hairpin turn in the tubule where most adjustment of the water balance of the body occurs.

Lower chordates: A group within the Chordata that shows chordate characteristics in the larvae but is separated from vertebrates by the lack of a skeleton.

Luciferase: One of a group of enzymes that catalyzes the oxidation of a luciferin.

Luciferin: One of a group of organic compounds that emits visible light when oxidized.

Lumen: The central opening through the digestive tract, which is continuous from the mouth to the anus.

Lung: A concave inpocketing of the body wall of an animal (in contrast with a gill); lungs occur in air-breathing animals.

Lymphatic vessels: Very thin tubes that carry water, proteins, and fats from the gut to the bloodstream.

Lymphocyte: White blood cell that produces either cell-mediated or humoral immunity in response to foreign antigens.

Macroevolution: Large-scale evolutionary processes that result in major changes in organisms and allow them to change rapidly, occupy new adaptive niches, or develop novel body plans.

Macrophage: An animal that feeds on whole plants or animals or their parts; these can be carnivores, herbivores, or omnivores; or, a mature phagocytic cell that works with lymphocytes in destroying foreign antigens.

Madreporite: A fine-meshed sieve that opens from the sea water into the water vascular system of the echinoderms.

Male: An organism that produces sperm, the smaller of two different types of gametes.

Malphigian tubule: The primitive excretory organ of insects.

Mammary glands: The milk-producing glands found in all mammals.

Mandible: The hard part of a jaw or beak.

Mane: Long, thick hair growing from the neck.

Mantle: The outermost living tissue of mollusks; it makes shells, mother-of-pearl, and pearls.

Marked: An individual animal that is identifiable by marks that may be either man-made, such as metal bands or tags, or natural, such as the pattern of a giraffe.

Marsupial: A pouched mammal that gives birth to embryonic young that complete development in a pouch, attached to the mother's nipples.

Marsupium: Abdominal pouch containing mammary glands, which shelters the offspring of marsupials until they are fully developed.

Mass extinction: An event in which a large number of organisms in many different taxa are eliminated; there have been five such events in the history of life that resulted in the disappearance of more than 75 percent of all species.

Master control gene: A gene that single-handedly triggers the formation of an organ or structure.

Mate choice: The tendency of members of one sex to mate with particular members of the other sex.

Mate competition: Competition among members of one sex for mating opportunities with members of the opposite sex.

Maternal: Referring to the female parent.

Matrilines: Several generations of adult females all related by common descent from one female ancestor.

Matrix: Proteins or protein-chitin polymers that act as nucleation sites for mineralization in exoskeletons, or for the production of nails or claws.

Matrotrophy: Nutrition of developing offspring directly from the mother.

Maxillary: Retaining to the upper jawbone.

Medulla: The innermost layer of a hair.

Medusa (*pl.* medusae): Adult umbrella- or bell-shaped forms of jellyfish, with mouth facing downward.

Meiosis: Reduction division of the genetic material in the nucleus to the haploid condition; it is the process used by animal cells to form the gametes, during which the genes from the two parents are mixed.

Melanistic: Having dark coloration of skin and hair.

Melanophore: A melanin-containing cell.

Menagerie: A French word for zoos first used in the early 1700's to describe the keeping (and later exhibition) of animals.

Menstrual cycle: A series of regularly occurring

changes in the uterine lining of a nonpregnant primate female that prepares the lining for pregnancy.

Mesenchyme: Embryonic undifferentiated connective tissue, derived from mesoderm, that will eventually give rise to all forms of connective tissue in the adult.

Mesoderm: The middle layer of the three fundamental tissue layers that appear during development; it will give rise to mesenchymal cells, muscles, bones, and blood.

Mesoglea: Gelatinous material lying between the inner and outer layers of a jellyfish.

Mesozoic era: A period of geologic time between 70 and 225 million years ago; subdivided into the Cretaceous, Triassic, and Jurassic periods.

Metabolic rate: The rate (expressed as calories per minute) at which an animal produces and consumes energy.

Metabolism: The conversion of carbohydrates, proteins, or fats into chemical energy that can be used to accomplish work and generate heat.

Metamorphosis: An abrupt change from a larval body form, accompanied by many physiological changes in the determination, differentiation, and distribution of cells, into an adult body form.

Metazoa: Organisms which are multicellular.

Microevolution: Small-scale evolutionary processes resulting from gradual substitution of genes and resulting in very subtle changes in organisms.

Microphages: Animals that feed on small microscopic particles suspended in water or deposited on bottom sediments.

Migration: The movement of individuals resulting in gene flow, changing the proportions of genotypes in a population.

Migratory animals: Animals that move from one place to another for feeding or breeding.

Milk ejection: Also known as milk letdown, this is the reflex response of the mammary gland to suckling of the nipple; the hormone oxytocin mediates this reflex.

Mimicry: The resemblance of one species (the model) by one or more other species (mimics), such that a predator cannot distinguish among them.

Mitochondria: Self-replicating units in a cell that are responsible for the metabolic generation of energy for cell processes; these structures are used to estimate the relationships between groups of organisms; the more similar the DNA, the more closely related the groups.

Mitosis: The process of cellular division in which the nuclear material, including the genes, is distributed equally to two identical daughter cells.

Modality: A specific type of sensory stimulus or perception, such as taste, vision, or hearing.

Molars: Flat, stout teeth used for grinding food.

Molecule: The smallest part of a chemical compound.

Mollusks: A phylum of aquatic invertebrates, usually shelled, such as clams, mussels, and squid.

Molting: The process of replacing exoskeletons or feathers.

Monogamy: A mating system in which one male pairs with one female.

Monophyletic: A group of species that is believed to have a common ancestor; thus, the species are all members of a clade.

Monotreme: A primitive mammal, such as the platypus and spiny anteater, which lays eggs and has other archaic features.

Morphogenesis: The development of form, including the overall form of the organism and the form of each organ and tissue.

Morphology: The scientific study of body shape, form, and composition; includes gross morphology, which examines entire structures or systems, such as muscles or bones; histology, which examines body tissues; and cytology, which focuses on cells and their components.

Mortality rate: The number of organisms in a population that die during a given time interval.

Morula: A solid ball or mass of cells resulting from early cleavage divisions of the zygote.

Mosaic development: The process whereby early embryonic cells are determined by the cytoplasm they receive from the egg; also called determinate development.

Motor neuron: A nerve cell that transmits impulses from the central nervous system to an effector such as a muscle cell.

Motor unit: A motor neuron together with the muscle cells it stimulates.

Mouth: The anterior part of the digestive system, used for ingesting food; it leads into the oral cavity, which opens into the esophagus.

Mucosa: The lining of the inner wall of the gut facing the lumen.

Mucus: A secretion of the salivary glands and other parts of the digestive system which lubricates passages; also covers the internal nasal structures to aid in humidification, warming, and particle filtration.

Müllerian ducts: The embryonic ducts that will become the female oviducts or Fallopian tubes, uterus, and vagina.

Multicellular organisms: Organisms consisting of more than one cell; there are diverse types of cells, specialized for different functions and generally organized into tissues and organs.

Musk: Bad-smelling liquids made to mark territory or for self defense.

Musth: Aggressive rutting behavior during elephant mating.

Mutation: A change in the nucleotide sequence of a gene or of a controlling site; changes in genes alter the protein, whereas changes in controlling sites determine where and how much of a protein is produced; usually refers to genetic change significant enough to alter the appearance or function of the organism.

Mutualism: A type of commensalism in which both symbiotes benefit from the association in terms of food, shelter, or protection.

Muzzle: The area around the nose and mouth of an animal.

Myelinated axon: An axon surrounded by a glistening sheath formed when a supporting cell has grown around the axon.

Myocarditus: Inflammation of the heart muscle.

Myoepithelial cells: The specialized cells within the mammary gland that surround the alveoli and contract to force milk into the ducts during milk ejection.

Myosin: One of the two major contractile proteins making up the thick myofibrils.

Myrmecology: The study of ants.

N: A standard abbreviation for the size of an actual population; if $\hat{n}$, it is an estimated value.

Nacre: Shiny, pearly lining of some mollusk shells; mother of pearl.

Nasohypophysial opening: An opening in the head of modern agnathans leading to a sac that aids in olfaction.

Natality rate: The number of individuals that are born into a population during a given time interval.

Natural selection: The process of differential survival and reproduction that leads to heritable characteristics that are best suited for a particular environment.

Navigation: To follow or control the course of movement from the place of origin to a specific destination.

Nekton: An aquatic organism that has the ability to swim.

Nematocyst: Poison sting cell.

Nematode: A long, cylindrical worm; some are parasitic.

Neoteny: Either the retention of immature characteristics in the adult form or the sexual maturation of larval stages; it results in new kinds of adult body plans.

Neotropical fauna: The geographic faunal region that includes Central and South America.

Nephridia: Excretory tubules specialized for excretion and osmoregulation, with an external opening and with or without an internal opening.

Nephron: Tubular structures in the kidneys that extract filtrate, reabsorb nutrients and other valuable substances, and secrete wastes.

Nerve: A cordlike bundle of sensory and/or motor axons in the peripheral nervous system.

Nest box programs: Construction and placement of nest boxes in suitable habitat to provide nesting platforms for specific birds of prey.

Nest-parasite: Also called brood-parasite; an individual (or species) that lays its eggs in the nest of another individual (or species) and does no parenting at all.

Neural integration: Continuous summation of the incoming signals acting on a neuron.

Neural spine: A projection extending off the upper side of a vertebra.

Neurobiology: The study of the biology of the brain.

Neuroethology: The study of behavior as it relates to brain functions.

Neurons: Complete nerve cells that respond to specific internal or external environmental stimuli, integrate incoming signals, and sometimes send signals to other cells.

Neurosecretory cells: Specialized neurons capable of manufacturing and releasing hormones (neurosecretions or neurosecretory hormones) and discharging them directly into circulation.

Neurotoxin: A substance that damages the nervous system, most often nerves that control breathing and heart action.

Neurotransmitter: A signaling molecule that provides neuron-to-neuron communication in animal nervous systems; some double as hormones.

Neurulation: The process by which the embryo develops a central nervous system; formation of a neural plate and subsequent closure of the plate to form a neural tube.

Niche: An organism's role in its habitat environment, such as food producer, decomposer, parasite, plant eater (herbivore), meat-eater (carnivore).

Nidifugous birds: Ground-nesting birds.

Nipple: The raised area on the surface of the skin over the mammary gland that contains the duct openings.

Nocturnal: Active at night.

Nomadic: Moving about from place to place according to the state of the habitat and food supply.

Nomenclature: The part of systematics that deals with establishing a valid name for a species, according to specific guidelines.

Nonrapid eye movement (NREM) sleep: Sleep characterized by relaxed muscles and slow brain waves.

Nonruminating: Digesting grasses without chewing cud.

Notochord: A dorsal, flexible, rodlike structure extending the length of a vertebrate's body; serves as an axis for muscle attachment.

Nuclear ribosomal deoxyribonucleic acid: Nuclear DNA that codes for the ribosomal DNAs.

Nuclear transfer: The insertion of genetic material from a donor cell to a recipient cell; in reproductive technologies the recipient cell is an egg cell from which the nucleus has been removed.

Nucleic acid: An organic acid chain or sequence of nucleotides, such as DNA or RNA.

Nucleus (pl. nuclei): A central cell structure that controls the activity of the cell because of the genetic material it contains; or, a cluster of neuron cell bodies within the central nervous system.

Numerical response: The abundance of predators dependent upon the abundance of prey in a habitat.

Nutrient: A nourishing food ingredient.

Nutrient cycle: A description of the pathways of a specific nutrient (such as carbon, nitrogen, or water) through the living and nonliving portions of an ecosystem.

Nymph: The sexually immature feeding stage in the development of those insects which undergo incomplete metamorphosis.

Occipital condyle: Ball-shaped bone that connected the back of the skull to the fused upper vertebrae of the spine.

Olfaction: The sense of smell.

Olfactory receptors: Receptor organs which have very high sensitivity and specificity and which are "distance" chemical receptors.

Ommatidium: Individual unit of the multifaceted compound eye.

Omnivore: An animal that eats both plant material and animal material.

Ontogenetic trajectory: A model of development represented as a graph on which the y axis is an abstract representation of morphology and the x axis is time; initiation of growth, rate of growth, and cessation of growth are shown.

Ontogeny: The successive stages during the development of an animal, primarily embryonic but also postnatal.

Oogenesis: Gamete formation in the female; it occurs in the female gonads, or ovaries.

Open circulation: An open-ended sinus or arterial vessel system in which the circulation system does not return blood or hemolymph directly back to the heart.

Opportunistic omnivore: An animal who includes a variety of plant and animal material in its diet, depending on the availability of different foodstuffs.

Opposable: Capable of rotation so that the fingerprint surface of the thumb or big toe approaches the corresponding surfaces of other fingers or toes.

Opsin: A membrane-bound protein or pigment, which absorbs light.

Optic nerve: The main nerve taking information from the eyes to higher processing areas.

Optimum temperature: The narrow temperature range within which the metabolic activity of an animal is most efficient.

Order: A group of closely related genera; in mammals, orders are the well-recognized major groups, such as the rodents, bats, whales, and carnivores.

Organelle: A subcellular structure found within the cytoplasm that has a specialized function.

Organism: Any form of life.

Orientation: An inherent sense of geographical location or place in time.

Orienting reflex: An unspecific reflex reaction caused by a change in the quantity or quality of a stimulus; it will disappear or decrease after repeated presentations of the stimulus.

Ornithischia: One of the two main dinosaur groups, characterized by a pelvis in which the pubis is swung backward.

Ornithology: The scientific study of birds.

Ornithopods: Bipedal, herbivorous dinosaurs.

Os: Bone.

Osculum: An opening through which a sponge ejects water.

Osmoconformer: An organism whose internal osmotic pressure approximates the osmotic pressure of its environment; such an organism is also referred to as poikilosmotic.

Osmoregulation: The regulation of the ratio between all dissolved particles, regardless of their chemical nature as ions or molecules, and water.

Osmoregulator: An organism that maintains its internal osmotic pressure despite changes in environmental osmotic pressure; such an organism is also referred to as euryosmotic.

Ossicles: The small bones of the middle ear.

Osteichthyes: The taxonomic class in which the bony fishes are placed; contains species related to the ancestors of higher vertebrates.

Osteoblast: A bone cell which makes collagen and causes calcium deposition.

Osteogenesis: The total biological process by which bone is formed within the body; the process involves the action of osteoblasts and is also called ossification.

Ostium: A surface pore through which water enters a sponge.

Otolith: A dense mineral frame supported by sensory hair cells immersed in aqueous fluid; used in the auditory system of lower animals to detect acceleration.

Oval gland: The poison gland in the scorpion telson.

Ovarian diverticulum: Used to house embryos and to obtain nutrients via absorptive cells among viviparous species.

Ovary: The female gonad that is the source of eggs to be fertilized and hormones to maintain pregnancy.

Oviduct: A narrow, hollow tube which takes the newly ovulated egg from the ovary, provides the site for fertilization, and transports the new embryo to the uterus.

Oviparity: Production of shelled eggs by females.

Oviphagy: A mode of reproduction in which embryo sharks develop in the maternal uterus and feed on unfertilized eggs produced by the mother.

Ovipositor: A tube that extends from the female's body for depositing eggs.

Ovum (pl. ova): The female reproductive cell (gamete); a mature egg cell.

Owl pellets: Compacted packets of undigested prey that is regurgitated; owl pellets may be used to determine food habits.

Oxytocin: Hormone involved with pleasure during bonding.

Ozone layer: The ozone-enriched layer of the upper atmosphere that filters out some of the sun's ultraviolet radiation, which causes skin and other types of cancer.

Pacemaker: A specialized group of cardiac muscle cells in the right atrium which initiates the heartbeat; also called the sinoatrial node.

Paedomorphosis: The appearance of youthful characters of ancestors in later ontogenetic stages of descendants.

Pair bond: Close relationship between a male and female for breeding purposes.

Pair-bonding: Prolonged and repeated mutual courtship display by a monogamous pair, serving to cement the pair bond and to synchronize reproductive hormones.

Pair-rule genes: Segmentation genes of fruit flies that divide the anterior-posterior axis into two-segment units.

Paleontology: The branch of geology that deals with prehistoric forms of life through the study of fossil animals, plants, and microorganisms.

Paleozoic era: Time period from 570 to 245 million years ago, which comprises the Cambrian, Ordovician, Silurian, Devonian, Carboniferous, and Permian periods.

Palp: Oral sensory organs of arthropods.

Palynology: The study of pollen and spores; also called paleopalynology.

Pancreas: An organ derived from the gut that secretes digestive enzymes; it is connected to the gut by a duct through which its secretions enter the gut.

Papillae: Sharp, curved projections on the tongue.

Paramyosin: A structural protein associated with myosin myofibrils and thought to support them.

Parapodia: Paired lateral projections on each side of most segments, variously modified for locomotion, respiration, or feeding.

Parasite: Any organism that lives on or in other living organisms and obtains its food from them.

Parasite-mix: All the individuals and species of symbiotes living in a host concurrently.

Parthenogenesis: Asexual reproduction from unfertilized gametes, producing female offspring only.

Partial pressure: The pressure exerted by a specific gas in a mixture of gases such as the atmosphere; it is analogous to concentration.

Patagium: A soft, flexible membrane.

Patch reef: A small, isolated reef that is typically found in lagoons.

Pathogens: Bacterium or viruses which cause diseases.

Pectines: Comblike structures on the ventral surface of the scorpion that are used in chemoreception.

Pectoral and pelvic girdles: Skeletal structures that form a structural base for attachment of the paired fins in fishes, connecting them to the rest of the body's skeleton.

Pectoral fin: One of a pair of fins just behind the head of a fish, where arms of terrestrial vertebrates are attached.

Pedicle: A small bone spur from which an antler grows (for example, in deer).

Pedipalps: Clawlike appendages that are used to catch and hold prey.

Pelage: A mammal's fur coat.

Pelagic: The area of open water in the oceans; organisms that occur in the water column.

Pelvic fins: Paired fins found either near the tail end of the fish body or below the pectoral fins; related to the hindlimbs of higher vertebrates.

Peptide: A chemical combination of certain amino acids.

Peramorphosis: A group of phenomena in which descendant morphologies exceed the ancestral morphologies either in the rate at which they develop or in the length of time during which development occurs.

Perineal: Located between scrotum and anus in males or the equivalent region in females.

Period: The length of one complete cycle of a rhythm; ultradian rhythms are shorter than twenty hours, circadian rhythms are about twenty-four hours (twenty to twenty-eight hours), and infradian rhythms are longer than twenty-eight hours.

Periodicity hypothesis: The proposal that mass extinctions have occurred approximately every 26 million years over the past 250 million years.

Periosteum: The fibrous membrane which covers all bones except at points of articulation, containing blood vessels and many connections to muscles.

Perissodactyls: Ungulates having an odd number of toes.

Permeability: The tendency to permit the movement of a gas across a membrane.

Persistent organic pollutants (POPs): Chemicals that remain in the environment for a very long time and can be found at long distances from where they are used or released; they are nearly all of human origin.

Perspiration: A watery fluid exuded by sweat glands that contains small quantities of salts, urea, and uric acid.

pH: The negative logarithm of the hydrogen ion concentration, with higher hydrogen ion concentrations indicating lower pH; the pH scale goes from 0 to 14, with a pH of 7 being neutral, values below 7 indicating acidity, and values above 7 indicating alkalinity.

Phagocytosis: Obtaining food by engulfing it.

Phalanges: The free toes of the foot; some can be modified to claws, hoofs, or nails.

Phanerozoic era: An era of geologic time beginning approximately 544 million years ago at the start of the Paleozoic era, when animals with mineralized skeletons became common.

Pharynx: Lower part of vocal tract, connecting the mouth and nasal cavities to the larynx.

Phasic receptors: Receptors that adapt quickly to a stimulus.

Phasmid: Sense organs located in the tail region of roundworms that are important for detecting chemical signals in the environment.

Phenotype: The visible or outward expression of the genetic makeup of an individual.

Pheromone: A chemical produced by one member of a species that influences the behavior or physiology of another member of the same species.

Photoperiod: The measure of the relative length of daylight as it relates to the potential physiological responses that exposure to daylight evokes.

Photoperiodism: The responses of an organism to seasonally changing day length that cause altered physiological states, such as flowering or nonflowering in plants and reproductive seasons in animals.

Photophore: A light-emitting organ consisting of a lens, reflector, and light-emitting photogenic cells.

Photoreceptor: A cell containing membranes that house light-sensitive pigments.

Photosynthesis: The process by which green

plants and algae use sunlight as energy to convert carbon dioxide and water into energy-rich compounds such as glucose.

Phylogenetics: The study of the developmental history of groups of animals.

Phylogeny: The evolutionary history of taxa, such as species or groups of species; order of descent and the relationships among the groups are depicted.

Phylum (pl. phyla): The taxonomic category of animals and animal-like protists that is contained within a kingdom and consists of related classes.

Physiology: The study of the functions, activities, and processes of living organisms.

Phytophagous: Animals, also referred to as herbivorous, that feed on plants.

Phytoplankton: Small, aquatic free-floating plants.

Pigments: A variety of colored substances which impart color to feathers.

Pinna (pl. pinnae): The external ear of an animal.

Pinnipeds: Flipper-footed marine mammals, such as sea lions, fur seals, true seals, walruses.

Pit viper: A poisonous snake, such as a rattlesnake, which detects its prey via paired heat-sensing pits in its head.

Pitch: The frequency of sound; the higher the frequency, the greater its pitch.

Placenta: Structure that connects a fetus to the mother's womb; indicative of internal gestation of young.

Placodermi: Extinct class of fishes characterized by dense armor plating made of dermal bone.

Placoid denticles: Toothlike scales found in sharks and rays.

Plankton: Floating or weakly swimming plants and animals, usually very small in size.

Plantigrade: Walking on the sole, with the heel touching the ground.

Planula: Free-swimming, ciliated jellyfish larva.

Plastron: The portion of the shell that protects a turtle's belly (venter).

Plate tectonics: Often referred to as continental drift; the modern theory assumes that the current position of continents and oceans are the consequence of dynamic forces that involve land masses "floating" on the earth's molten core.

Pleistocene era: A period from about 1.8 million years ago to the last 100,000 years, characterized by alternating glacial and interglacial cycles; during the glacial periods ice sheets many miles deep covered as much as a third of the earth's surface.

Pleurocoel: Chamber formed in the centrum of a vertebra to reduce its weight.

Plexus: A group of nerve cells and their connections to one another.

Plumage: The feathers of birds.

Podites: The parts of the jointed appendages of arachnids.

Poikilotherm: Cold-blooded or ectothermic; any organism having a body temperature that varies with its surroundings; in general, reptiles, amphibians, fish, and invertebrates.

Polyestrous: Multiple sexual heats or periods of sexual excitement.

Polygamy: A mating system in which one male mates with several females (polygyny) or one female mates with several males (polyandry).

Polymorphism: The occurrence of two or more structurally or behaviorally different individuals within a species.

Polyp: Immature cylindrical forms of jellyfish with mouth facing upward.

Polyphyletic: A group of species believed not to have a common ancestor; thus, they are not members of a clade but are probably a product of convergent evolution.

Population: A group of individuals of the same species that live in the same location at the same time.

Population analysis: The study of factors that influence growth of biological populations.

Population density: The number of individuals in a population per unit area or volume.

Population regulation: Stabilization of population size by factors such as predation and competition, the relative impact of which depends on abundance of the population in a habitat.

Pore canals: Sites that house the cytoplasmic extensions of the crustacean hypodermis.

Positional information: A concept by which differentiating cells organize themselves to produce a particular tissue type based upon cell-to-cell interactions.

Postdisplacement: A form of paedomorphosis in which the initiation of growth in a descendant occurs later than in the ancestor, ensues at the ancestral rate, and ceases at the ancestral point.

Postnuptial molt: Replacement of feathers following the mating season.

Precambrian era: The earliest chapter of the earth's history, covering the time interval between the formation of the earth, about 4.6 billion years ago, and the beginning of the Cambrian period, about 570 million years ago.

Precocial: The condition of being strong and relatively well developed at birth or hatching, and thus not particularly dependent upon parental care.

Predation: The act of killing and consuming another organism.

Predentary bone: Keeled and pointed bone that terminated the lower jaw.

Predisplacement: A form of peramorphosis in which the initiation of growth in a descendant occurs earlier than in the ancestor, ensues at the ancestral rate, and ceases at the ancestral point.

Preening: A bird's act of grooming itself by cleaning and straightening its feathers.

Prehensile: Adapted for seizing or grasping.

Primary consumer: An organism that get its nourishment from eating primary producers, which are mostly green plants and algae.

Primary emotions: Emotions related to innate motivations.

Primary feathers: Flight feathers on the outer joint of the wing.

Primary production: The energy assimilated by green plants and stored as organic tissue.

Primates: A group of mammals including apes, chimpanzees, monkeys, humans, lemurs, and tarsiers.

Primer pheromone: A chemical substance that affects behavior by altering physiology and is therefore not rapid in its effects.

Primitive: Referring to a feature that reflects an ancestral condition, rather than those that represent more recent evolutionary changes (derived).

Principle of antithesis: The observation that signals communicating opposite meaning tend to be expressed using displays having opposite characteristics.

Prismatic layer: The outer crystalline layer of the molluscan shell.

Proboscis: A coiled, springlike sucking tube or "tongue" used to drink nectar.

Producers: Green plants and chemosynthetic organisms that can produce food from inorganic materials.

Progenesis: A form of paedomorphosis in which the initiation of growth in a descendant occurs at the same time as in the ancestor, ensues at the ancestral rate, but ceases earlier than at the ancestral point.

Progesterone: The hormone essential for maintenance of pregnancy that is secreted by the corpus luteum.

Proglottid: A body segment of a tapeworm that contains a set of reproductive organs, usually both ovaries and testes.

Prokaryote: An organism that has no internal membranes in its cells; the only membrane in a prokaryote is the cell membrane.

Prolactin: A hormone responsible for secretions of milk from the mammary glands of mammals and from the crops of birds.

Promiscuity: A mating system in which sexual partners do not form lasting pair bonds, where their relationship does not persist beyond the time needed for copulation and its preliminaries.

Prosimians: A group of primates that retain some primitive characteristics absent in higher primates.

Protandry: The condition of starting out male with the potential to become female.

Protective mimicry: Use of both color and form to mimic an inanimate feature of the environment.

Protein: An organic molecule containing carbon, hydrogen, oxygen, nitrogen, and sulfur and composed of large polypeptides in which over a hundred amino acids are linked together; proteins are the chief building blocks of cellular structures.

Protein hormone: A hormone type composed of protein, a long chain of amino acids encoded by a gene.

Prothoracic gland: The gland where ecdysone is made in insects.

Prothoracicotropic hormone (PTTH): A hormone made in the brain of insects which stimulates the prothoracic gland to make ecdysone.

Protists: Members of the kingdom Protista; the greater number are unicellular but, unlike bacteria, possess a nucleus and chromosomes.

Protogrammar: Word coined to signify the early foundation for grammar development found in primates.

Protogyny: The condition of starting out female with the potential to become male.

Protostomes: Annelids, mollusks, flatworms, and arthropods, a group linked by features of cell development including retention of the blastopore as the mouth.

Protozoa: Single-celled, animal-like organisms; all are eukaryotic (the cells have nuclei and other internal structures).

Proximal: Occurring near the base end of a limb.

Pseudocoel: A fluid-filled body cavity that is bounded by mesodermal muscle on the outside and endodermal epithelium on the internal boundary.

Pseudopods: Cytoplasmic extensions of a protozoan's body, used for locomotion and the engulfing of food.

Pubis: One of the three bones that make up the pelvis (the others are the ischium and ilium).

Pugmarks: Pawprints.

Punctuated equilibrium: The idea that new species form during relatively short (a few generations) speciation events and then persist for millions of years unchanged until they go extinct.

Pupa: Intermediate stage between the larval and adult stages of the life cycle, during which metamorphosis occurs.

Pyramid of numbers: A graph that plots the size of animals against the number of individuals, generally yielding a pyramidal shape when the X axis is shifted to the middle of the figure.

Quadrat: A sample plot of a specific size and shape used in one method of determining population size or species diversity.

Quadruped: Animal with four feet.

Quills: Sometimes referred to as spines; modified guard hairs; quills have barbed tips which can work themselves deeper into flesh once they have penetrated.

r strategy: A reproductive strategy involving high reproductive output; found often in unstable or previously unoccupied areas.

Rabies: Viral disease of the nervous system.

Radial sesamoid: Forefoot wrist bone.

Radial symmetry: An arrangement of body parts of an organism like the pieces of a pie around an imaginary central axis.

Radiochronometry: The determination of the age of an object using radioactive isotope decay rates.

Radioisotopes: Unstable elements that decay into stable forms at a constant rate, which are used to determine how long ago volcanic rocks solidified.

Radula: A tonguelike, toothed organ used to grind food or drill holes in shells of prey.

Rami: The branches of an arthropod limb or appendage.

Random genetic drift: The random change of gene frequencies because of chance, especially in small populations.

Random mating: The assumption that any two individuals in a population are equally likely to mate, independent of the genotype of either individual; this is equivalent to saying that all the gametes of all the individuals in a population are placed into a large pool, from which gametes are paired at random.

Rapid eye movement (REM) sleep: Sleep characterized by fast brain waves, during which dreaming typically occurs.

Recapitulation: The repetition of phylogeny in ontogeny or of the ancestral adult stages in the embryonic stages of descendants.

Recaptured: A previously marked animal that is either seen, trapped, or collected again after its initial marking.

Receptive field: The area upon or surrounding the body of an animal that, when stimulated, results in the generation of a response in the sense organ.

Receptor: A nerve ending specialized for the reception of stimuli; or, a protein molecule on or in a cell that responds to the hormone by binding to it and initiating a series of events that form the response.

Receptor cells: Sensory cells within sense organs that are directly responsible for detecting stimuli.

Receptor molecule: A molecule on the cell membranes of target tissues that binds to the hormone molecule and initiates the action of the hormone.

Receptor potential: A change in the distribution of electric charge across the membrane of a receptor cell in response to the presentation of a stimulus.

Recessive: Requiring two copies of a gene for the trait to be expressed.

Reciprocal relationship: Any type of coevolved, highly interdependent relationship between two or more species.

Reciprocal sacrifice: One explanation for acts of altruism among unrelated animals; an individual sacrifice is made under the assumption that a similar sacrifice may in turn aid the individual in the future.

Rectrices: Tail feathers.

Reflex arc: The entire nerve path involved in a reflex action.

Regulative development: The process whereby early embryonic cells are determined by their interactions with other cells; also called indeterminate development.

Releaser: A stimulus that releases a sequence of reflexes that always occur in the same order and manner.

Releaser pheromone: A chemical that generates an immediate and more or less predictable behavioral response in another member of the same species.

Remnants: Nesting material not deliberately placed in nest by birds.

Reproductive mode: A combination of life-history characteristics, including egg-deposition site and type of parental care.

Reproductive strategy: A set of traits that characterizes the successful reproductive habits of a group of organisms.

Reproductive success: The number of offspring produced by one individual relative to other individuals in the same population.

Reproductives: Sexually mature males and females.

Reptile: A class of vertebrates characterized by cleidoic eggs, dry scaly skin, a single bone in the middle ear, and several bones in each jaw.

Reservoir host: A host species other than the one of primary interest in a given research study.

Resilience stability: Stability exhibited by a community that changes its structure when disturbed but returns to its original structure when the disturbance ends.

Resource: A requirement for life, such as space for living, food (for animals), or light (for plants), not including conditions such as temperature or salinity.

Resource defense: The control of a resource indirectly or directly.

Resource-holding potential: The ability of an individual to control a needed resource relative to other members of the same species.

Respiration: The utilization of oxygen; in air-breathing vertebrates, the inhalation of oxygen and the exhalation of carbon dioxide.

Respiratory medium: The water or air that contains the oxygen used by an animal to carry out biochemical reactions.

Respiratory pigment: A protein that "supercharges" the body fluid (blood) with oxygen; the oxygen can bind to the pigment and then be released.

Respiratory surface: The gill, lung, or skin site at which oxygen is taken up from the air or water into the animal, with the release of carbon dioxide at the same time and site.

Restriction: Reduction of the developmental potency of a cell.

Retardation: The appearance of an organ later in the development of a descendant than in the ancestor as a result of a slowing of development.

Reticulate: Netlike covering of scales on legs.

Retina: The light-sensitive membrane at the back of the eye.

Reynolds numbers: The results of a formula that takes into account the velocity of an object, its characteristic length divided by the dynamic viscosity of the fluid.

Ribonucleic acid (RNA): A long, single-stranded molecule that amplifies, transports, and expresses the coded information in DNA.

Ribosomes: Small cytoplasmic particles that function in protein synthesis.

Ritualization: An evolutionary process that formalizes the context and performance of a display so that its meaning is clear and straightforward.

RNA. *See* **Ribonucleic acid (RNA)**.

Rostral bone: Keeled and pointed bone that terminated the upper jaw.

Royal jelly: The protein- and hormone-rich food that worker bees feed queen larvae.

Rudimentary: Short or small.

Ruminant: A herbivore that chews and swallows food which enters its stomach, is partly digested, is regurgitated and chewed again, and reenters the stomach for more digestion.

Rut: Activities associated with cervid mating behaviors.

Saggital: Bony skull top.

Salinity: A measure of the quantity of dissolved salts in seawater.

Saliva: The liquid containing enzymes secreted by the salivary glands that is injected into the host when the adult mosquito feeds.

Sampling: The process of collecting data, usually in such a manner that a statistically valid set of data can be acquired.

Saprovore: Any organism that consumes dead or decaying plant or animal matter.

Sarcomere: The fundamental unit of contraction within a muscle cell; repeating bands of actin and myosin.

Sarcoplasmic reticulum: Membrane-bound sacs that surround the myofibrils of muscle cells and which store and release calcium ions.

Saurischia: One of the two main dinosaur groups, characterized by a pelvis in which the pubis points forward.

Sauropods: Large, quadrupedal herbivorous dinosaurs that lived from the Early Jurassic to the end of the Cretaceous.

Scale of being: An arrangement of life-forms in a single linear sequence from lower to higher; also called a chain of being.

Scavenger: An animal that feeds on the dead carcasses of other animals.

Schema (pl. schemata): An innate releasing mechanism; a neural process that programs an animal for receiving a particular sign stimulus and causes a specific behavioral response.

Sclerotin: A hard, horny protein constituent of the exocuticle found in arthropods such as insects; it is superficially similar to vertebrate horn or keratin.

Scutellate: Overlapping of platelike or shieldlike scales.

Secondary emotions: Emotions with a strong social component.

Secondary feathers: Flight feathers on the inner wing.

Secondary metabolite: A biochemical that is not involved in basic metabolism, often of unique chemical structure and capable of serving a defensive role for the organism.

Secondary palate: The palate forming a shelf in the mouth that places the internal nares far back in the throat.

Segment polarity genes: Segmentation genes of fruit flies that divide the anterior-posterior axis into individual segments.

Segmentation: The division of a structure into linearly arranged segments; it can lead to the formation of somites, or it can lead to the formation of separate chondrification centers in a developing digit.

Segmentation genes: These include the gap rule, pair-rule, and segment polarity genes; they establish anterior-posterior segmentation in fruit flies without specifying segment identity.

Selection: A process that prevents some individuals from surviving and propagating while allowing others to do so.

Selective pressure: Evolutionary factors that favor or disfavor the genetic inheritance of various characteristics of a species.

Semen: Fluid produced by the male reproductive system that contains the sperm.

Semiochemical: A chemical messenger that carries information between individual organisms of the same species or of different species; pheromones and allelochemics are semiochemicals, but hormones are not.

Sensilla: Hairlike structures associated with nerves that act as mechanoreceptors and chemoreceptors.

Sensitive period: A period during which a given event produces a stronger effect on development or in which a given effect can be produced more rapidly than it can earlier or later.

Sensitization: An arousal or an alerting reaction which increases the likelihood that an organism will react; also, a synonym for loss of habituation with increased intensity of response.

Septa: A thin, muscular tissue used to separate segments into a series of ringlike cavities.

Septum: The bony structure that divides the nose into two sections.

Sequential hermaphrodite: Species or individual with the potential to change from one sex to the other.

Sequester: To store a material derived from elsewhere. In defenses, some predators sequester defensive properties from their prey to defend themselves from their own predators.

Sessile: An organism incapable of moving from its point of origin.

Setae: A needlelike, chitinous structure of the integument, used for locomotion; also, hairlike organs arising from the cuticle that are typically sens.

Sex hormones: Hormones (androgens in males, estrogens in females) associated with sex characteristics and sexual behavior.

Sex-limited traits: Features that are only expressed in one sex.

Sex pheromone: A volatile chemical released into the air by females to attract males.

Sex-role reversal: Generally used to refer to species in which the male does most of the parenting.

Sexual dimorphism: A difference in structure or behavior between males and females.

Sexual reproduction: Reproduction in which genes are exchanged between individuals.

Sexual selection: The process that occurs when inherited physical or behavioral differences among individuals cause some individuals to obtain more matings than others.

Sexual swelling: An estrogen-induced water retention that causes reddening and swelling in the perineal region and around the buttocks.

Shaft: Long, central spine of the feather; or, the main hair part, made of dead cells arranged in a complex fashion.

Shedding: A process through which organisms lose and replace their external covering.

Siblicide: Infanticide committed by the siblings of the individual killed.

Signal: Information transmitted through sound, such as bird calls, or through sight, such as body posture.

Signal pheromone: Nearly a synonym of releaser pheromone, but used with mammals to remove the suggestion of a programmed response and to indicate a more complex response.

Silurian era: A geological period from about 440 million years ago to about 400 million years ago; the first jawed fishes appeared during this period.

Simultaneous hermaphroditism: The condition of being simultaneously male and female.

Sinuses: Larger spaces, thought to represent through channels, for hemolymph in open circulatory systems, sometimes bound by membranes.

Siphon tube: Tube that extends from the rear of the larval abdomen at the air water interface and allows the larva to breath air.

Sit-and-wait foraging: Sitting in one place, waiting, and attacking prey as they move.

Slotting: The separation of primary feathers at the tip of the wings.

Slow muscle: Muscle cells that respond slowly to nervous impulses; in invertebrates, these muscle fibers have long sarcomeres and a high ratio of thin to thick myofibrils.

Snailing: The process in which the free-swimming larva (miracidium) of flukes utilizes the tissue of a snail as an intermediate host.

Social grooming: An activity maintaining social interaction, whereby debris is removed from a primate's hair.

Social wasp: A wasp living in a large colony.

Sociality: The tendency to form and maintain stable groups.

Sociobiology: The study of the biological basis of the social behavior of animals.

Soldiers: Large workers who defend the colony and often raid other colonies.

Solitary: Lives alone.

Somite: A condensation of mesoderm on either side of the notochord (a rod extending the length of the embryo), whose three different cell populations will differentiate during cellular migration into various types of connective tissue, muscles, and other structures and tissues.

Sound frequency: The distances between crests of sound waves measured in hertz.

Sound intensity: The loudness of a sound directly related to the amplitude of the sound waves measured in decibels.

Speciation: The formation of new species as a result of geographic, physiological, anatomical, or behavioral factors that prevent previously interbreeding natural populations from breeding with each other any longer.

Species: A group of animals capable of interbreeding under normal natural conditions; the smallest major taxonomic category.

Species selection: The idea that species are independent entities with their own properties, such as birth (speciation) and death (extinction); a higher level of selection above that of natural selection is postulated to take place on the species level.

Species-specific: A behavior or trait that characterizes members of a species, is innate, and is exclusive to that species.

Specific metabolic rate: The rate of metabolism per unit body mass (calories per gram per hour).

Spectacle: Transparent scale covering the eye as a replacement for the eyelid; occurs in some lizards and all snakes.

Spermaceti: A waxy substance in heads of sperm whales; usable for candles, ointments, lubricants, and cosmetics.

Spermatogenesis: Gamete formation in the male; it occurs in the male gonads, or testes.

Spermatophore: A tiny, mushroomlike structure deposited by a male salamander for transferring sperm to a female during courtship.

Spermiogenesis: The structural and functional

changes of a spermatid that lead to the formation of a mature sperm cell.

Sphincter: A ring of muscle that can close off a portion of the gut.

Spicule: A needlelike structure that is part of a sponge skeleton.

Spindle: In a muscle, the bundle of nuclear fibers formed during one stage of mitosis.

Spiracles: Openings on the outside of the insect abdomen that lead to breathing tubes.

Spongin: A fibrous skeletal material in soft sponges.

Stasis: The long-term stability and lack of change in fossil species, often spanning millions of years of geologic time.

Status badge: A visual feature that, based on its size or color or some other variation, indicates the social status of the bearer.

Stegosaurs: Quadrupedal, herbivorous dinosaurs with vertical bony plates along their backbones.

Stem cell: A determined, undifferentiated cell that is hormonally activated and changes into a specific cell type.

Stenohaline: The inability of an organism to tolerate wide ranges of salinity.

Stereoscopic vision: Vision with good depth perception.

Stereotyped behavior: An unlearned and unchanging behavior pattern that is unique to a species.

Steroid: A hormone that is made from cholesterol.

Stimulus: Any environmental cue that is detected by a sensory receptor and can potentially modify an animal's behavior.

Stomach: The part of the digestive system where mechanical breakdown of food is completed and chemical digestion begins.

Strategy: A behavioral action that exists because natural selection favored it in the past (rather than because an individual has consciously decided to do it).

Stratigraphy: In geology, a sequence of sedimentary or volcanic layers, or the study of them; indispensable for dating specimens.

Striated muscle: Voluntary or skeletal muscle, capable of conscious ennervation.

Strong acid: An acid that dissociates almost completely into its component ions; hydrochloric acid, for example, dissociates almost completely into hydrogen ions and chloride ions.

Studbook: A record-keeping system that provides information on an animal's lineage.

Submersible: A vessel, like a submarine, that can operate with or without occupants.

Subspecies: A group or groups of interbreeding organisms that are distinct and separated from similar related groups but not fully reproductively isolated.

Subungulates: Nonhoofed mammals that support their weight on more than the terminal phalanges; some, such as elephants and hyraxes, have pads under their metatarsals, others, such as the sirenians, have forelimbs modified into flippers.

Succession: Change in a plant or animal community over time, with one kind of organism or plant being replaced by others in a more or less predictable pattern.

Survivorship: The pattern of survival exhibited by a cohort throughout its life span.

Sustained annual yield: The harvest of no more animals than are produced, so that the total population remains the same.

Swim bladder: An internal organ evolved from the gut that allows a fish to regulate its vertical position in the water column (maintain its balance); also called an air bladder.

Symbiosis: A type of coevolved relationship between two species in which both participants benefit; a type of mutualism.

Sympatric: Populations of organisms living in the same place, not separated by a barrier that would prevent interbreeding.

Synapse: The point of contact between adjacent neurons, where nerve impulses are transmitted from one to the other.

Synchronization: Causing events to occur simultaneously.

Syrinx: The vocal mechanism of birds, consisting of one or more membranous structures at the lower end of the trachea, where the windpipe divides into two bronchial tubes leading to the lungs; the membranes vibrate due to pressure differences when air streams across their surfaces.

Systematics: The subdivision of biology that deals with the identification, naming, and classification of organisms and with understanding the evolutionary relationships among them.

Systemic: Referring to a group of organs that function in a coordinated and controlled manner to accomplish some end, such as respiration.

Systolic pressure: The pressure in blood vessels when the ventricles of the heart contract.

Tadpole: The larval stage of frogs and toads.

Tagmatization: Functional specialization of groups of segments.

Talons: The long, curved, and sharply pointed claws of a bird of prey; used for slashing and killing, holding and carrying of prey, and for defense.

Tapetum: Membrane layers.

Taphonomy: The study of the processes that lead to fossilization.

Target cells: Cells that contain hormone receptors.

Target organ: A specific body part that a particular hormone directly affects.

Taxon (*pl.* taxa): The basic unit of taxonomy; any of the categories of classification to which an organism may be assigned.

Taxonomy: A classification scheme for organisms based primarily on structural similarities; taxonomic groups consist of genetically related animals.

Teat: An elongated form of nipple that contains one duct opening.

Tectonic plate: Tectonic plate theory suggests that the earth's surface is composed of a number of oceanic and continental plates which have the ability to move slowly across the earth's asthenosphere.

Teleosts: Members of the infraclass Teleostei, the most advanced of the ray-finned fishes; they compose the vast majority of living bony fish species.

Tentacles: A long flexible arm or projection.

Terrestrial: Living on land.

Territorial behavior: The combination of methods and actions through which an animal or group of animals protects its territory from invasion by other species.

Tetrapods: Four-legged vertebrates (amphibians, reptiles, birds, and mammals).

Thanatocoenosis: An assemblage of fossil species from a particular environment; a fossilized community; compare to "biocoenosis," an assemblage of living species.

Thermogenesis: The generation of heat in endotherms by shivering or increased oxidation of fats.

Thermoregulation: The process by which animals maintain body temperatures within a certain range.

Theropoda: "Beast foot"; dinosaurs that lived from the Late Triassic to the terminal Cretaceous extinction event; most predatory and carnivorous dinosaurs belong to this group.

Thoracic: Related to the middle section of the body of arthropods, which are generally subdivided into a head, thorax, and abdomen; both thorax and abdomen may have several segments.

Thorax: The middle section of insects from which the legs and wings protrude.

Threat display: A territorial behavior exhibited by animals during defense of a territory, such as charging, showing bright colors, and exaggerating body size.

Threatened species: Animals or plants so few in number that they may soon be endangered and then extinct.

Thrust: The forward force that is developed by engines, rotors, or moving wings, which pushes airplanes, helicopters, or flying animals and which opposes drag.

Tissue: A group of cells with a similar structure and their intercellular matrix, which work together to accomplish a specific function.

Tonic receptors: Receptors that typically show little or no adaptation to a continuously applied stimulus.

Torpid: Dormant, numb, sluggish in action.

Totipalmate: Having all four toes fully webbed.

Totipotent: The ability of a cell to develop into any kind of cell in the body.

Toxin: Any substance, such as the venom in snakes or spiders, that is toxic to an animal.

Tracheal system: The respiratory system of insects and other terrestrial invertebrates; it consists of numerous air-filled tubes with branches extending into tiny channels in direct contact with body cells.

Tract: A cordlike bundle of parallel axons within the central nervous system.

Trait: A genetically inherited characteristic.

Transdetermination: An event by which a determined, undifferentiated cell changes its determination, thereby giving rise to a different tissue type.

Transduction: The translation of a stimulus's energy into the electrical and chemical signals that are meaningful to the nervous system.

Triploid: Having three of each chromosome; an abnormal state which is unable to produce normal haploid gametes.

Trophallaxis: The exchange of bodily fluids between nestmates, either by regurgitation or by feeding on secreted or excreted material.

Trophic hormones: Hormones that stimulate another endocrine gland.

Trophic level: A level in a food chain or food web at which all organisms consume the same general types of food.

Tropomyosin: A double-stranded protein that lies in the grooves of actin myofibrils, blocking actin from attachment to myosin.

Troponin: A globular protein composed of three subunits; one subunit binds calcium ions, and another draws tropomyosin away from actin, which allows myosin to form crossbridges constituting the third subunit.

True horn: The permanent horns found in animals such as cattle, sheep, and goats.

Tubule: The long, slender part of the nephron that is the location of almost all kidney function.

Turbinates: Bony structures that define the internal nasal anatomy.

Turbulence: Flow that is chaotic and may create stall conditions through the loss of lift.

Twitch: A rapid muscular contraction followed by relaxation that occurs in response to a single stimulus.

Tympanic membrane: Eardrum or other surface which serves the purpose of converting sound waves into mechanical vibration.

Ungulate: A hoofed mammal from the order Artiodactyla (pigs, cattle, antelope, and their relatives) or from the order Perissodactyla (horses, rhinoceroses, tapirs, and their relatives).

Unguligrade: Walking on the tips of the toes.

Unicellular: Consisting of only one cell.

Uniformitarianism: The belief that the earth and its features are the result of gradual biological and geological processes similar to the processes that exist today.

Uniramous: Having only one rami; antennae appear as a single, nonbifurcated structure.

Upper Paleolithic era: The era from 30,000 B.C.E. to 3500 B.C.E., when humans first began to affect European wildlife populations.

Urea: A substance formed from by-products of protein metabolism and excreted by the kidney.

Urogenital groove: A slitlike opening behind the genital tubercle that will become enclosed in the penis but remain open in females.

Uterus: The hollow, thick-walled organ in the pelvic region of females that is the site of menstruation, implantation, development of the fetus, and labor.

Valves: Specialized, thickened groups of muscle cells in the heart chambers, major arterial trunks, arterioles, and veins which prevent backflow of blood.

Vane: Flat, broad web emerging from opposing sides of the feather shaft.

Vascular papilla: A protuberance having a blood supply.

Vector: Transmits pathogens from one host to another.

Vein: A tubelike, elastic channel with thin walls enforced with smooth muscles and valves that transport oxygen-poor blood; veins start as fine venules in tissues connected with capillary beds.

Velvet: A hairy skin richly endowed with blood vessels that covers developing antlers.

Venom: A toxic substance that must be injected (instead of ingested) to immobilize or kill prey.

Ventilation: The movement of the respiratory medium to and across the site of gas exchange.

Ventricle: The right and left chambers of the heart, which pump blood into pulmonary and systemic circulation, respectively.

Vertebrates: A subphylum of the phylum Chordata that includes all animals with backbones, and thus all animals possessing endoskeletons.

Vestigial: An organ that is no longer biologically useful.

Vibrissae: Stiff hairs, whiskers, or bristles projecting as feelers from the nose and the head, used to provide positional information to an animal.

Vicariance: An event involving the erection of a barrier to dispersal that splits the distribution of an organism and facilitates the differential evolution of the geographically separated descendants.

Viscera: Any internal body organ, such as intestines or entrails.

Viscosity: The stickiness of a fluid created by internal forces as molecular attractions.

Visual cortex: The part of the cerebral cortex concerned with vision.

Visual predation: Catching prey (such as insects) by sighting them visually, judging their exact position and distance, and pouncing on them.

Vitamin: An organic nutrient that an organism requires in very small amounts and which generally functions as a coenzyme.

Vitelline envelope: The protective layers that form around the egg while it is still in the ovary.

Vivaria: A Latin word for a structure housing living animals, first used by the Romans to describe the places holding elephants and other animals for their shows.

Viviparous: Characterized by live birth (as opposed to egg-laying).

Vocal folds: Small, laminated sheets of muscle which meet at the front of the larynx; they are open for breathing, and are brought together to vibrate for voiced sounds.

Volatiles: Chemical compounds that are vapor or gas at environmental temperatures or that readily release many of their molecules to the vapor phase.

Voluntary: Capable of being consciously controlled.

Warm-blooded: Referring to animals whose body temperatures are maintained at a constant level by their own metabolisms.

Warning coloration: The bright colors seen on many dangerous and unpalatable organisms that warn predators to stay away.

Weak acid: An acid that does not dissociate to a great extent; carbonic acid, for example, dissociates to produce some ions, but most of the molecules remain in their original forms.

White matter: The part of the central nervous system primarily containing myelinated axon tracts.

Wildlife: All living organisms; traditionally, the term included only mammals and birds that were hunted or considered economically important.

Wildness: Characteristics that define the biological and behavioral life of a species in the wild.

Wing loading: Ratio of weight to lifting area of wing; birds with large wings and light weight, such as the swallow-tailed kite, have a low wing loading, while birds with small wings and large, heavy bodies, such as pheasant or turkey, have a high wing loading.

Wolffian ducts: An embryonic duct system that becomes the internal accessory male structures that carry the sperm.

Workers: Sterile, wingless female social insects.

Xenotransplantation: The transplantation of organs from one species to another.

Zeitgeber: "Time giver" in German, also referred to as a synchronizer or entraining agent; most often the light-dark cycle, noise, feeding, or, for humans, societal factors.

Zona pellucida: Mammalian protective layer analogous to the vitelline envelope.

Zone fossils: Fossils that characterize a period of time and can be used to provide a relative date for a rock.

Zone of polarizing activity (ZPA): A region at the posterior base of the limb bud that seems to influence the distal development of pattern in a developing limb.

Zoo: An English abbreviation of the term "zoological garden"; first used in the early 1800's to describe the early British zoos, it tended to replace the word menagerie in the 1900's.

Zoogeography: The study of the distribution of animals over the earth.

Zoonoses: Diseases transmissible from animals to humans.

Zygodactyl: Having two toes pointing backward and two toes pointing forward.

Zygote: The single cell formed when gametes from the parents (ova and sperm) unite; a one-celled embryo.

ANIMAL TERMINOLOGY

Animal	*Male*	*Female*	*Young*	*Group of Animals*
aardvark	—	—	pup	—
alligator	bull	cow	hatchling	congregation, pod, bask
alpaca	—	—	cria	herd, flock
American pronghorn	buck	doe	fawn	herd
anteater	—	—	pup	—
antelope	buck	doe	calf	herd
ape	male	female	baby	shrewdness
armadillo	male	female	pup	—
ass	jack	jenny	foal	herd, band, pace
baboon	male	female	infant	tribe
badger	boar	sow	kit, cub	cete
bat	male	female	pup, battling	colony
bear	boar	sow	cub	sloth, sleuth
beaver	male	female	pup, kitten	lodge, colony
bee	drone	queen, worker	larva	hive, swarm, bike, drift, grist
bird	cock	hen	hatchling, nestling, chick	dissimulation, fleet, flight, flock, parcel, pod, volery, drove, brace
boar	boar	sow	piglet, shoat, farrow, boarlet	singular, sounder
buffalo	bull	cow	calf	gang, herd
butterfly	male	female	caterpillar, larva, pupa, chrysalis	swarm, rabble, army (caterpillars)
camel	bull	cow	calf	flock
cat	tomcat	queen	kitten, kit, kittling, cub	clutter, clowder, litter, kindle, pounce
cheetah	male	female	cub	—
chicken	rooster	hen	chick, pullet (female), cockrell (male)	flock, brood (hens), clutch or peep (chicks)
chinchilla	—	—	—	herd
clam	—	—	larva	bed
cockroach	—	—	nymph	—
codfish	male	female	codling, hake, sprag, sprat	school
cormorant	cock	hen	chick	gulp
cow	bull	cow, heifer	calf	herd, drove, drift, mob
coyote	dog	bitch	pup, whelp	pack, rout

Animal	*Male*	*Female*	*Young*	*Group of Animals*
crane	cock	hen	chick, craneling	sedge, siege, herd
crocodile	bull	cow	crocklet	bask, congregation
crow	cock	hen	chick	murder, muster, horde
deer	buck, stag, hart	doe, hind	fawn, knobber, brockett	herd, mob, bevy
dinosaur	bull	cow	hatchling, juvenile	herd (plant-eaters), pack (meat-eaters)
dog	dog	bitch	pup	kennel, pack (wild dogs), litter
dolphin	bull	cow	pup, calf	pod, herd, school
donkey	jack, jackass	jennet, jenny	colt, foal	drove, herd
dove	cock	hen	chick	flight, piteousness
duck	drake	duck	duckling	badelynge, brace, bunch, flock, paddling, raft, team, flush
eagle	male	female	fledgling, eaglet	aerie, convocation, jubilee
echidna	—	—	puggle	—
eel	male	female	elver	swarm
elephant	bull	cow	calf	herd, parade, memory
elk	bull	cow	calf	herd, gang
falcon	tercel, terzel	falcon	chick	—
ferret	hob	jill	kit	business
finch	cock	hen	chick	charm
fish	male	female	fry, fingerling, alevin	draft, run, school, shoal, nest
flamingo	—	—	—	stand
fly	—	—	maggot	cloud, swarm
fox	reynard, dog	vixen	kit, cub, pup	skulk, leash, earth
frog	male	female	tadpole, pollywog, froglet	army
gerbil	buck	doe	pup	horde
giraffe	bull	doe, cow	calf	herd, corps, tower, group
gnat	male	female	larva	cloud, horde
gnu	bull	cow	calf	herd
goat	buck, billy	doe, nanny	kid, billy	trip, tribe, herd, flock
goldfish	male	female	—	charm
goose	gander	goose	gosling	flock, gaggle, skein, wedge, brood
gorilla	male	female	infant	band
grasshopper	male	female	nymph	swarm
grouse	cock	hen	chick, cheeper	covey, pack
guinea pig	buck	doe	pup	group
gull	cock	hen	chick	colony
hamster	buck	doe	pup	horde

Animal	*Male*	*Female*	*Young*	*Group of Animals*
hare	buck	doe	leveret	down, husk, warren
hawk	tiercel	hen	eyas	aerie, cast, kettle, boil
hedgehog	boar	sow	piglet, pup	array
heron	cock	hen	chick	sedge, siege
herring	male	female	sprat	army
hippopotamus	bull	cow	calf	herd, bloat
hog	boar	sow	shoat, farrow	drove, herd, litter
hornet	male	female	larva	nest
horse	stallion, stud	mare, dam	foal, colt (male), filly (female)	stable, harras, herd, team, string, field
hound	dog	bitch	pup	cry, mute, pack, kennel
hummingbird	cock	hen	chick	charm
hyena	male, dog	female, bitch	cub, pup	clan, crackle
hyrax	buck	doe	bunny	bury
jay	cock	hen	chick	band, party, scold
jellyfish	—	—	ephyna	smack
kangaroo	buck, boomer, jack	doe, flyer, jill, roo	joey	troop, mob, herd
koala	male	female	joey	—
lark	cock	hen	chick	exaltation
lemur	male	female	infant	group
leopard	leopard	leopardess	cub	leap, prowl
lion	lion	lioness	cub	pride
llama	male	female	cria	herd
locust	—	—	—	host, swarm
louse	male	female	nymph	colony, infestation
magpie	cock	hen	chick	tiding, tribe, charm, gulp, flock, murder
mallard	drake	duck	duckling	flush, sord, brace
mole	male	female	pup	labor
monkey	male	female	infant	troop, cartload
moose	bull	cow	calf	herd
mosquito	male	female	nymph	swarm
mouse	buck	doe	pup, pinkie, kitten	horde, mischief
mule	jack	hinney	foal	barren, pack, span
nightingale	cock	hen	chick	watch
opossum	jack	jill	joey	passel
ostrich	cock	hen	chick	flock
otter	male, dog	female, bitch	whelp, pup, cub	family, raft, bevy, romp
owl	male	female	owlet	parliament

Animal	*Male*	*Female*	*Young*	*Group of Animals*
ox	steer, bull	cow	stot, calf	drove, herd, yoke, team
oyster	—	—	spat	bed
panda	boar	sow	cub	—
parrot	cock	hen	chick	company, flock
partridge	cock	hen	chick, cheeper	covey
peafowl	peacock	peahen	peachick	muster, ostentation, pride
penguin	male	female	chick	rookery, colony, parcel, huddle
pheasant	cock	hen	chick	brood, nye, bouquet, nide, nest
pig	boar	sow	piglet, shoat, farrow, suckling	drove, drift, herd, litter, sounder
pigeon	cock	hen	squab, squeaker	flock, kit
polar bear	boar	sow	cub	pack
porcupine	boar	sow	pup, porcupette	prickle
porpoise	bull	cow	calf	herd, pod, school
prairie dog	boar	sow	pup	coterie, town
quail	cock	hen	chick	bevy, covey, drift
rabbit	buck	doe	kitten, bunny, kit	colony, drove, leash, nest, trace, warren, down, husk, trip, bury
raccoon	boar	sow	cub	nursery
rat	buck	doe	pup, pinkie, kitten	horde, mischief
raven	cock	hen	chick	congress, unkindness
reindeer	bull	cow	calf	herd
rhinoceros	bull	cow	calf	crash
salmon	—	—	parr, smolt, grilse	school, run, shoal
sand dollar	male	female	larva, pluteus, juvenile	—
sea urchin	male	female	larva, pluteus, juvenile	—
seal, sea lion	bull	cow	pup	herd, pod, rookery, harem, colony, crash
shark	bull	female	pup, cub	school, shiver
sheep	buck, ram	ewe, dam	lamb, lambkin, lambling, cosset	drift, drove, flock, herd, mob, trip
skunk	—	—	kit	surfeit
snake	male	female	snakelet, neonate, hatchling	bed, nest, pit
snipe	cock	hen	chick	wisp, walk
sparrow	cock	hen	chick	host
spider	male	female	spiderling	—
squirrel	buck	doe	pup, kit, kitten	dray, scurry
starfish	male	female	larva, pluteus, juvenile	—
stork	cock	hen	chick	mustering
swallow	cock	hen	chick	flight
swan	cob	pen	cygnet, flapper	bevy, game, herd, team, wedge

Animal	*Male*	*Female*	*Young*	*Group of Animals*
tapir	male	female	calf	—
termite	male	female	larva	swarm, colony
tiger	tiger	tigress	cub, whelp	ambush, streak
toad	male	female	tadpole, toadlet	knot
trout	male	female	fry, fingerling	hover
turkey	tom	hen	poult	rafter, gang
turtle	male	female	hatchling	bale, nest
turtle dove	—	—	chick	pitying
viper	male	female	snakelet	nest
vulture	cock	hen	chick	colony
wallaby	jack	jill	joey	mob
walrus	bull	cow	cub, pup	pod, herd
wasp	drone	queen, worker	larva	colony
weasel	dog, buck, jack, hob	bitch, doe, jill	kit	gang
whale	bull	cow	calf	gam, grind, herd, pod, school
wolf	dog	bitch	pup, whelp	pack, rout
wombat	jack	jill	joey	mob, warren
woodpecker	male	female	chick	descent
wren	cock	jenny	chick	flock, herd
yak	bull	cow	calf	herd
yellow jacket	drone	queen, worker	larva	colony
zebra	stallion	mare	colt, foal	herd, crossing

TIME LINE

The evolutionary diagram presented below, based on fossils and taxonomic studies, indicates a branch-like bush. All currently living groups are at the tips of the branches. The arrangement of animal groups in classifications on the evolutionary tree thus indicates not only how the groups are related to one another but also when the linkages occurred.

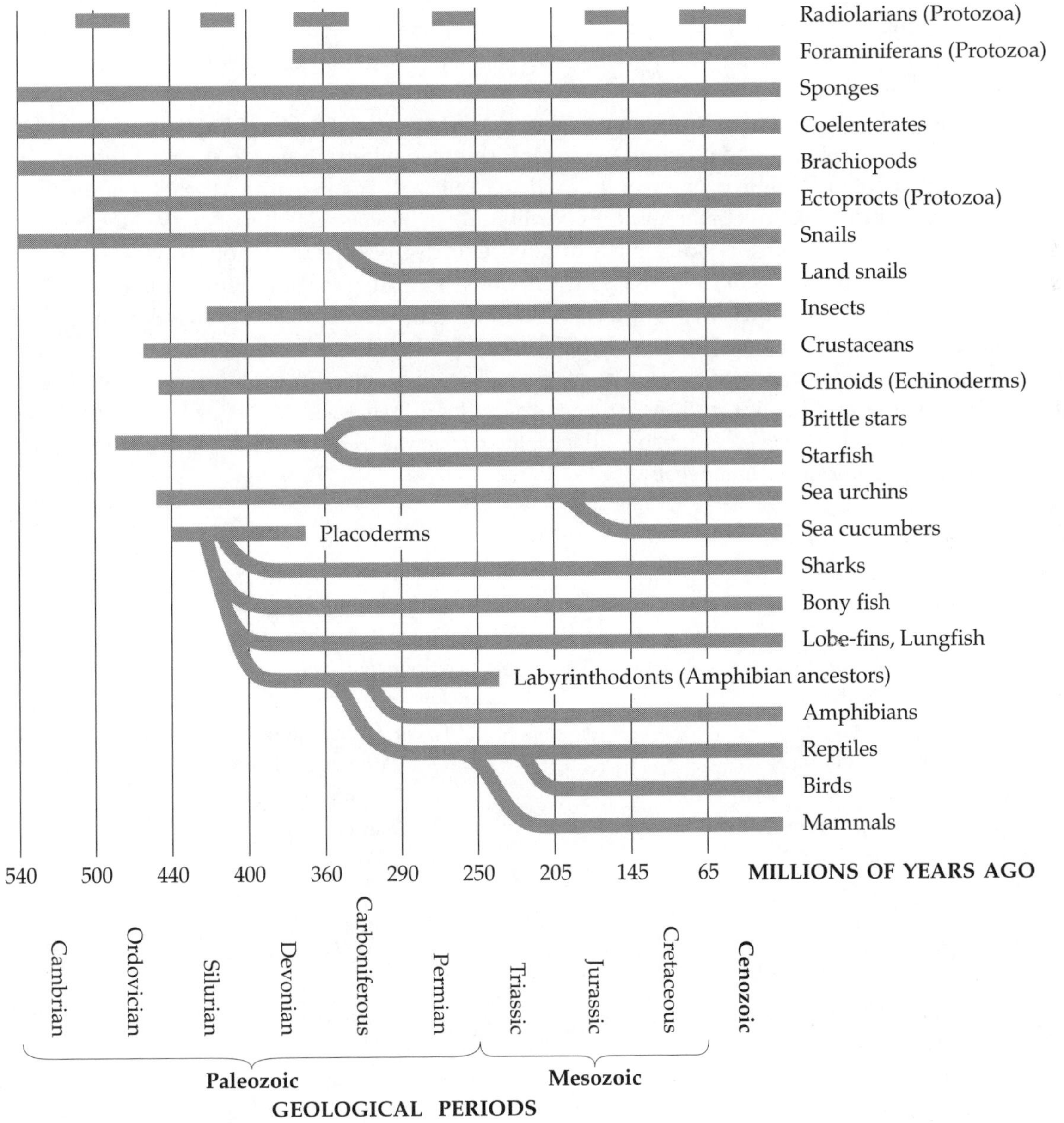

CLASSIFICATION TABLE

The scientific groupings of the animal world on which microbiologists and zoologists more or less agree are shown in their Latin, not vernacular, designations. They indicate both evolutionary history—how the animal kingdom developed over time—as well as the relationships among groups. In the seventeenth and eighteenth centuries, when these classifications were first compiled by Carolus Linnaeus (1707-1778) and others, structure and appearance were used as the most important basis. After Charles Darwin (1809-1882) propounded his theory of natural selection, animals were understood to have evolved and changed during long periods of geological time, with one kind of animal giving rise to another. Thus, it was only after the mid-nineteenth century that the arrangement of animal groups in a classification was expected to indicate how the groups were related to one another in this historical and evolutionary sense.

The following list, taken from *The New Larousse Encyclopedia of Animal Life* (1980), begins with the oldest animal form, phylum Protozoa, and proceeds to the most recent, phylum Chordata. Whether the phylum includes a large number of members or fewer than a dozen, all share many features that both unite them and mark them as distinct from all other fauna. Subordination in the classification of phyla—the primary divisions of the animal kingdom—is indicated by differences in degrees of indentation. Thus, where applicable, each phylum in the kingdom Animalia is broken down according to subphylum, superclass, class, subclass, superorder, order, infraclass, suborder, superfamily, family, section, and group. Since the number of animal species, from microscopic one-celled protozoans to humans, is estimated to be well over one million and perhaps over two million, and new ones are still being discovered, the lowest levels of classification, namely, genus and species, as well as subgenus and subspecies, are necessarily omitted. Thus, *Canis lupus* for wolf or *Canis familiaris* for dog (all varieties), from the family Canidae, will not appear because the numbers would be unmanageable.

—*Peter B. Heller*

PHYLUM PROTOZOA

- Class Mastigophora
  - Subclass Phytomastigophora
    - Order Phytomonadida
    - Order Euglenoidida
    - Order Cryptomonadida
    - Order Chrysomonadida
    - Order Dinoflagellata
  - Subclass Zoomastigophora
    - Order Protomonadida
    - Order Polymastigida
    - Order Trichomonadida
    - Order Opalinida
- Class Sarcodina
  - Subclass Rhizopoda
    - Order Rhizomastigida
    - Order Amoebina
    - Order Testacida
    - Order Foraminifera
    - Order Heliozoida
    - Order Radiolarida
- Class Sporozoa
  - Subclass Gregarinomorpha
    - Order Archigregarinida
    - Order Eugregarinida
      - Suborder Cephalina
      - Suborder Acephalina
  - Subclass Coccidiomorpha
    - Order Eucocciida
      - Suborder Adeleidea
      - Suborder Eimeriidea
      - Suborder Haemosporidia
- Class Cnidosporidea
- Class Ciliata
  - Subclass Holotricha
    - Order Gymnostomatida

Suborder Rhabdophorina
Suborder Cyrtophorina
Order Suctorida
Order Trichostomatida
Order Hymenostomatida
Order Peritrichida
Order Astomatida
Subclass Spirotricha
Order Heterotrichida
Order Hypotrichida
Order Entodiniomorphida

PHYLUM MESOZOA
Order Dicyemida
Order Orthonectida

PHYLUM PORIFERA
Class Calcarea
Class Hexactinellida
Class Demospongiae
Class Sclerospongiae

PHYLUM CNIDARIA
Class Hydrozoa
Order Athecata
Family Tubulariidae
Family Clavidae
Family Corynidae
Family Bougainvilliidae
Family Hydridae
Order Thecata
Family Campanulariidae
Family Lafoeidae
Family Sertulariidae
Order Limnomedusae
Order Trachymedusae
Order Narcomedusae
Order Siphonophora
Order Hydrocorallinae
Suborder Milleporina
Suborder Stylasterina
Class Scyphozoa
Order Semaeostomeae
Order Rhizostomeae
Order Coronatae
Order Cubomedusae
Order Stauromedusae
Class Anthozoa
Subclass Octocorallia
Order Alcyonacea
Order Gorgonacea
Order Pennatulacea
Subclass Zoantharia
Order Actiniaria
Order Corallimorpharia
Order Scleractinia
Order Zoanthiniaria
Subclass Ceriantipatharia
Order Antipatharia
Order Cerianthidea

PHYLUM CTENOPHORA
Class Tentaculata
Order Cydippida
Order Lobata
Order Cestida
Order Platyctenea
Class Nuda
Order Beroida

PHYLUM PLATYHELMINTHES
Class Turbellaria
Order Polycladida
Suborder Cotylea
Suborder Acotylca
Order Tricladida
Suborder Maricola
Suborder Paludicola
Suborder Terricola
Order Protricladida
Suborder Crossocoela
Suborder Cyclocoela
Order Eulecithophora
Order Perilecithophora
Order Archoophora
Order Temnocephala
Class Monogenea
Class Digenea
Class Cestoda
Subclass Cestodaria (= Cestoda Monozoa)
Subclass Eucestoda (= Cestoda Merozoa)

PHYLUM NEMERTINA (= RHYNCHOCOELA)
- Class Anopla
  - Order Palaeonemertini
  - Order Heteronemertini
- Class Enopla
  - Order Hoplonemertini
  - Order Bdellonemertini

PHYLUM ASCHELMINTHES
- Class Nematoda
  - Subclass Aphasmida
    - Order Chromadorida
    - Order Enoplida
  - Subclass Phasmida
    - Order Rhabditida
    - Order Strongylida
    - Order Ascaridida
    - Order Tylenchida
    - Order Spirurida
- Class Rotifera
  - Order Seisonidea
  - Order Bdelloidea
  - Order Monogononta
    - Suborder Ploima
    - Suborder Flosculariacea
    - Suborder Collothecacea
- Class Gastrotricha
  - Order Chaetonotoidea
  - Order Macrodasyoidea
- Class Kinorhyncha
- Class Priapuloidea
- Class Nematomorpha
  - Order Nectonematoidea
  - Order Gordioidea

PHYLUM ACANTHOCEPHALA

PHYLUM ENTOPROCTA

PHYLUM BRYOZOA
- Class Phylactolaemata
- Class Gymnolaemata
  - Order Cyclostomata
  - Order Cheilostomata
  - Order Ctenostomata

PHYLUM PHORONIDA

PHYLUM BRACHIOPODA
- Class Inarticulata
- Class Articulata

PHYLUM MOLLUSCA
- Class Monoplacophora
- Class Amphineura
  - Subclass Polyplacophorea (= Loricata)
  - Subclass Aplacophorea
- Class Gastropoda
  - Subclass Prosobranchea
    - Order Archaeogastropoda
    - Order Neritoidea
    - Order Mesogastropoda
    - Order Neogastropoda
  - Subclass Opisthobranchia
    - Order Cephalaspidea (= Bullomorpha)
    - Order Aplysiacea (= Anaspidea)
    - Order Thecosomata
    - Order Gymnosomata
    - Order Notaspidea(= Pleurobranchomorpha)
    - Order Acochlidiacea
    - Order Sacoglossa
    - Order Acoela (= Nudibranchia)
  - Subclass Pulmonata
    - Order Basommatophora
    - Order Geophila (= Stylommatophora)
- Class Scaphopoda
- Class Bivalvia (= Lamellibrancha = Pelecypoda)
  - Subclass Protobranchia
  - Subclass Lamellibranchia
- Class Cephalopoda
  - Subclass Nautiloidea
  - Subclass Coleoidea
    - Order Sepioidea
    - Order Teuthoidea
    - Order Octopoda
    - Order Vampyromorpha

PHYLUM SIPUNCULA

PHYLUM ECHIUROIDEA

PHYLUM ANNELIDA
- Class Polychaeta
  - Family Phyllodocidae
  - Family Tomopteridae
  - Family Nephtyidae
  - Family Glyceridae
  - Family Aphroditidae
  - Family Polynoidae
  - Family Nereidae
  - Family Syllidae
  - Family Eunicidae
  - Family Amphinomidae
  - Family Ariciidae
  - Family Cirratulidae
  - Family Magelonidae
  - Family Arenicolidae
  - Family Capitellidae
  - Family Maldanidae
  - Family Spionidae
  - Family Chaetopteridae
  - Family Sabellariidae
  - Family Oweniidae
  - Family Pectinariidae (= Amphictenidae)
  - Family Ampharetidae
  - Family Terebellidae
  - Family Sabellidae
  - Family Serpulidae
- Class Oligochaeta
- Class Hirudinea

PHYLUM ARTHROPODA
- Class Onychophora
- Class Pauropoda
- Class Diplopoda
- Class Chilopoda
- Class Symphyla
- Class Insecta
  - Subclass Apterygota
    - Order Thysanura
      - Family Lepismatidae
      - Family Machilidae
    - Order Diplura
      - Family Campodeidae
      - Family Japygidae
      - Family Projapygidae
    - Order Protura
    - Order Collembola
      - Suborder Arthropleona
        - Superfamily Poduroidea
        - Superfamily Entomobryoidea
      - Suborder Symphypleona
        - Family Sminthuridae
  - Subclass Exopterygota
    - Order Ephemeroptera
      - Family Baetidae
      - Family Siphlonuridae
      - Family Caenidae
      - Family Ecdyonuridae
    - Order Odonata
      - Suborder Zygoptera
        - Family Agriidae
        - Family Coenagriidae
      - Suborder Anisozygoptera
      - Suborder Anisoptera
        - Family Aeshnidae
        - Family Libellulidae
        - Family Gomphidae
        - Family Petaluridae
    - Order Plecoptera
      - Family Eustheniidae
      - Family Pteronarcidae
      - Family Leuctridae
      - Family Capniidae
      - Family Nemouridae
      - Family Perlidae
    - Order Grylloblattodea
    - Order Orthoptera
      - Family Tettigoniidae
      - Family Stenopelmatidae
      - Family Gryllidae
      - Family Gryllotalpidae
      - Family Acrididae
      - Family Pneumoridae
      - Family Tetrigidae
      - Family Tridactylidae
      - Family Cylindrachetidae
    - Order Phasmida
      - Family Phylliidae
    - Order Dermaptera
      - Suborder Forficulina
        - Family Forficulidae
        - Family Labiidae

Family Labiduridae
Suborder Hemimerina
Suborder Arixeniina
Order Embioptera
Family Clothodidae
Family Embiidae
Order Dictyoptera
Suborder Blattodea
Suborder Mantodea
Order Isoptera
Family Mastotermitidae
Family Kalotermitidae
Family Hodotermitidae
Family Rhinotermitidae
Family Termitidae
Order Zoraptera
Order Psocoptera
Suborder Eupsocida
Family Psocidae
Family Mesopsocidae
Family Pseudocaeciliidae
Suborder Trogiomorpha
Suborder Troctomorpha
Order Mallophaga
Suborder Amblycera
Family Menoponidae
Suborder Ischnocera
Family Philopteridae
Family Trichodectidae
Suborder Rhynchophthirina
Order Siphunculata
Family Echinophthiriidae
Family Hoplopleuridae
Family Linognathidae
Order Hemiptera
Suborder Heteroptera
Section Geocorisae
Family Pentatomidae
Family Coreidae
Family Pyrrhocoridae
Family Lygaeidae
Family Tingidae
Family Reduviidae
Family Nabidae
Family Anthocoridae
Family Cimicidae
Family Miridae
Family Saldidae
Section Amphibicorisae
Family Gerridae
Family Veliidae
Family Hydrometridae
Section Hydrocorisae
Family Naucoridae
Family Belostomatidae
Family Nepidae
Family Notonectidae
Family Corixidae
Suborder Homoptera
Section Auchenorrhyncha
Family Cicadidae
Family Cicadellidae
Family Membracidae
Family Cercopidae
Group Fulgoroidea
Family Delphacidae
Section Sternorrhyncha
Group Aphidoidea
Family Psyllidae
Family Coccoidae
Family Pseudococcidae
Family Aleyrodidae
Order Thysanoptera
Suborder Terebrantia
Suborder Tubulifera
Subclass Endopterygota
Order Neuroptera
Suborder Megaloptera
Family Sialidae
Family Corydalidae
Family Raphidiidae
Suborder Plannipennia
Family Sisyridae
Family Hemerobiidae
Family Chrysopidae
Family Mantispidae
Family Myrmeleontidae
Family Ascalaphidae
Order Mecoptera
Family Panorpidae
Family Bittacidae
Order Lepidoptera

Suborder Zeugloptera
Family Micropterygidae
Suborder Monotrysia
Family Eriocraniidae
Family Hepialidae
Family Incurvariidae
Suborder Ditrysia
Family Sesiidae
Family Tinaeidae
Family Gracillariidae
Family Plutellidae
Family Orneodidae
Family Cossidae
Family Psychidae
Family Zygaenidae
Family Tortricidae
Family Eucosmidae
Family Olethreutidae
Superfamily Pyralidoidea
Family Galleriinae
Family Crambinae
Family Phycitinae
Family Pyralidae
Family Pyraustinae
Family Lasiocampidae
Family Saturniidae
Family Bombycidae
Family Nymphalidae
Family Lycaenidae
Family Pieridae
Family Papilionidae
Family Hesperiidae
Family Geometridae
Family Sphingidae
Family Noctuidae
Family Notodontidae
Family Lymantriidae
Family Arctiidae
Order Trichoptera
Family Rhyacophilidae
Family Hydroptilidae
Family Hydropsychidae
Family Phryganeidae
Family Limnephilidae
Family Leptoceridae
Order Diptera
Suborder Nematocera
Family Tipulidae
Family Psychodidae
Family Culicidae
Family Cecidomyiidae
Family Bibionidae
Family Mycetophilidae
Family Simuliidae
Family Chironomidae
Suborder Brachycera
Family Stratiomyidae
Family Rhagionidae
Family Tabanidae
Family Asilidae
Family Bombyliidae
Family Empididae
Family Dolichopodidae
Suborder Cyclorrhapha
Family Syrphidae
Family Phoridae
Section Acalyptratae
Family Agromyzidae
Family Psilidae
Family Tephritidae
Family Chloropidáe
Section Calyptratae
Family Oestridae
Family Calliphoridae
Family Tachinidae
Family Muscidae
Family Hippoboscidae
Order Siphonaptera
Order Hymenoptera
Suborder Symphyta
Family Xyelidae
Family Siricidae
Family Diprionidae
Family Pergidae
Family Orussidae
Family Tenthredinidae
Suborder Apocrita
Section Parasitica
Family Ichneumonidae
Family Braconidae
Family Cynipidae
Superfamily Chalcidoidea

Family Trichogrammatidae
Family Mymaridae
Family Agaontidae
Superfamily Proctotrupoidea
Family Scelionidae
Family Platygasteridae
Section Aculeata
Family Dryinidae
Family Chrysididae
Family Scoliidae
Family Tiphiidae
Family Formicidae
Family Pompilidae
Family Vespidae
Family Sphecidae
Family Prosopidae
Family Andrenidae
Family Megachilidae
Family Apidae
Order Coleoptera
Suborder Adephaga
Family Carabidae
Family Cicindelidae
Family Paussinae
Family Haliplidae
Family Dytiscidae
Family Gyrinidae
Suborder Archostemata
Suborder Myxopbaga
Family Cupedidae
Suborder Polyphaga
Family Hydrophilidae
Family Histeridae
Family Silphidae
Family Staphylinidae
Family Passalidae
Family Lucanidae
Family Geotrupidae
Family Scarabaeidae
Family Elateridae
Family Buprestidae
Family Cantharidae
Family Lampyridae
Family Dermestidae
Family Anobiidae
Family Cleridae
Family Nitidulidae
Family Coccinellidae
Family Tenebrionidae
Family Meloidae
Family Cerambycidae
Family Chrysomelidae
Family Curculionidae
Family Scolytidae
Order Strepsiptera
Class Crustacea
Subclass Cephalocarida
Family Hutchinsoniellidae
Subclass Branchiopoda
Order Anostraca
Order Notostraca
Order Conchostraca
Order Cladocera
Family Daphniidae
Subclass Mystacocarida
Subclass Copepoda
Order Calanoida
Family Diaptomidae
Family Centropagidae
Order Cyclopoida
Family Cyclopidae
Family Notodelphyidae
Order Harpacticoida
Order Caligoida
Order Monstrilloida
Order Lernaeoida
Subclass Branchiura
Subclass Ostracoda
Order Myodocopa
Order Cladocopa
Order Platycopa
Order Podocopa
Subclass Cirripedia
Order Thoracica
Order Rhizocephala
Order Ascothoracica
Subclass Malacostraca
Superorder Phyllocarida
Order Leptostraca
Superorder Hoplocarida
Order Stomatopoda
Superorder Syncarida

Order Anaspidacea
Order Stygocaridacea
Order Bathynellacea
Superorder Peracarida
Order Spelaeogriphacea
Order Thermosbaenacea
Order Mysidacea
Suborder Lophogastrida
Suborder Mysida
Order Tanaidacea
Order Isopoda
Suborder Asellota
Family Asellidae
Suborder Flabellifera
Family Limnoriidae
Family Sphaeromidae
Family Anthuridae
Suborder Gnathiidea
Suborder Valvifera
Suborder Phreatoicidea
Suborder Oniscoidea
Family Ligiidae
Family Armadillidiidae
Suborder Epicaridea
Family Entoniscidae
Family Bopyridae
Order Amphipoda
Suborder Hyperiidea
Suborder Gammaridea
Family Gammaridae
Family Talitridae
Suborder Caprellidea
Family Caprellidae
Family Cyamidae
Suborder Ingolfiellida
Order Cumacea
Family Pseudocumidae
Order Euphausiacea
Order Decapoda
Suborder Natantia
Section Penaeidea
Family Penaeidae
Family Sergestidae
Family Leuciferidae
Section Caridea
Family Atyidae
Family Alpheidae
Section Stenopodidea
Family Stenopodidae
Suborder Reptantia
Section Palinura
Section Astacura
Family Homaridae
Family Astacidae
Family Parastacidae
Family Austroastacidae
Section Anomura
Family Paguridae
Family Coenobitidae
Family Lithodidae
Family Galatheidae
Family Porcellanidae
Section Brachyura
Family Dromiidae
Family Calappidae
Family Portunidae
Family Potamonidae
Family Xanthidae
Family Pinnotheridae
Family Grapsidae
Family Oxypodidae
Family Majidae
Class Arachnida
Order Scorpiones
Order Pseudoscorpiones
Order Opiliones
Order Acari
Family Eriophyidae
Order Palpigradi
Order Uropygi
Order Schizomida
Order Amblypygi
Order Araneae
Suborder Orthognatha (= Mygalomorpha)
Suborder Labidognatha (= Araneomorpha)
Family Araneidae (= Argiopidae)
Family Theridiidae
Family Agelenidae
Family Thomisidae
Family Lycosidae
Family Salticidae
Order Solifugae (= Solpugida)

Order Ricinulei
Class Merostomata
Class Pycnogonida

PHYLUM PENTASTOMIDA

PHYLUM TARDIGRADA

PHYLUM CHAETOGNATHA

PHYLUM POGONOPHORA

PHYLUM ECHINODERMATA
Class Asteroidea
Class Ophiuroidea
Class Echinoidea
Class Holothuroidea
Class Crinoidea

PHYLUM CHORDATA
Subphylum Hemichordata
Class Enteropneusta
Class Pterobranchia
Subphylum Urochordata
Class Ascidiacea
Class Thaliacea
Class Larvacea
Subphylum Cephalochordata
Subphylum Vertebrata
Superclass Agnatha
Class Cephalaspidomorphi
Order Petromyzoniformes
Class Pteraspidomorphia
Order Myxiniformes
Superclass Gnathostomata
Class Chondrichthyes
Superorder Selachimorpha
Order Heterodontiformes
Order Hexanchiformes
Order Lamniformes
Family Rhincodontidae
Family Orectolobidae
Family Odontaspidae
Family Lamnidae
Family Scyliorhinidae
Family Carcharinidae
Family Sphyrnidae
Order Squaliformes
Family Squalidae
Family Pristiophoridae
Family Squatinidae
Superorder Batoidimorpha
Family Pristidae
Family Rhinobatidae
Family Torpedinidae
Family Rajidae
Family Dasyatidae
Family Potamotrygonidae
Family Myliobatidae
Family Mobulidae
Subclass Holocephali
Order Chimaeriformes
Family Chimaeridae
Family Rhinochimaeridae
Family Callorhynchidae
Class Osteichthyes
Subclass Dipneusti
Order Ceratodiformes
Order Lepidosireniformes
Family Lepidosirenidae
Family Protopteridae
Subclass Crossopterygii
Order Coelacanthiformes
Family Latimeriidae
Subclass Brachioptergyii
Family Polypteridae
Subclass Actinopterygii
Order Acipenseriformes
Family Acipenseridae
Family Polyodontidae
Order Semionotiformes
Family Lepisosteidae
Order Amiiformes
Family Amiidae
Order Osteoglossiformes
Order Mormyriformes
Order Clupeiformes
Family Clupeidae
Family Engraulidae
Order Elopiformes
Order Anguilliformes
Order Notacanthiformes

- Order Salmoniformes
  - Suborder Esocoidei
  - Suborder Salmonoidei
  - Suborder Argentinoidei
  - Suborder Stomiatoidei
- Order Gonorynchiformes
- Order Cypriniformes
  - Suborder Characoidei
  - Suborder Cyprinoidei
- Order Siluriformes
- Order Myctophiformes
- Order Polymixiiformes
- Order Percopsiformes
- Order Gadiformes
- Order Batrachoidiformes
- Order Lophiiformes
- Order Indostomiformes
- Order Atheriniformes
  - Suborder Exocoetidei
  - Suborder Cyprinodontoidei
  - Suborder Atherinoidei
- Order Lampridiformes
- Order Beryciformes
- Order Zeiformes
- Order Syngnathiformes
- Order Gasterosteiformes
- Order Synbranchiformes
- Order Scorpaeniformes
  - Suborder Scorpaenoidei
- Order Dactylopteriformes
- Order Perciformes
  - Suborder Percoidei
  - Suborder Mugiloidei
  - Suborder Sphyraenoidei
  - Suborder Polynemoidei
  - Suborder Labroidei
  - Suborder Blennioidei
  - Suborder Icosteoidei
  - Suborder Ammodytoidei
  - Suborder Gobioidei
  - Suborder Acanthuroidei
  - Suborder Scombroidei
  - Suborder Stromateoidei
  - Suborder Anabantoidei
  - Suborder Channoidei
  - Suborder Mastacembeloidei
- Order Gobiesociformes
- Order Pleuronectiformes
  - Family Psettodidae
  - Family Citharidae
  - Family Bothidae
  - Family Pleuronectidae
  - Family Soleidae
  - Family Cynoglossidae
- Order Tetraodontiformes
  - Family Balistidae
  - Family Ostraciontidae
  - Family Tetraodontidae
  - Family Diodontidae
  - Family Molidae

Class Amphibia

- Order Apoda [sometimes Gymnophiona]
  - Family Caeciliidae
- Order Caudata (= Urodela)
  - Suborder Cryptobranchoidea
    - Family Hynobiidae
    - Family Cryptobranchidae
  - Suborder Ambystomatoidea
    - Family Ambystomatidae
  - Suborder Salamandroidea
    - Family Salamandridae
    - Family Amphiumidae
    - Family Plethodontidae
    - Family Proteidae
  - Suborder Sirenoidea
    - Family Sirenidae
- Order Anura (= Salientia)
  - Suborder Amphicoela
    - Family Ascaphidae
  - Suborder Opisthocoela
    - Family Pipidae
    - Family Discoglossidae
    - Family Rhinophrynidae
  - Suborder Anomocoela
    - Family Pelobatidae
    - Family Pelodytidae
  - Suborder Diplasiocoela
    - Family Ranidae
    - Family Rhacophoridae
    - Family Microhylidae
    - Family Phrynomeridae
  - Suborder Procoela

Family Pseudidae
Family Bufonidae
Family Atelopidae
Family Hylidae
Family Leptodactylidae
Family Centrolenidae

Class Reptilia
- Order Rhynchocephalia
- Order Testudines (= Chelonia)
  - Family Testudinidae
  - Family Emydidae
  - Family Kinosternidae
  - Family Platysternidae
  - Family Chelydridae
  - Family Chelidae
  - Family Trionychidae
  - Family Chelonidae
  - Family Dermochelidae
  - Family Pelomedusidae
  - Family Carrettochelyidae
- Order Crocodylia (= Loricata)
  - Family Gavialidae
  - Family Crocodylidae
  - Family Alligatoridae
- Order Squamata
  - Suborder Sauria (= Lacertilia)
    - Family Gekkonidae
    - Family Agamidae
    - Family Iguanidae
    - Family Lacertidae
    - Family Teiidae
    - Family Helodermatidae
    - Family Varanidae
    - Family Chamaeleonidae
    - Family Scincidae
    - Family Cordylidae
    - Family Anguidae
    - Family Amphisbaenidae
  - Suborder Serpentes (= Ophidia)
    - Family Boidae
    - Family Typhlopidae
    - Family Colubridae
    - Family Elapidae
    - Family Viperidae
    - Family Crotalidae
    - Family Hydrophidae

Class Aves
- Order Struthioniformes
- Order Rheiformes
- Order Casuariiformes
  - Family Dromaiidae
  - Family Casuariidae
- Order Apterygiformes
- Order Tinamiformes
- Order Sphenisciformes
- Order Gaviiformes
- Order Podicipediformes
- Order Procellariiformes
  - Family Diomedeidae
  - Family Procellariidae
  - Family Hydrobatidae
  - Family Pelecanoididae
- Order Pelecaniformes
  - Family Phaethontidae
  - Family Pelecanidae
  - Family Sulidae
  - Family Phalacrocoracidae
  - Family Anhingidae
  - Family Fregatidae
- Order Ciconiiformes
  - Family Ardeidae
  - Family Cochleariidae
  - Family Balaenicipitidae
  - Family Scopidae
  - Family Ciconiidae
  - Family Threskiornithidae
  - Family Phoenicopteridae
- Order Anseriformes
  - Family Anhimidae
  - Family Anatidae
- Order Falconiformes
  - Family Cathartidae
  - Family Accipitridae
  - Family Pandionidae
  - Family Falconidae
  - Family Sagittariidae
- Order Galliformes
  - Family Megapodiidae
  - Family Cracidae
  - Family Tetraonidae
  - Family Phasianidae
  - Family Numididae

Family Meleagrididae
Family Opisthocomidae
Order Gruiformes
Family Mesitornithidae
Family Turnicidae
Family Pedionomidae
Family Gruidae
Family Aramidae
Family Psophiidae
Family Rallidae
Family Heliornithidae
Family Rhynochetidae
Family Eurypygidae
Family Cariamidae
Family Otididae
Order Charadriiformes
Family Jacanidae
Family Rostratulidae
Family Haematopodidae
Family Charadriidae
Family Scolopacidae
Family Recurvirostridae
Family Phalaropodidae
Family Dromadidae
Family Burhinidae
Family Glareolidae
Family Thinocondae
Family Chionididae
Family Stercorariidae
Family Laridae
Family Rynchopidae
Family Alcidae
Order Columbiformes
Family Pteroclidae
Family Columbidae
Order Psittaciformes
Family Psittacidae
Order Cuculiformes
Family Musophagidae
Family Cucilidae
Order Strigiformes
Family Tytonidae
Family Strigidae
Order Caprimulgiformes
Family Steatornithidae
Family Podargidae
Family Nyctibiidae
Family Aegothelidae
Family Caprimulgidae
Order Apodiformes
Family Apodidae
Family Hemiprocnidae
Family Trochilidae
Order Coliiformes
Order Trogoniformes
Order Coraciiformes
Family Alcedinidae
Family Todidae
Family Momotidae
Family Meropidae
Family Leptosomatidae
Family Coraciidae
Family Upupidae
Family Phoeniculidae
Family Bucerotidae
Order Piciformes
Family Galbulidae
Family Bucconidae
Family Capitonidae
Family Indicatoridae
Family Ramphastidae
Family Picidae
Order Passeriformes
Suborder Eurylaimi
Family Eurylaimidae
Suborder Tyranni
Family Dendrocolaptidae
Family Furnariidae
Family Formicariidae
Family Conopophagidae
Family Rhinocryptidae
Family Pittidae
Family Philepittidae
Family Acanthisittidae (= Xenicidae)
Family Tyrannidae
Family Oxyruncidae
Family Pipridae
Family Cotingidae
Family Phytotomidae
Suborder Menurae
Family Menuridae
Family Atrichornithidae

Suborder Passeres (= Oscines)
- Family Alaudidae
- Family Hirundinidae
- Family Motacillidae
- Family Campephagidae
- Family Pycnonotidae
- Family Irenidae
- Family Laniidae
- Family Vangidae
- Family Bombycillidae
- Family Dulidae
- Family Cinclidae
- Family Troglodytidae
- Family Mimidae
- Family Prunellidae
- Family Muscicapidae
- Family Paridae
- Family Certhiidae
- Family Sittidae
- Family Climacteridae
- Family Dicaeidae
- Family Nectariniidae
- Family Zosteropidae
- Family Meliphagidae
- Family Emberizidae
- Family Parulidae
- Family Drepanididae
- Family Vireonidae
- Family Icteridae
- Family Fringillidae
- Family Estrildidae
- Family Ploceidae
- Family Sturnidae
- Family Oriolidae
- Family Dicruridae
- Family Callaeidae
- Family Grallinidae
- Family Artamidae
- Family Cracticidae
- Family Ptilonorhynchidae
- Family Paradisaeidae
- Family Corvidae

Class Mammalia
- Subclass Prototheria
  - Order Monotremata
  - Infraclass Metatheria

Order Marsupialia
- Family Didelphidae
- Family Dasyuridae
- Family Notoryctidae
- Family Peramelidae
- Family Phalangeridae
- Family Phascolarctidae
- Family Phascolomidae
- Family Macropodidae

Infraclass Theria

Order Insectivora
- Family Erinaceidae
- Family Soricidae
- Family Tenrecidae
- Family Solenodontidae
- Family Talpidae
- Family Chrysochloridae
- Family Potamogalidae
- Family Macroscelididae
- Family Tupaiidae

Order Dermoptera

Order Chiroptera
- Suborder Megachiroptera
- Suborder Microchiroptera
  - Family Megadermatidae
  - Family Rhinolophidae
  - Family Vespertilionidae
  - Family Phyllostomatidae
  - Family Noctilionidae
  - Family Molossidae
  - Family Desmodontidae

Order Primates
- Suborder Prosimii
  - Family Lemuridae
  - Family Cheirogaleidae
  - Family Lorisidae
  - Family Indriidae
  - Family Daubentoniidae
  - Family Tarsiidae
- Suborder Anthropoidea (= Simiae)
  - Family Cebidae
  - Family Callithricidae
  - Family Cercopithecidae
  - Family Pongidae

Order Edentata
- Family Myrmecophagidae

- Family Bradypodidae
- Family Dasypodidae

Order Pholidota

Order Lagomorpha

- Family Leporidae
- Family Ochotonidae

Order Rodentia

- Suborder Bathyergomorpha
- Suborder Hystricomorpha
    - Family Echimyidae
    - Family Dasyproctidae
    - Family Erethizontidae
    - Family Hystricidae
    - Family Cuniculidae
    - Family Dinomyidae
    - Family Chinchillidae
    - Family Caviidae
- Suborder Sciuromorpha
    - Family Sciuridae
    - Family Aplodontidae
    - Family Castoridae
    - Family Geomyidae
    - Family Heteromyidae
- Suborder Myomorpha
    - Family Anomaluridae
    - Family Pedetidae
    - Family Ctenodactylidae
    - Family Dipodidae
    - Family Zapodidae
    - Family Muscardinidae
    - Family Lophiomyidae
    - Family Spalacidae
    - Family Rhizomyidae
    - Family Muridae

Order Cetacea

- Suborder Mysticeti
    - Family Balaenidae
    - Family Balaenopteridae
    - Family Eschrichtiidae
- Suborder Odontoceti
    - Family Physeteridae
    - Family Kogiidae
    - Family Phocaenidae
    - Family Delphinidae
    - Family Zephiidae
    - Family Monodontidae
    - Family Platanistidae

Order Carnivora

- Family Canidae
- Family Ursidae
- Family Procyonidae
- Family Mustelidae
- Family Viverridae
- Family Hyaenidae
- Family Protelidae
- Family Felidae

Order Pinnipedia

- Family Phocidae
- Family Otariidae
- Family Odobenidae

Order Tubulidentata

Order Proboscidea

Order Hyracoidea

Order Sirenia

Order Perissodactyla

- Family Equidae
- Family Tapiridae
- Family Rhinocerotidae

Order Artiodactyla

- Suborder Suiformes
    - Family Suidae
    - Family Tayassuidae
    - Family Hippopotamidae
- Suborder Tylopoda
    - Family Camelidae
- Suborder Ruminantia
    - Family Tragulidae
    - Family Cervidae
    - Family Giraffidae
    - Family Antilocapridae
    - Family Bovidae

GEOGRAPHICAL LIST OF ANIMALS

All Continents (except Antarctica)
Ants
Arachnids
Bats
Bees
Beetles
Butterflies
Cats
Cattle
Centipedes and millipedes
Chickens
Cockroaches
Dogs
Ducks
Eagles
Finches
Flatworms
Flies
Foxes
Frogs
Grasshoppers
Hawks
Horses
Lizards
Mice
Mollusks
Mosquitoes
Moths
Owls
Pelicans
Pigs and hogs
Quail
Rats
Roundworms
Scorpions
Sheep
Snails
Snakes
Sparrows
Spiders
Storks
Swans
Termites
Toads
Turtles and tortoises
Worms

All Oceans
Clams
Coral
Crabs
Eels
Elephant seals
Horseshoe crabs
Jellyfish
Lobsters
Mollusks
Octopuses
Otters
Oysters
Rays and skates
Sand dollars
Sea cucumbers
Sea urchins
Seahorses
Seals
Sharks
Squid
Starfish
Whale sharks
Whales, baleen
Whales, toothed
White sharks
Zooplankton

Africa
Aardvarks
Antelope
Baboons
Buffalo
Camels
Chameleons
Cheetahs
Chimpanzees
Civets and genets
Cranes
Crocodiles
Deer
Donkeys
Elephants
Flamingos
Giraffes
Goats
Gorillas
Hippopotamuses
Hyenas
Hyraxes
Leopards
Lions
Lungfish
Manatees
Meerkats
Mongooses
Monkeys
Mules
Ostriches
Parrots
Penguins
Pheasant
Porcupines
Rabbits and hares
Rhinoceroses
Shrews
Squirrels
Vultures
Weasels
Woodpeckers
Zebras

Antarctica
Penguins
Seals

Arctic
Bears
Caribou
Dolphins
Ermines
Geese
Lemmings
Mink
Moose
Polar bears
Reindeer
Sables
Shrews
Squirrels
Voles
Walruses
Wolverines
Wolves

Asia
Antelope
Bears
Buffalo
Camels
Chameleons
Cobras
Cranes
Crocodiles
Deer
Donkeys
Elephants
Elk
Ermines
Flamingos
Gibbons
Goats
Hyenas
Jerboas
Komodo dragons
Lemmings
Leopards
Lions
Marmots
Mink
Moles
Monkeys
Moose
Mules
Orangutans
Pandas
Parrots
Pheasant
Pikas
Porcupines
Pythons
Rabbits and hares
Reindeer
Rhinoceroses
Sables
Salmon
Shrews
Snow leopards
Squirrels
Tapirs
Tigers
Trout
Vultures
Wasps and hornets
Wolverines
Wolves
Woodpeckers
Yaks

Atlantic Ocean
Manatees
Porpoises
Salmon
Sponges
Trout

Australia
Camels
Cassowaries
Cranes
Crocodiles
Dingoes
Emus
Goannas
Goats
Kangaroos
Koalas
Kookaburras
Lungfish
Moles
Monitor lizards
Parrots
Penguins
Platypuses
Tasmanian devils
Wallabies
Wombats

Caribbean
Coral
Crocodiles
Flamingos
Iguanas
Manatees
Sponges

Central America
Anteaters
Armadillos
Boa constrictors
Caimans
Coatimundis
Coyotes
Cranes
Crocodiles
Deer
Donkeys
Gophers
Hummingbirds
Jaguars
Lungfish
Macaws
Manatees
Monkeys
Mules
Ocelots
Opossums
Otters
Parrots
Porcupines

Quetzals
Rabbits and hares
Raccoons
Salamanders and newts
Shrews
Skunks
Sloths
Squirrels
Tapirs
Toucans
Vultures
Wasps and hornets
Wolves
Woodpeckers

Europe
Badgers
Bears
Chameleons
Cranes
Deer
Donkeys
Elk
Ermines
Flamingos
Geese
Goats
Jerboas
Lynx
Marmots
Martens
Mink
Moles
Moose
Mules
Otters
Pheasant
Pikas
Rabbits and hares
Reindeer
Sables
Salmon
Shrews
Squirrels
Trout
Vultures
Wasps and hornets
Weasels
Wolverines
Wolves
Woodpeckers

Galápagos Islands
Finches
Iguanas
Penguins
Turtles and tortoises

Madagascar
Boa constrictors
Chameleons
Civets and genets
Lemurs
Mongooses
Parrots
Wasps and hornets

Middle East
Antelope
Baboons
Bears
Camels
Cheetahs
Cobras
Donkeys
Geese
Goats
Hyenas
Hyraxes
Jerboas
Leopards
Martens
Mules
Vultures
Wolves

North America
American pronghorns
Antelope
Armadillos
Badgers
Bears
Beavers
Bison
Caribou
Chipmunks
Condors
Coyotes
Cranes
Crocodiles
Deer
Donkeys
Elk
Fishers
Geese
Goats
Gophers
Grizzly bears
Groundhogs (woodchucks)
Hummingbirds
Iguanas
Kangaroo rats
Lynx (bobcats)
Manatees
Marmots
Martens
Mink
Moles
Moose
Mountain lions (cougars)
Mules
Muskrats
Opossums
Otters
Parrots
Pikas
Porcupines
Prairie dogs
Rabbits and hares
Raccoons
Rattlesnakes
Salamanders and newts
Salmon
Shrews
Skunks
Squirrels
Trout
Turkeys

Vultures
Wasps and hornets
Weasels
Wolverines
Wolves
Woodpeckers

Pacific Islands
Cassowaries
Coral
Crocodiles
Echidnas
Elephant seals
Goats
Iguanas
Kiwis
Mongooses
Monitor lizards
Parrots
Penguins
Pythons
Wallabies

Pacific Ocean
Porpoises
Salmon
Trout

South America
Alpacas
Anacondas
Anteaters
Armadillos
Bears
Boa constrictors
Caimans
Capybaras
Chinchillas
Coatimundis
Condors
Crocodiles
Deer
Dolphins
Donkeys
Flamingos
Hummingbirds
Iguanas
Jaguars
Llamas
Lungfish
Macaws
Manatees
Monkeys
Mules
Ocelots
Opossums
Parrots
Peccaries
Penguins
Porcupines
Pumas (cougars)
Quetzals
Rabbits and hares
Raccoons
Rheas
Shrews
Sloths
Squirrels
Tapirs
Toucans
Vicuñas
Vultures
Wasps and hornets
Wolves
Woodpeckers

BIOGRAPHICAL LIST OF SCIENTISTS

Alexander Agassiz (1835-1910). Born in Neuchâtel, Switzerland, Agassiz came to the United States in 1849 to join his father, Louis Agassiz, and was graduated from Harvard nine years later. He became a specialist in marine ichthyology. In 1865, he published *Seaside Studies in Natural History*, which he coauthored with his stepmother, Elizabeth Cabot Agassiz. He was curator of the Museum of Comparative Zoology at Harvard from 1874 to 1885, and most of his writing about marine zoology appeared in the publications of that museum.

Jean Louis Rodolphe Agassiz (1807-1873). As a student at the universities of Zürich, Heidelberg, and Munich, Agassiz, who was born on the Murtensee in Switzerland, studied natural history, giving special attention to botany. He received the Ph.D. from the University of Erlangen in 1829 and the following year received an M.D. from the University of Munich. Agassiz became an ichthyologist almost accidentally. When J. B. Spix, renowned for the collection of Amazon fish that he brought home from Brazil, died in 1826, Agassiz was chosen to complete the classification of these specimens. Besides becoming a well-respected ichthyologist, Agassiz did considerable work in tracking the course of recent glaciers. Agassiz did not accept Charles Darwin's theory of evolution, clinging to his own more conventional notion of independent creations.

John James Audubon (1785-1851). Born on the island of Saint-Domingue (now Haiti), Audubon was educated in Paris, studying art under the tutelage of Jacques-Louis David. Returning to the United States to avoid conscription, he came to live on a farm near Philadelphia. There he became intrigued by natural history and began a comprehensive study of birds, which he began to draw in remarkably accurate detail. His *Birds of America* (4 vols., 1827-1838), sold by subscription, consisted of 435 plates in color that contained 1,055 life-sized figures of birds. Between 1831 and 1839, he published, with William MacGillivray, five descriptive volumes to accompany *Birds of America* under the composite title *Ornithological Biography*. He remains the best-known American avian artist.

Henry Walter Bates (1825-1892). Usually thought of in connection with Batesian mimicry in the animal kingdom, Bates was born and spent much of his life in Britain, although he made fruitful expeditions to Brazil and other venues to observe animal behavior and to gather specimens. It was Bates who first observed in the Amazon basin that the black-banded orange-brown monarch butterfly (*Danaus plexippus*) of North America had a mimic, the viceroy (*Limentitis archippus*). The monarch develops a natural protection against predators by ingesting cardenolide toxins from the milkweed it consumes as a caterpillar, but the viceroy lacks this protection.

Comte de Buffon (Georges-Louis Leclerc; 1707-1788). A leading French naturalist of the eighteenth century, Buffon's greatest contribution to the life sciences was in establishing the field of biogeography. Practitioners in this field study not only the current geographical distribution of plants and animals but, viewing them over long periods of time, consider changes that have occurred in their geographical distribution.

This field has provoked substantial thought among scientists. Biogeography is generally credited with having been the most prominent factor in Charles Darwin's formulation of his theory of natural selection.

Rachel Carson (1907-1964). Best known for her books about the sea and pesticides, Carson, an American biologist and environmentalist who was born in Pennsylvania, gained renown with the publication of *The Sea Around Us* (1951) and *The Silent Spring* (1962), both of which warn about the effects of pesticides on the environment. Trained at The Johns Hopkins University and Woods Hole Marine Biological Laboratory, Carson became an aquatic biologist for the United States Bureau of Fisheries in 1936. In this position she learned a great deal about the environmental fragility of the sea.

Dennis Hubert Chitty (1912-). A professor emeritus at the University of British Columbia, Chitty is perhaps the world's leading authority on the fluctuating lemming populations of the world. His book, *Do Lemmings Commit Suicide? Beautiful Hypotheses and Ugly Facts* (1996), raises interesting questions about the widely held belief that lemmings are at heart suicidal.

Eugenie Clark (1922-). A native New Yorker, Clark received the Ph.D. from New York University in 1950, having done graduate study at both the Scripps Institution of Oceanography and Woods Hole Marine Biological Station prior to completing her doctorate. She served as an oceanographic chemist on an expedition to the Philippines in 1946. She recounted some of her oceanographic research in *Lady with a Spear* (1953), which was a Book-of-the-Month Club selection. Continuing to pursue her work with sharks, she contributed to P. W. Gilbert's *Sharks and Survival* (1963) and published *The Lady and the Sharks* (1969). She has made nearly twenty trips to the Red Sea to pursue her research.

Georges Cuvier (1769-1832). French comparative anatomist Cuvier is the founder of modern comparative anatomy. Regarded as the most renowned person in his field of study, he gained his reputation by departing from the natural philosophers, who tried to shape their facts to fit their preconceived notions. Cuvier permitted his generalizations to proceed from an objective consideration of the facts with which he was dealing. His *Leçons d'anatomie comparée* (5 vols., 1800-1805; lessons on comparative anatomy) brought all known anatomical data into a well-ordered system of knowledge. His later work, *Le Règne animal distribué d'après son organisation* (1817; *The Animal Kingdom, Arranged in Conformity with Its Organization*, 1827-1832), protested the idea of the scale of all living things, establishing in its place a series of types based on objective studies of the anatomy of various forms of life.

Charles Darwin (1809-1882). Possibly the most influential scientist of the nineteenth century, British-born Darwin made false starts in medicine and theology before scientists he met at Cambridge University sparked his interest in natural sciences. In 1831, an A.B. from Cambridge in hand, he sailed aboard the HMS *Beagle* on a scientific voyage around the world. On this trip, Darwin moved slowly away from the prevalent view that all living things are immutable, products of divine creation. He began to believe that species evolve by descent from other species. He expressed some such views early in *Notebooks on the Transmutation of Species* (1837), but articulated them most fully in his *On the Origin of Species* (1859).

Theodosius Dobzhansky (1900-1975). A native of Ukraine, Dobzhansky came to the United States in 1927 to work with Thomas Hunt

Morgan, an American geneticist who studied the genetics of fruit flies (*Drosophila*). In his research with fruit flies, Dobzhansky discovered that there is considerable genetic diversity among individual specimens. He demonstrated that most organisms possess one or more abnormal genes, so-called wild genes. These genes are of less immediate use than normal genes, but they are important because, as Dobzhansky's research reveals, species with large genetic loads are more adaptable than those with low genetic loads. His *Genetics and the Origin of Species* (1937) presents in detail his conclusions about genetic overload.

Gerald Malcolm Durrell (1925-1995). Born in India to British parents, Durrell, a novelist and writer of children's books, devoted much of his life to the study of nature. His book, *The Amateur Naturalist* (1982), which was made into a television series for the British Broadcasting Corporation, shows Durrell seeking out animal life in many parts of the world, while *My Family and Other Animals* (1956) is a classic memoir of family life and zoological exploration. In 1959, he founded a zoological park and shelter on the island of Jersey to care for endangered species. He is remembered for the genuine concern and respect with which he always treated animals.

Edward Forbes (1815-1854). Renowned for his research in ocean life, British zoologist Forbes advanced the field of paleontology through analyzing specimens he collected when he served as the naturalist on a voyage of HMS *Beacon* to Asia Minor. He systematically catalogued much of the sea life he encountered on this trip. His most notable books are his *History of British Starfishes* (1841) and his four-volume collaboration with Sylvanus Hanley, *History of British Mollusca* (1853).

Dian Fossey (1932-1985). Born in San Francisco, Fossey went to Rwanda in 1967 to found the Karisoke Research Centre, where she served as project coordinator from 1980 to 1983. On her first visit to Tanzania in 1963, she had impressed Louis Leakey, who offered her a position as one of his "ape girls," women who carefully observed the behavior of various primates. Fossey was assigned to observe mountain gorillas and spent much of the rest of her life doing so. She wrote about it in such books as *Gorillas in the Mist* (1983), which was made into a film by Warner Brothers and Universal Pictures in 1988. Fossey, who received a Ph.D. from Cambridge University in 1974, was active in organizing antipoaching patrols in Rwanda. Native resentment generated by this activity is presumed to have led to her murder in 1985.

Karl von Frisch (1886-1982). An Austrian zoologist, Frisch determined the means by which bees know direction and are able to communicate with other bees. As early as 1910, Frisch proved that fish can ascertain colors, shattering the long-held notion that fish and other lower animals are color-blind. His experiments with bees revealed that they can see all colors except red, as well as ultraviolet light. Observing bees through glass partitions, he discovered that they communicate by dancing and that their dances have very specific meanings. Frisch shared the 1973 Nobel Prize in Physiology or Medicine with Konrad Lorenz and Nikolaas Tinbergen.

Biruté Galdikas (1946-). Focusing her scholarly activity on the orangutans of Borneo and Sumatra, Galdikas, a lecturer in primatology at Simon Fraser University in Vancouver, Canada, spends most of her time on the Malay peninsula studying every aspect of orangutan life and behavior. Raised in Canada by her Lithuanian parents,

Galdikas in 1971 went to Borneo, where she founded the Orangutan Research and Conservation Project in Tanjun Puting National Park. In 1986, Galdikas established the Orangutan Foundation International.

Étienne Geoffroy Saint-Hilaire (1772-1844). A French anatomist, Geoffroy Saint-Hilaire was one of the leading comparative anatomists of the nineteenth century. He sought to reduce all animal forms to a single ideal plan. He brought considerable condemnation upon himself in 1830 by contending that cephalopod mollusks developed from the same basic pattern as vertebrates, a concept that was publicly deplored by Georges Cuvier, then the most celebrated figure in the field of comparative anatomy. Geoffroy Saint-Hilaire's arguments are believed to have helped pave the way for acceptance of Charles Darwin's theories.

Jane Goodall (1934-). Born in London, Goodall had no formal higher education and no experience in studying animal behavior, ethology, when Louis Leakey encouraged her to observe the behavior of chimpanzees at the Gombe Stream Reserve in Tanzania in 1960. Her careful and detailed observations led to her receiving a doctorate from Cambridge University in 1965 on the basis of a thesis on chimpanzee behavior. She was the first person to report authoritatively that chimpanzees, contrary to previous speculation, are not strictly vegetarian but will track down, kill, and eat small animals. Even more important, she observed that chimpanzees can make simple tools, until then thought to be a strictly human activity. Among her books are *My Friends the Wild Chimpanzees* (1967), *In the Shadow of Man* (1971), and *The Chimpanzees of Gombe: Patterns of Behavior* (1986), her most important book.

Stephen Jay Gould (1941-). An American paleontologist, Gould, working from his study of fossil remains, developed his theory of punctuated equilibrium. According to Gould, fossil evidence suggests that there is seldom gradual evolution from one species to another. Rather, according to this theory, fossils show periods of relatively little change, but then a more evolved species arises quite quickly. Which species proliferate and which species become extinct is as much a matter of chance as of the ability to adapt.

Ernst Haeckel (1834-1919). This German biologist was a spirited supporter of Charles Darwin's theory of evolution espoused in 1859 in *On the Origin of Species.* Trained as a physician, Haeckel soon abandoned his medical practice and, during a soul-searching period spent in Italy, joined a scientific expedition to Messina. His publication in 1862 of the results of this expedition helped secure him a position at the University of Jena, where he served as chair of zoology from 1865 until he retired in 1909. He is best known for his recapitulation theory, now largely discredited, espousing the notion that a recapitulation of phylogeny is found in ontogeny. In simpler terms, he contends that animals, as they develop, ape the characteristics of their lineage.

Baron Alexander von Humboldt (1769-1859). This German scientist and explorer was among the most versatile naturalists of his time. Born in Berlin, Humboldt studied mining and geology at the Freiberg School of Mines. An inheritance in 1796 enabled him to pursue his scientific studies uninterruptedly. In South America, he sailed the Orinoco and the Rio Negro, systematically collecting botanical and zoological specimens. He studied the current in the Pacific Ocean that now bears his name. His major publication was the five-volume *Kosmos: Entwurf einer physischen Weltbeschreibung* (1845-1862; *Cosmos: A Sketch of a Physical Description of the Universe*, 1848-1858), which sought to explain human beings' place in the overall natural scheme.

Thomas Henry Huxley (1825-1895). Until Huxley, a British biologist, embarked on his landmark studies in paleontology, evolution, and comparative anatomy, comparative studies were pursued primarily through the deductive method. Huxley changed this practice by observing as carefully and objectively as he could and permitting his conclusions to proceed from his data. The inductive method, with his help, has become the prevailing method of scientific research.

Libbie Henrietta Hyman (1888-1969). A famed American invertebrate zoologist, Hyman challenged some of the most entrenched theories of metazoan evolution, examining closely the types of organisms that first developed a mesoderm and in flatworms formed a coelom. Hyman contended that mesodermal tissue originally appeared in a group of flatworms, suggesting, according to Hyman, that they were derived directly from the planula larvae of some other animal, giving them an ancestral relationship to all the higher animals, a revolutionary evolutionary concept.

Donald Johansen (1943-). Johansen's specialty is paleoanthropology, which has taken him to excavations all over the world. In 1974, he discovered the *Australopithecus afarensis* skeleton known as Lucy, the oldest, most complete skeleton of a human ancestor. Johansen received the Ph.D. from the University of Chicago. He has attempted to inform the general public about his field through hosting many Public Broadcasting System (PBS) series, including *In Search of Human Origins* (1995), and writing popular books such as *Lucy: The Beginnings of Humankind* (1981).

Ernest Everett Just (1883-1941). A native of Charleston, South Carolina, Just was a biologist who was particularly concerned with cell life and metabolism. He became expert in egg fertilization, artificial parthenogenesis, and cell division. He was instrumental in training many young biologists at Woods Hole Marine Biology Laboratory in Massachusetts. In 1939, his major book, *Biology of the Cell Surface*, was published. Just was the first African American zoologist to have his image featured on a United States postage stamp. Disheartened by restrictions on African Americans during the 1930's, he resided in Europe during most of that decade.

Charles J. Krebs (1936-). An American teaching at the University of British Columbia, Krebs is known for his research on the fluctuating deer mouse (*Peromyscus maniculatus*) and Townsend's vole (*Microtus oregoni*) populations near Vancouver. In 1976, Krebs began a ten-year study of snowshoe hares at Kluane Lake. Eleven years later, he embarked on a study of the dynamics of the lemming population on the Arctic coast of Canada's Northwest Territory.

Edwin Gerhard Krebs (1918-). Best known for his work on the regulation of enzyme activity, which he undertook with Edmond Fischer, Krebs was born in Lansing, Iowa. He earned an M.D. from the University of Washington in 1943, but became a professor of biochemistry rather than a practicing physician. For their work on the regulation of enzyme activity, Krebs and Fischer were awarded the 1992 Nobel Prize in Physiology or Medicine.

Schack August Steenberg Krogh (1874-1949). Born in Jutland, this Danish physiologist and zoologist is remembered mostly for his pioneering work focusing on human circulation and respiration. His strong background in physics enabled him to design the mechanical devices necessary to carry out his research. He showed how the lungs absorb oxygen and expel carbon dioxide solely by diffusion. Instruments he devised to measure blood flow and respiration during exercise led him to the important discovery that during exercise capillaries expand, permitting an increased flow of oxygenated blood to stressed muscles. For this discovery, he received the Nobel Prize in Physiology or Medicine in 1920.

Jean-Baptiste-Pierre-Antoine de Monet, chevalier de Lamarck (1744-1829). Born in Picardy, the French naturalist Jean-Baptiste Lamarck articulated the first systematic evolutionary theory in biology. He was a highly competent field botanist, but his interest in zoology grew after he was appointed professor of lower vertebrate animals at the newly founded Jardin des Plantes in 1793. He made a major contribution to the field with the publication of *Histoire naturelle des animaux sans vertèbres* (1815; natural history of invertebrate animals), which remained a standard work for half a century.

Sir Edwin Ray Lankester (1847-1929). A British embryologist in the post-Darwinian period, Lankester was active in comparative anatomical studies. Lankester was centrally concerned with the nature and development of the body cavities, notably the coelom. His pioneering work in this field led to his being knighted by King George V. He was a prolific writer and editor, whose works included *Diversions of a Naturalist* (1915) and *Great and Small Things* (1923).

Louis Leakey (1903-1972). Leakey, born in Kenya to British missionaries, spent his life in Kenya and was initiated into the Kikuyu tribe of that nation. He went to England to study at Cambridge University, from which he was graduated with first class honors in modern languages, archaeology, and anthropology. He is credited with the discovery of the first large-toothed australopithecine at Olduvai Gorge in 1955. He uncovered the first fossils of *Homo habilis* at Olduvai Gorge in 1960, claiming them as direct ancestors of modern humans. Among his many books, perhaps the most significant are *Adam's Ancestors* (1934; rev. ed. 1960), *Stone Age Races in Kenya* (1935; 2d ed. 1970), and *Olduvai Gorge, 1951-1961* (1965).

Mary Douglas Leakey (1913-1996). Born in London, Mary Leakey studied prehistoric archaeology in France before taking part in several British excavations between 1930 and 1934. Louis Leakey, impressed by her drawings of stone artifacts, sought her out. In 1936, they married, and Leakey later moved to Kenya, where she introduced modern archaeological techniques to the excavation of late Pleistocene sites at Hyrax Hill and Njoro River Cave. In 1960, she established a base camp at Olduvai Gorge, from which she directed the excavations in that area. In 1976, with the discovery of several sets of bipedal footprints at the site, Leakey found substantiation of her theory that Laetoli hominids were the earliest conclusive hominid samples found up to that time. Her *Africa's Vanishing Art: The Rock Paintings of Tanzania* (1983) is a highly significant archaeological study.

Richard Erskine Frere Leakey (1944-). The son of Louis and Mary Leakey, Richard Leakey was born in Nairobi and holds Kenyan citizenship. In 1968, he was appointed administrative director of the National Museums of Kenya, advancing to the directorship in 1974. He founded and obtained funding for the International Louis Leakey Memorial Institute for African Prehistory. An active paleontologist, Leakey has found some amazing human fossils in

Kenya, but his chief skill is as an administrator who has organized and directed international teams bent on recovering human fossils. His analyses of these fossils have been exceptionally valuable. Among his significant books are *The Making of Mankind* (1981) and *The Origins of Humankind* (1994).

Carolus Linnaeus (1707-1778). A Swedish naturalist and physician, Linnaeus devised a descriptive process for biology using standard terminology and nomenclature. Although his most renowned contributions were in botany, the methods he established and advanced are used in zoology and mineralogy as well. Linnaeus elaborated his binomial nomenclature for plants in *Species Plantarum* (1753). The tenth edition of his *Systema Naturae* (1758) introduced a similar system for animals. He sought to make his method of classification as consistent and objective as possible. A versatile man with far-ranging interests, he also wrote significantly in ethnology and geography, basing much of this writing on his travels to underdeveloped parts of Scandinavia.

Konrad Zacharias Lorenz (1903-1989). Viennese zoologist Lorenz spent most of his adult life promoting the comparative study of animal and human behavior. The father of modern ethology, the study of animal behavior, Lorenz had doctorates in both philosophy and medicine from the University of Vienna. He studied comparatively the behavior of birds, dogs, and fish, developing his theory of action-specific energy. Lorenz contended that animals' behavioral patterns are often genetically based. He studied animal behavior in relation to its adaptive survival value for the species. For this work, he shared the Nobel Prize in Physiology or Medicine in 1973.

Patrick Matthew (1790-1874). Like many British naturalists of his day, Matthew was an inveterate traveler who collected specimens in his travels and also philosophized about the origins of life and its evolution. Considered a precursor of Charles Darwin, Matthew questioned the prevailing notions about how human beings came to exist in their present form and how other forms of life adapted, seemingly in response to changing conditions. Matthew's two most noted books, both published in 1839, were *Emigration Fields, North America, the Cape, Australia, and New Zealand* and *Two Addresses to the Men of Perthshire and Fifeshire*.

Gregor Mendel (1822-1884). A Roman Catholic priest, Austrian scientist Mendel is noted for his work in heredity that began when he raised and hybridized peas in his monastery's garden, keeping careful records of them over several generations. Through his experiments, he concluded that the fertilization of an egg involved only one male sex cell. Although his vision was failing, Mendel continued his research on hawkeyes, four o'clocks, and bees. He generally did not publish his findings, although his noted article, *Versuche über Pflanzenhybriden* (1856; *Experiments with Plant Hybrids*), was published as a pamphlet and later in William Bateson's *Mendel's Principles of Heredity* (1913).

Maria Sibylla Merian (1647-1717). In 1660, Merian became the first known person to study systematically the life cycle of silkworms from caterpillars to moths. A gifted painter of flowers, Merian had no trouble selling her art work. Soon she began to add insects to her paintings. Always fascinated by insects, she began to collect caterpillars in 1674, keeping them in containers where she could watch their transformation into moths. She painted what she observed. In 1699, Merian and her daughter departed for two years in Surinam, where they both closely observed insects and other animals, painting them in immense and accurate detail.

Thomas Hunt Morgan (1866-1945). One of the leading founders of the science of modern genetics, Morgan, a native of Lexington, Kentucky, is renowned for his discovery of how chromosomes function in the transmission of heredity, a discovery for which he was awarded the 1933 Nobel Prize in Physiology or Medicine. After doctoral studies in embryology at The Johns Hopkins University, Morgan began a long career as a professor of biology, first at Columbia University, then at the California Institute of Technology. His studies centered on the fruit fly (*Drosophila melanogaster*). His collaborative study, *The Mechanism of Mendelian Heredity* (1915), contained evidence that genes too small to be seen transmit heredity and exist in the chromosomes.

Sir Richard Owen (1804-1892). Superintendent of the natural history department of the British Museum from 1856 until 1884, Owen, born in Lancaster, England, was among the great comparative anatomists of his day. His interest in vertebrate paleontology was piqued when Charles Darwin asked him in 1837 to describe a fossil specimen. Owen espoused the notion that there was an archetype or plan of organization in nature, attributing the plan of unity within diversity of animals to a creator. Although he did not completely oppose Darwin's theory of evolution, he objected to Darwin's theory of natural selection.

John Ray (1627-1705). This early British naturalist devoted himself to observing and writing about nature, trying always to do so in a systematic way. He set out to devise a taxonomy that he could apply to the classification of plants and animals. He classified plants on the basis of seed vessels, whereas he classified animals according to anatomy and habitat.

René-Antoine Ferchault de Réaumur (1683-1757). A French physicist and naturalist, Réaumur in 1731 invented a thermometer on whose temperature scale the freezing point of water was zero, as in the Celsius thermometer, and the boiling point was 80 degrees. His scientific research included work on the regeneration of crayfish and digestion in birds, particularly the role of gastric juices that he isolated. Between 1734 and 1742, he published six volumes that grew out of his extensive research on insects.

Michael Sars (1809-1869). A Norwegian biologist, Sars, professor of zoology at the Royal Frederic's University in Christiana (now the University of Oslo), carried out pioneering marine research. He observed marine mollusks carefully and finally discovered their metamorphosis. He also established the relation of crinoids to similar fossil groups. His research helped to make clear the knowledge of the alternation of generations.

Thomas Say (1787-1834). Born in Philadelphia, American naturalist Say made numerous expeditions into the Rocky Mountains as well as to Mexico, Florida, Georgia, and Minnesota in quest of insect specimens. Curator of the American Philosophical Society from 1821 to 1827, he published his major multivolume work, *American Entomology*, between 1824 and 1828. In 1824, he eloped to Robert Owen's utopian community at New Harmony, Indiana. His wife illustrated his *American Conchology* (6 vols., 1830-1834).

George B. Schaller (1933-). Born in Berlin, Germany, Schaller emigrated to the United States and received a B.S. in zoology and an

A.B. in anthropology from the University of Alaska in 1955. His Ph.D. is from the University of Wisconsin. Schaller is best known for studies of the mountain gorilla, the deer, and the tiger. His monumental study, *The Serengeti Lion: A Study of Predator-Prey Relations* (1972), was recipient of the 1973 National Book Award in nonfiction. Schaller has been praised for his ability to write clearly and engagingly about technical subjects. He is among the most accessible writers in field biology.

David Takayoshi Suzuki (1936-). Born in Vancouver, British Columbia, Suzuki received his education in the United States, graduating from Amherst College with an A.B. degree, and from the University of Chicago with a Ph.D. in zoology in 1961. A leading geneticist, his major research interests have been in the regulation and development of behavior, in the genetic organization of chromosomes, and in developmental and behavioral genetics. Suzuki is much concerned in his writing with the implications of his discipline for the future, as evidenced in such books as *Inventing the Future* (1989), *It's a Matter of Survival* (with Anita Gordon, 1990), and *The Sacred Balance: Rediscovering Our Place in Nature* (with A. McConnell,1997).

Nikolaas Tinbergen (1907-1988). A British zoologist born in the Netherlands, Tinbergen, along with German zoologists Konrad Lorenz and Karl von Frisch, is considered the cofounder of ethology, the study of animal behavior based on observing animals in their natural habitats. The three shared the Nobel Prize in Physiology or Medicine in 1973. Tinbergen is best known for his analyses of stimuli that cause specific behavioral responses in animals. His *The Study of Instinct* (1951) is the first real handbook of ethology. He is also known for his *Social Behavior in Animals* (1953).

Charles Henry Turner (1867-1923). Biologist Turner was born in Cincinnati, Ohio, and received a B.S. degree in 1891 and an M.S. in 1892, both from the University of Cincinnati. There he came under the influence of Professor Clarence Luther Herrick, a leader in the emerging field of psychobiology. Herrick encouraged Turner to do research and to publish. In 1907, Turner received the Ph.D. in zoology from the University of Chicago, one of the first African Americans to achieve this distinction. Turner is best known for his work on insect behavior, most notably the unique gyrations that ants make as they return to their nests. These gyrations are referred to as "Turner's circling." Turner also studied the behavior of bees, wasps, and cockroaches, determining that the latter learn by trial and error.

Alfred Russel Wallace (1823-1913). Although Charles Darwin receives much of the credit for espousing the theories of evolution and natural selection, Wallace, quite independently, embraced such notions earlier than Darwin. Born in England of Scottish parents, Wallace became interested in natural science in grammar school. With the encouragement of Henry Walter Bates, he collected and studied beetles. He accompanied Bates to Brazil in 1848, remaining there for four years. In 1854, he went to the Malay Archipelago, where he remained for eight years. There he devised an evolutionary theory. In 1855, he wrote an essay on natural selection, which he sent to Darwin. He expanded this essay, entitled "On the Tendency of Varieties to Depart Indefinitely from the Original Type," and published it in conjunction with Darwin as a joint effort, titled "On the Tendency of Species to Form Varieties," in 1858. His *Contributions to the Theory of Natural Selection* (1870) and Darwin's *On the Origin of Species* (1859) are the two most significant early books on evolution.

Edward Osborne Wilson, Jr. (1929-). Born in Birmingham, Alabama, Wilson attended the University of Alabama before pursuing doctoral studies at Harvard University, where he received a Ph.D. in 1955. A professor of zoology at Harvard, Wilson has served as curator of entomology at the university's Museum of Comparative Zoology. He is also a trustee of the Marine Biological Laboratory at Woods Hole, Massachusetts. He received the Pulitzer Prize for general nonfiction in 1979 for *On Human Nature* (1978) and again for *The Ants* (with Bert Hölldobler, 1990). His *The Insect Societies* (1971) remains an important book in its field. He is one of the most significant figures in the controversial field of sociobiology, which examines the biological bases of behavior.

Roger Arliner Young (1889-1964). The first African American woman to earn a Ph.D. in zoology, which she received from the University of Pennsylvania in 1940. Young was also the first African American woman to publish extensively in her field. Born in Clifton Forge, Virginia, she conducted productive early research on the biological structures that control salt concentration in paramecia. She also published cogent research on the effect of direct and indirect radiation on sea urchin eggs. Burdened with a staggering teaching load and little support for her research, she finally was overcome by the pressures upon her and died in 1964.

—*R. Baird Shuman*

GENERAL BIBLIOGRAPHY

General Works, Almanacs, and Resources

Allaby, Michael. *A Dictionary of Zoology*. 2d ed. New York: Oxford University Press, 1999. A comprehensive, authoritative, and easy-to-understand resource on diverse aspects of animal life. Includes newly discovered organisms and animals, taxonomic revisions, and reclassification of some species.

Angier, Natalie. *The Beauty of the Beastly: New Views on the Nature of Life*. Boston: Houghton Mifflin, 1995. Analyzes and describes the beauty, power, and meaning of the natural world. Includes discussions on cockroaches, dolphins, hyenas, bees, birds, sheep, beetles, lemurs, and many others.

Boitani, Luigi, and Todd K. Fuller, eds. *Research Techniques in Animal Ecology*. New York: Columbia University Press, 2000. Presents common issues and challenges in the study of animal ecology, surveys past and present research techniques and their limitations and misuses, and examines new perspectives and strategies.

Carter, David. *Butterflies and Moths*. New York: Dorling Kindersley, 1992. This Eyewitness handbook, with photographs by Frank Greenaway, is a detailed yet easy-to-use visual guide to more than five hundred species of moths and butterflies throughout the world.

Chinery, Michael, ed. *The Grolier Illustrated Encyclopedia of Animals*. Danbury, Conn.: Grolier, 1994. Revised and updated, these four volumes provide detailed information about animals "from aardvark to zorille." Features an endangerment status code, glossary, alternative names, and cross-references.

Dalton, Stephen. *The Miracle of Flight*. Buffalo, N.Y.: Firefly Books, 1999. Discusses the evolution of wings and the ability to fly in birds and insects. Looks at human attempts, failures, and achievements. Provides striking illustrations and photographs.

Dean, Loral. *Animals of North America*. Secaucus, N.J.: Chartwell Books, 1984. Provides information about the rich and teeming wildlife of the continent. Magnificent photographs display animals in their natural habitats. Among these are bison, sea lions, seals, owls, badgers, ermines, bears, pelicans, gulls, beavers, deer, lynxes, hares, and many others.

Garber, Steven D. *The Urban Naturalist*. New York: John Wiley & Sons, 1987. Highlights the natural plant and animal life to be found in cities and suburbs. Discusses transformations in the ecosystem caused by urban growth and urban sprawl, and suggests ways for the urban naturalist to observe and study animals in these areas. Animal life includes insects and invertebrates, fish, amphibians, reptiles, birds, and mammals.

Greenwood, Pippa, Andrew Halstead, A. R. Chase, and Daniel Griffin. *Pests and Diseases: The Complete Guide to Preventing, Identifying, and Treating Plant Problems*. New York: Dorling Kindersley, 2000. A practical guide detailing and describing pests and diseases that attack plants. Includes an A-to-Z directory of pests, such as varieties of mites, beetles, ants, oakworms, maggots, wasps, and more. Discusses symptoms, causes, and means of controlling disease, and provides a glossary, colorful photographs, and black-and-white drawings.

Hairston, Nelson G. *Vertebrate Zoology: An Experimental Field Approach*. New York: Cambridge University Press, 1994. Analyzes the origins and adaptations of the major living vertebrate groups, their senses, classification, ecology, geographical distribution, migration patterns, behaviors, and behavior interactions between groups. Includes field studies and continuing questions about vertebrate life and behaviors.

Katz, Cathie. *Nature: A Day at a Time.* San Francisco: Sierra Club Books, 2000. An almanac celebrating an animal, bird, insect, or organism every day of the year. Enriched by black-and-white pen-and-ink drawings, as well as quotations from poets and writers.

Leopold, Aldo. *A Sand County Almanac.* New York: Oxford University Press, 1949. This special edition commemorates the one-hundredth anniversary of the birth of conservationist Leopold. The author relates his experiences on a farm on the Wisconsin river, describes the land and wildlife in the area, and considers the relationship of humans to nature. Black-and-white sketches by Charles W. Schwartz enhance the work "here and there," as the author says.

Mattison, Chris. *The Encyclopedia of Snakes.* New York: Facts on File, 1995. Details all aspects of a snake's life, including biology, evolution, classification, habits, and behaviors. Myths, superstitions, and attitudes toward snakes are also examined.

Mehrtens, John M. *Living Snakes of the World.* New York: Sterling, 1987. A comprehensive pictorial reference presenting information on classification, nomenclature, habitats, geographic range, and natural history of different snakes throughout the world. Includes magnificent full-color photographs.

Sirch, Willow Ann, and the Humane Society of the United States. *Careers with Animals.* Golden, Colo.: Fulcrum Resources, 2000. Examines a variety of careers with animals. These include animal shelter workers, veterinarians, pet sitters, wildlife specialists, animal photographers, lawyers, and lobbyists for animal welfare and rights. Vignettes of individuals in these and other animal-related careers are presented.

Stivins, Dal. *The Incredible Egg: A Billion Year Journey.* New York: Weybright and Talley, 1974. Explores patterns of reproductive behavior and species adaptation for survival. Examines the process of evolution, classification of animals, early animal life, eggs and fishes, amphibians and reptiles, and mammals that lay eggs.

Tanner, Ogden. *Animal Defenses.* New York: Time-Life Films, 1978. Based on the television series *Wild Wild World of Animals.* Provides a study, along with intriguing photographs, of how animals of all types have developed attributes and strategies for primary and secondary defense. Explains evolutionary adaptations, and examines humans in this pattern of defense. Types of defenses explored are cover and concealment, camouflage and disguise, warning signals, bluffs and threats, groups and coalitions, and fight or flight.

Thompson, Earnest. *Anatomy of Animals: Studies in the Forms of Mammals and Birds.* London: Bracken Books, 1996. First published in 1896, the book is still considered a classic, and is of interest as a general study of animal anatomy. Provides detailed illustrations.

Wildlife Journal. Published four times a year by Ennis Communications and Creative Street. A weekly half-hour series, *Game Warden Wildlife Journal,* appears on television. Provides information on all aspects of animal life in nature.

Witherspoon, James D. *Two Hundred Life Science Experiments for the Amateur Biologist.* Blue Ridge Summit, Pa.: TAB Books, 1993. Provides suggestions and instructions for experiments for high school students or other amateur scientists. Experiments include ways to research eggs, small organisms and cells, fruit flies, invertebrates of many kinds, fish, birds, and mammals.

Woodward, Fred. *Shells.* Secaucus, N.J.: Chartwell Books, 1993. Contains information on molluscan biology, and the classification and nomenclature of a number of species. Provides facts about each shell's general description, geographical distribution, habitat depth, and rarity. Explains cleaning and storing of shells for collectors.

Animal Emotional Life

Beckoff, Mark, ed. *The Smile of a Dolphin: Remarkable Accounts of Animal Emotions.* New York: Discovery Books, 2000. Experts on elephants, pigs, iguanas, hyenas, dolphins, cats, dogs, camels, ravens, chimpanzees, wolves, and guppies report and describe emotions displayed by these and other animals. Emotions and related behaviors include love, fear, aggression, anger, joy, grief, and friendship. Enriched by attractive and poignant color photographs.

Guglielmo, Anthony, and Carl Lynn. *The Walrus on My Table: Touching True Stories of Animal Healing.* New York: St. Martin's Press, 2000. Describes the true experiences of a massage therapist as he utilizes his skills to heal sick or stressed creatures such as dolphins, racehorses, penguins, walruses, cats, and sharks.

Masson, Jeffrey Moussaieff. *The Emperor's Embrace: Reflections on Animal Families and Fatherhood.* New York: Pocket Books, 1999. Explores positive and negative paternal behaviors and parenting styles of animals and birds such as penguins, seahorses, marmosets, beavers, lions, bears, and humans.

Masson, Jeffrey Moussaieff, and Susan McCarthy. *When Elephants Weep: The Emotional Lives of Animals.* New York: Dell, 1995. Looks into the hearts and feelings of animals, and gathers evidence from numerous sources to document animal emotions and emotional behaviors.

Arthropods and the Insect World

Berenbaum, May. *Bugs in the System: Insects and Their Impact on Human Affairs.* Reading, Mass.: Addison-Wesley, 1995. Provides extensive information on insect life, behaviors, and ecological impact.

Browning, John G. *Tarantulas.* Neptune City, N.J.: T. F. H., 1989. Contains detailed information on tarantula anatomy, varieties, behaviors, health, food, breeding, and needs. Has more than seventy full-color photographs.

Evans, Howard Ensign. *The Pleasures of Entomology: Portraits of Insects and the People Who Study Them.* Washington, D.C.: Smithsonian Institution Press, 1985. Examines insect life in all its beauty, power, and fragility, and shows how fascinating the study of fleas, crickets, killer bees, boll weevils, beetles, flies, and other insects can be. Reviews the work of European and American entomologists, among them Carl Linnaeus, Thomas Jefferson, and Benjamin Franklin, from the seventeenth century onward. Illustrated by Peter Eades.

Gordon, David George. *The Complete Cockroach: A Comprehensive Guide to the Most Despised (and Least Understood) Creature on Earth.* Berkeley, Calif.: Ten Speed Press, 1996. Provides an absorbing look at the ancestors, life cycle, social life and behavior patterns of these insects, which have survived for over 340 million years on earth. Discusses efforts to control and eradicate them, and describes their surprising presence in song, music, and literature.

Hubbell, Sue. *A Book of Bees . . . and How to Keep Them.* New York: Random House, 1988. Takes the reader on an informative journey through the year of the bee and beekeeper, from autumn through summer. Includes a glossary. Drawings by Sam Pothoff.

Imes, Rick. *The Practical Entomologist.* New York: Simon & Schuster, 1992. A comprehensive and colorful guide to the anatomy, morphology, and life cycle of various insect groups. Describes insect senses, behaviors, social lives, "minor insect orders," and enemies and allies. Documents entomology in action.

Jolivet, Pierre. *Interrelationship Between Insects and Plants.* Boca Raton, Fla.: CRC Press, 1998. Examines the way animal and plant life coexists in nature, how animals and plants have coevolved, and how they support and adapt to each other.

Style, Sue. *Honey from Hive to Honeypot: A Celebration of Bees and Their Bounty.* San Francisco: Chronicle Books, 1992. Provides an interesting history of honey and the bees that make it, the important features of the beekeeper's year, types of bees, their health and their hives, and bees and honey in literature and culture. Presents a variety of recipes that use honey. Illustrated by Graham Evernden.

Taber, Stephen Welton. *The World of the Harvester Ants.* College Station: Texas A & M University Press, 1998. Provides a comprehensive view of the evolution, geographical distribution, behavior patterns, communication techniques, and tribal and social lives of these insects. Discusses identification, name meanings, and past status. Features black-and-white photographs and drawings.

Birds

Allen, Hayward. *The Great Blue Heron.* Minocqua, Wis.: Northword Press, 1991. Portrays the natural history, behaviors, family life, and habitats of this magnificent bird. Provides full color photographs, as well as stories from different cultures.

Barth, Kelly L. *Birds of Prey.* San Diego, Calif.: Lucent Books, 2000. This volume in the Endangered Animals and Habitats series takes a close look at endangered birds of prey such as northern goshawks, some kinds of owls, and various eagles. Discusses causes of endangerment, details worldwide rescue and recovery efforts, and provides names of organizations to contact.

Blaugrund, Annette, and Theodore Stebbins, Jr. *John James Audubon: The Watercolors for "The Birds of America."* New York: Random House and the New York Historical Society, 1993. Offers a visual feast of Audubon's numerous and brilliant watercolors celebrating the rich and complex bird life of North America. Essays examine the naturalist's techniques, creativity, and entrepreneurship, and provide background information on each painting.

Discovery Channel. *Birds.* New York: Discovery Books, 1999. This detailed, well-organized, and practical handbook provides information on bird life, anatomy, and behavior. Explains bird identification, and looks at birdwatching in one's own backyard.

Forshaw, Joseph M, ed. *Birds.* San Francisco: Fog City Press, 2000. This well-illustrated guide explores the origins of birds, their habitats, anatomy, plumage cycles, characteristics, and feeding habits. Explains how taxonomists classify and name birds.

Hyman, Susan. *Edward Lear's Birds.* Stamford, Conn.: Longmeadow Press, 1989. Presents and critically analyzes the detailed, astonishing bird paintings of Lear, who is famous for his limericks and nonsense rhymes. Includes caricatures and colorplates of parrots, pigeons, kestrels, herons, cockatoos, toucans, peacocks, owls, cranes, storks, pheasants, and a host of other birds.

Johnsgard, Paul A. *North American Owls: Biology and Natural History.* Washington, D.C.: Smithsonian Institution Press, 1988. Offers a comparative biology of owls, including their evolution, classification, ecology, morphology, physiology, and behavior. Examines owls in myth and legend. Provides a glossary, useful appendices, range maps, and attractive paintings and drawings by Louis Agassiz Fuertes.

Katz, Barbara. *So Cranes May Dance.* Chicago: University of Chicago Press, 1993. Describes the efforts of two ornithologists to rescue this endangered species.

Lynch, Wayne. *Penguins of the World.* Willowdale, Ontario: Firefly Books, 1997. A careful and enlightening study of the complex lives, behaviors, and adaptation techniques of these birds. Useful appendices provide information on geographical distribution, breeding, species status and endangerment, and the relationship between humans and penguins.

Sibley, David Allen. *The Sibley Guide to Birds.* New York: Alfred A. Knopf, 2000. Magnificently illustrated by the author, this is the National Audubon Society's comprehensive, practical, and easy-to-use guide to 810 species of North American birds.

Singer, Arthur, Alan Singer, and Virginia Buckley. *State Birds.* New York: Lodestar Books/E. P. Dutton, 1986. Endorsed by the National Audubon Society and the National Wildlife Federation. Provides information on the various state birds of America, their origins, habitats, and characteristics, and gives reasons and dates for selection. Colorfully illustrated.

Fish and Sea Animals

Cahill, Tim. *Dolphins.* Washington, D.C.: National Geographic Society, 2000. Features spectacular photographs from the documentary film of the same name, accompanied by a readable, scientific, and detailed text.

Cleave, Andrew. *Giants of the Sea.* Stamford, Conn.: Longmeadow Press, 1993. Provides detailed information and remarkable perspectives on the lives and habitats of the enormous residents of the deep, such as whales, dolphins, sharks, marine turtles, and others. Beautiful and breathtaking photographs enrich the work.

Cousteau, Jacques. *Whales.* New York: Harry N. Abrams, 1986. An exhaustive study of all aspects of whales, including their history and relationship to humans, characteristics, anatomy, geographical range, behaviors, social life and interaction, means of communication, body language, and place in literature, legend, and art. Enhanced by striking and detailed photographs and drawings.

Cousteau, Jacques, and Philippe Cousteau. *The Shark: Splendid Savage of the Sea.* Garden City, N.Y.: Doubleday, 1970. Takes the reader into the dangerous and exotic world of the shark.

Ellis, Richard. *The Search for the Giant Squid.* New York: Penguin Books, 1998. The author, a marine biologist, explorer, and artist, surveys science, myth, and literature to provide a cultural and scientific account of one of the largest sea creatures, about which little is known.

Ford, John K. B., M. Ellis Graeme, and Kenneth C. Balcomb. *Killer Whales.* 2d ed. Vancouver, Canada: University of British Columbia Press, 2000. Investigates the changing relationship between humans and *Orcinus orca*, or the killer whale. Discusses the habitats and social structure of whale societies, whale watching, whale identification and classification, conservation concerns, and the future of whales. Black-and-white and color photographs, glossary, list of resources, and interesting sidebars.

Knop, Daniel. *Giant Clams: A Comprehensive Guide to the Identification and Care of Tridacnid Clams.* Translated by Eva Hert and Sebastian Holzberg. Ettlingen, Germany: Dähne Verlag, 1996. Explains species identification of this clam family, and examines ecological status, reproduction, anatomy, and diseases caused by other marine life. Discusses aquariums, and export and trade regulations.

Lichatowich, Jim. *Salmon Without Rivers.* Washington, D.C.: Island Press, 1999. Presents the evolutionary and environmental history of the salmon of the Pacific Northwest, detailing transformations as the region changes and becomes more populated. Investigates efforts to protect and restore the salmon, and suggests that society's worldview regarding the links between nature and humans must change to make these efforts successful.

Pinkguni, Manolito. *Piranhas: Keeping and Breeding Them in Captivity.* Philadelphia: Chelsea House, 1999. Explains the physical characteristics and needs of piranhas of various types. Discusses care, feeding, and aquarium maintenance. Glossy, colorful photographs enhance this guide.

Ripple, Jeff. *Manatees and Dugongs of the World.* Stillwater, Minn.: Voyageur Press, 1999. Focuses on the life, behaviors, and endangered situations of these aquatic mammals, and documents them through riveting and beautiful photographs by Doug Perrine. Describes associated myths and traditions.

Primates

De Waal, Frans. *Bonobo: The Forgotten Ape.* Berkeley: University of California Press, 1997. Explores the world, nature, and social life of the peaceable, gentle, and sensitive bonobo, a relatively unknown member of the ape family. Investigates bonobos in zoos and natural habitats, and reflects upon bonobo and human relations. Provides appealing and colorful photographs by Frans Lanting.

Fossey, Dian. *Gorillas in the Mist.* Boston: Houghton Mifflin, 1983. Presents the moving story of the author's life and research among the endangered mountain gorillas of Zaire, Rwanda, and Uganda. Describes Fossey's attempts to preserve the disappearing rain forests, which are the natural habitat of the gorillas.

Fouts, Roger, and Stephen Tukel Mills. *Next of Kin: My Conversations with Chimpanzees.* New York: Avon Books, 1997. Provides authentic, humorous, and moving documentation of the author's work with chimpanzees. This research leads to a greater understanding of intelligence, communication strategies, and use of language in both chimpanzees and humans.

Goodall, Jane. *The Chimpanzees of Gombe: Patterns of Behavior.* Cambridge, Mass.: Belknap Press of Harvard University Press, 1986. Documents the author's research on the complex social behavior, relationships, character, and perceptual world of chimpanzees. The engrossing and scientific text is enlivened by photographs revealing the animals' lives and activities.

_______. *In the Shadow of Man.* Boston: Houghton Mifflin, 1971. This classic describes Goodall's early years among the chimpanzees of Gombe, and is a record of her observations on chimpanzee life, emotions, habitats, and relationship to humans. Photographs by Hugo Van Lawick.

_______. *Through a Window: My Thirty Years with the Chimpanzees of Gombe.* Boston: Houghton Mifflin, 1990. Describes the author's life and scientific research, and analyzes the many facets and complexities of chimpanzee emotions and behavior.

Levine, Stuart P. *The Orangutan.* San Diego, Calif.: Lucent Books, 2000. This volume in the Endangered Animals and Habitats series describes past and present history and habitats of the orangutan, decreasing populations, and the impact of research, captivity, habitat loss, and human population growth. Provides names of organizations to contact, black-and-white photographs, and a glossary.

Other Mammals

Alexander, Shana. *The Astonishing Elephant.* New York: Random House, 2000. Gives a detailed account of the unique qualities of the elephant, including behavior patterns, communication, and emotional depths. Discusses genocide of elephants in American zoos and circuses, the march toward extinction of both the Asian and African elephant, and the struggle of scientists, biologists, and zoologists to understand and protect the animal. Also examines the role of elephants in religion, war, and entertainment.

Bass, Rick. *The New Wolves.* New York: Lyons Press, 1998. Chronicles the reintroduction of Mexican wolves, or lobos, to the American southwest. Presents concerns of environmentalists, ranchers, and others, and examines the conflict between modern life and nature. Describes, in a sensitive and dramatic manner, the dangers and wonders faced by wolves released on the mesa.

Bolgiano, Chris. *Mountain Lion.* Mechanicsburg, Pa.: Stackpole Books, 1995. Records seven years of extensive research on the place of mountain lions—cougars, panthers, and pumas—in North American history, folklore, and ecology. Describes controversial efforts, in laboratories and captivity, to save these endangered animals.

Caras, Roger. *Animals in Their Places.* San Francisco: Sierra Club Books, 1987. Presents absorbing accounts of wolves, elephants, bears, squirrels, panthers, condors, and many other animals in their natural surroundings. Indicates threats to this natural world, especially from humans.

Clyne, Densey. *The Best of Wildlife in the Suburbs.* South Melbourne, Australia: Oxford University Press, 1993. Detailed, amusing essays examine, on a month by month basis, the rich array of insects, birds, and animals in a suburban Australian garden. Behaviors, diets, and characteristics are described. Simple but charming drawings by Martyn Robinson enliven the text.

Fenton, M. Brock. *Bats.* New York: Facts on File, 1992. Investigates the world of different bats, their origins, anatomy, echolocation skills, food habits, social organization, family life, ecological value, health, and diversity. Describes attitudes toward bats, and their image in various cultures.

Gauthier-Pilters, Hilde, and Anne Innis Dagg. *The Camel: Its Evolution, Ecology, Behavior, and Relationship to Man.* Chicago: University of Chicago Press, 1981. Describes the life, traits, behaviors, and habitats of the camel, and discusses the interdependency of humans and camels in the desert.

Geist, Valerius. *Moose.* Stillwater, Minn.: Voyageur Press, 1999. Using striking color photographs by Michael H. Francis, this work explores all aspects of the life and world of the moose.

Hall, Tarquin. *To the Elephant Graveyard.* New York: Atlantic Monthly Press, 2000. Describes in a dramatic and vivid manner the search for a killer elephant in Northeast India. Reflects on the changing environment and its impact on human and elephant relationships.

Harrison, Kit, and George Harrison. *America's Favorite Backyard Wildlife.* New York: Simon & Schuster, 1985. Provides facts and details about wildlife that might be observed in American backyards, such as the box turtle, gray squirrel, opossum, woodchuck, sparrow, finch, and others. Accompanied by black-and-white photographs and a glossary.

Himsel, Carol A. *Rats.* Hauppage, N.Y.: Barron's Educational Series, 1991. Written as a manual for those raising rodent pets, the book has a chapter on understanding rats, as well as practical information on care, habitats and housing, nutrition, training, and health. Includes a glossary and interesting photographs and helpful drawings by Karin Skogstad, Fritz W. Kohler, and Michele Earle-Bridges.

Hoffman, Matthew, ed. *Dogs: The Ultimate Care Guide.* Emmaus, Pa.: Rodale Press, 1998. Covers numerous topics concerning dog characteristics, behaviors, and care. Includes information on dog breeds, health and longevity, training, feeding, emotions, communication, and grooming.

Kanze, Edward. *Kangaroo Dreaming: An Australian Wildlife Odyssey.* San Francisco: Sierra Club Books, 2000. Describes the experiences of two naturalists as they travel across Australia observing its unique wildlife: the kookaburra and other unusual birds, giant lizards, kangaroos, koalas, platypuses, crocodiles, wombats, and others. Provides interesting information on habitats and characteristic behaviors and gives humorous accounts of encounters with different animals.

Kirk, Mildred. *The Everlasting Cat.* New York: Galahad Books, 1977. Provides a history of the cat in literature, folklore, and religion, and discusses the relationship of people and cats through the ages.

Long, Kim. *Squirrels.* Boulder, Colo.: Johnson Books, 1995. Presents a comprehensive look at squirrel species, taxonomy, anatomy, diet, geographical range, behaviors, endangerment status, relationship to humans, and place in folklore. Includes useful illustrations, a list of wildlife and squirrel organizations, online resources, and products for squirrels.

McNamee, Thomas. *The Return of the Wolf to Yellowstone.* New York: Henry Holt, 1997. Chronicles the reintroduction of the gray wolf to Yellowstone National Park. Portrays conflicts and consequences surrounding this event. Politics, conservation, wolf biology and nature, and the relationship of wolves and humans are some of the topics studied.

Matthiessen, Peter. *Tigers in the Snow.* New York: North Point Press, 2000. A detailed and moving look at the Siberian tiger, its origins, history, place in mythology, and struggle for survival. Tells the story of the Siberian Tiger Project. Provides appealing and colorful photographs by Maurice Hornocker.

_______. *Wildlife in America.* New York: Viking, 1987. A classic history of the continent's rich wildlife, its exploitation and destruction by early European settlers and growing populations, and modern efforts to protect disappearing species. Provides photographs, maps, drawings, appendices, and information on wildlife legislation.

Mitchell, Hayley R. *The Wolf.* San Diego, Calif.: Lucent Books, 1998. This volume in the

Endangered Animals and Habitats series demystifies the image of the wolf by detailing its history, discussing its loss of habitat, and describing efforts to save it from extinction and provide new homes for the gray, red, Mexican, and Ethiopian wolf. Provides black-and-white photographs, a glossary, and the names of organizations to contact.

Mowat, Farley. *Never Cry Wolf.* New York: Bantam Books, 1983. Poignantly narrates the story of two summers and a winter spent on the frozen tundra studying wolves and caribou of the region. Debunks the image of the wolf as savage killer and enemy of humans, and pleads for understanding and preservation of this misunderstood animal.

Natural History New Zealand. *Wild Asia.* Gretna, La.: Pelican, 1999. Enriched by over 250 spellbinding photographs, the book is based on an award-winning international television series. Presents the deserts, rain forests, mountains, rivers, and woodlands of Asia, which are home to the teeming and varied wildlife of the massive continent.

Neary, John, and Time-Life Television Books. *Wild Herds.* Time-Life Films/Vineyard Books, 1977. Based on the television series *Wild Wild World of Animals*, the book provides information on what constitutes a herd, herds of the past and present, reasons for herding, and hooves and other anatomical features of herd animals. Among the wild herds are antelope, zebras, horses, camels, pigs, bison, buffalo, deer, sheep, goats, and horses. Presents attractive and informative photographs and drawings.

Penny, Malcolm. *Rhinos: Endangered Species.* New York: Facts on File, 1988. Explores reasons for the swift depletion of the rhinoceroses population, including beliefs and myths that support the hunting of these beleaguered animals. Investigates the evolution, breeding habits, social behaviors, and diets of Asian and African rhinoceroses, and discusses ways of saving them. Provides addresses of organizations.

Quinn, John R. *Wildlife Survivors: The Flora and Fauna of Tomorrow.* Blue Ridge Summit, Pa.: TAB Books, 1994. Examines major environmental changes and their impact on different animals. Considers, and documents through black-and-white photographs, the habits, characteristics, and habitats of reptiles, amphibians, insects, fish, birds, and mammals which are likely to survive. Reflects on why others will not. Provides a list of environmental organizations.

Reddish, Paul. *Spirits of the Jaguar: The Natural History and Ancient Civilizations of the Caribbean and Central America.* London: BBC Books, 1996. Serves as a companion to the BBC television series of the same name. Provides brilliant photographs and animated discussions of the region's natural history, peoples, and teeming wildlife.

Roth, Sara. *The Complete Pig: An Entertaining History of Pigs.* Stillwater, Minn.: Voyageur Press, 2000. Offers, through amusing photographs, drawings, and extensive research, a whimsical account of pigs in history, folklore, and literature.

Rundquist, Eric M. *Reptile and Amphibian Parasites.* Philadelphia: Chelsea House, 1999. Gives detailed information about the different parasites preying on reptiles and amphibians, the illnesses they can cause, and treatments that can be provided. Contains bright, glossy, and detailed photographs.

Schaller, George B. *The Last Panda.* Chicago: The University of Chicago Press, 1993. Describes the author's years in China studying the life of pandas in their natural habitat. Emphasizes the urgent need to fight to save pandas from extinction, and suggests strategies to do so.

Scherr, Lynn. *Tall Blondes: A Book About Giraffes.* Kansas City: Andrews McMeel, 1997.

Examines the physical traits, personalities, and current status of giraffes. Surveys their place in history, and their impact on culture and literature.

Souder, William. *A Plague of Frogs*. New York: Hyperion, 2000. Describes the race to identify and understand the reasons for mysterious deformities discovered in frogs in a Minnesota pond in 1995. Similar abnormalities have shown up elsewhere in the world. Considers frog physiology, environmental changes, and government and scientific efforts to solve the problem.

Steinhart, Peter. *The Company of Wolves*. New York: Alfred A. Knopf, 1995. Explores facts and myths about wolves and their relationship with humans, and presents points of view of biologists, ranchers, trappers, people who love or hate wolves, and people who study them.

Sterry, Paul. *Beavers and Other Rodents*. New York: Todtri, 1998. Gives a detailed account of the characteristics and behaviors of beavers and rodents both loved and hated by humans. Enriched by attractive and colorful photographs.

Stirling, Ian. *Polar Bears*. Ann Arbor: University of Michigan Press, 1988. Brilliant photographs by Dan Guravich depict the life and habitat of polar bears. Introduces readers to the study of these animals, their life cycle and behaviors, conflicts with humans, and conservationists' concerns for their survival.

Thomas, Elizabeth Marshall. *The Hidden Life of Dogs*. Boston: Houghton Mifflin, 1993. Identified by the author as "a book about dog consciousness," this work chronicles thirty years of observing and sharing life with various kinds of dogs, including dingoes and wolves. Throws light on the need of dogs to have a social life, create rituals and patterns of behavior, and display thoughts and feelings.

_______. *The Tribe of the Tiger: Cats and Their Culture*. New York: Simon & Schuster, 1994. Studies the behaviors, social lives, and individuality of both domestic and wild cats. Illustrated by Jared Taylor Williams.

Tuttle, Merlin. *America's Neighborhood Bats*. Austin: University of Texas Press, 1988. Dispels existing myths about bats, and explains their value to humans. Examines bat species, origins, behaviors, feeding habits, health, habitats, and colonies. Provides a glossary, photographs, and drawings.

Watson, Mary Gordon, Russell Lym, and Sue Montgomery. *Horse: The Complete Guide*. New York: Barnes & Noble Books, 1999. Examines in detail all the physiological, psychological, and behavioral aspects of horses. Reviews care of horses, equestrian sports, different breeds, and the impact of the horse on the human imagination.

Winston, Mark L. *Nature Wars: People vs. Pests*. Cambridge, Mass.: Harvard University Press, 1997. Explains how humans put their world at risk by indiscriminately attempting to control and destroy pests and organisms that are a necessary part of the chain of existence and balance of nature. Provides a brief history of pests and pesticides, and discusses the escalation of the battle between people and pests in modern times. Presents scientific and thought-provoking data, and pleads to manage rather than control, and reduce rather than eradicate.

Zimmer, Carl. *Parasite Rex: Inside the Bizarre World of Nature's Most Dangerous Creatures*. New York: The Free Press, 2000. Examines in minute detail the evolution, growth, and activities of parasites. Discusses parasite hosts, including humans, and argues that human beings are themselves parasites of other living things on earth.

—Nillofur Zobairi

JOURNALS

The Anatomical Record
Roger R. Markwald, Ph.D., Editor
Department of Anatomy and Cell Biology
Medical University of South Carolina
173 Ashley Avenue, Suite 601
P.O. Box 250508
Charleston, SC 29425
Ph.: 843-792-7658
Fax: 843-792-7611
E-mail: markwald@musc.edu
www.anatomy.org/anatomy
The official publication of the American Association of Anatomists (AAA), which advances the science and art of anatomy. It encourages research and publication in the field and maintaining high standards in the teaching of anatomy. Subscription and author guidelines are available online.

The Auk
Kimberly G. Smith, Editor
Department of Biological Sciences
WAAX 19
University of Arkansas
Fayetteville, AK 72701
E-mail: auk@comp.uark.edu
A quarterly journal of ornithology, published by the American Ornithologists' Union. Founded in 1883, the American Ornithologists' Union is the oldest and largest organization in the New World devoted to the scientific study of birds. Although the AOU primarily is a professional organization, its membership of about four thousand includes many amateurs dedicated to the advancement of ornithological science.

Australian Camel News
PMB 118 William Creek
Via Port Augusta 5710, Australia
Ph.: Australia 8-8670 7846
Fax: Australia 8-8672 3268
E-mail: austcamel@bigpond.com
www.austcamel.com.au/inform'n.htm
A quarterly journal for camel owners. *Camel News* is Australia's leading journal dedicated to the dromedary camel, since 1996. It closely follows the meteoric growth of the Australian camel industry and keeps subscribers informed and up-to-date on Australian and foreign developments as they occur, especially with regard to camel racing and camel management. It is specifically targeted toward the camel owner, camel pastoralist, and camel hobbyist, providing that essential link between the cutting edge of camelid research and the hands-on practical camel owner.

Bee Craft
Alison Mouser
79 Strathcona Avenue
Bookham
Leatherhead
Surrey, KT23 4HR, UK
E-mail: secretary@bee-craft.com
A monthly British journal that aims to provide the latest beekeeping ideas and scientific research for beginners and seasoned apiarists alike.

Bioacoustics
Professor A. N. Popper, Coeditor
Department of Zoology
University of Maryland
College Park, MD 20742-4415
www.zi.ku.dk/zi/bioacoustics/title.gif
International peer-reviewed journal devoted to the study and recording of animal sounds. Subscription and author guidelines are available online.

Biological Journals and Abbreviations
arachne.prl.msu.edu/journams
Web site contains the abbreviations, full titles, and links to Web pages for a large variety of biological and medical journals.

The Chameleon Journals
www.chameleonjournals.com
Web site offers stories and information pertaining to Old World chameleons. The publication has a featured article section and welcomes submissions for articles concerning anything from interesting stories about chameleons, product reviews, and serious scientific papers. The journals will continue to chronicle "Life with My Chameleons"—the starting block for this Web site. Also included are growth charts for the animals.

Clinical Anatomy
Stephen W. Carmichael, Ph.D., D.Sc., Editor in Chief
Department of Anatomy
Mayo Clinic
Stabile Building 9
Rochester, MN 55905
Ph.: 507-284-3743
Fax: 507-284-2707
E-mail: carmichael.stephen@mayo.edu
www.clinicalanatomy.org
One of several official publications of the American Association of Clinical Anatomists (AACA). Subscription and author guidelines are available online.

Communique
American Zoo and Aquarium Association
8403 Colesville Road, Suite 710
Silver Spring, MD 20910-3314
Ph.: 301-562-0777
Fax: 301-562-0888
www.aza.org
An official publication of the American Zoo and Aquarium Association.

Development
Chris Wylie, Editor in Chief
Division of Developmental Biology
Children's Hospital Medical Center
3333 Burnet Avenue
Cincinnati, OH 45229-3039
Ph.: 513-636-2090
Fax: 513-636-4317
E-mail: Development.Journal@chmcc.org
www.biologists.com/images/devtitle.gif
Provides insights into mechanisms of plant and animal development, from molecular and cellular to tissue levels.

Ecotoxicology and Environmental Safety
Editorial and Production Offices
525 B Street, Suite 1900
San Diego, CA 92101-4495
www.apnet.com/www/journal/es.htm
Publishes studies that examine the biologic and toxic effects of natural or synthetic chemical pollutants on animal, plant, or microbial ecosystems and their routes into the affected organisms. Research Areas include health problems and biological effects in humans arising from discharges into surface waters, meteorological factors, industrial effluents, industrial products, radiation, and fuels. Subscription and author guidelines are available online.

Electronic Journals in Biology
mcb.harvard.edu/Admin_Res/Library/edjbio.htm
Web site offer access to journals focusing on basic biology, including molecular biology but not medicine, and has tables of contents available on the World Wide Web. Some of the journals also have abstracts; some have full articles including graphics. Web pages usually include information on subscriptions and instructions to authors.

Environmental Research
www.apnet.com/www/journal/gfx/ercurrent.gif
Publishes original reports describing studies of the toxic effects of environmental agents on humans and animals. The principal aims of the journal are to define the etiology of environmentally induced illness and to increase understanding of the mechanisms by which environmental agents cause disease. It emphasizes multidisciplinary studies as well as studies employing biological markers of toxic exposure and effect. Occasional critical reviews and selected book reviews are included. Other emphasized research areas include biochemistry, cancer research, environmental and occupational medicine, epidemiology and risk analysis, immunology, mineral and organic agents, molecular and cellular biology, neuroscience, pathology, pharmacology, reproductive biology, and toxicology. Subscription and author guidelines are available online.

Folia Primatologica: International Journal of Primatology
Dr. R. H. Crompton
S. Karger AG
Editorial Office 'Folia Primatologica'
CH–4009 Basel (Switzerland)
194.209.48.2/journals/fpr/images/l_fpr1.gif
Official journal of the European Federation for Primatology.

Ibis
Dr. A. G. Gosler, Editor
c/o Edward Grey Institute of Field Ornithology
Department of Zoology
South Parks Road
Oxford, OX1 3PS, UK
www.bou.org.uk/pubibisc.html
An international journal of avian science.

International Journal of Veterinary Medicine
Stephen J. Baines, M.A. VetMB CertVR CertSAS MRCVS, Editor
vet@priory.com
www.priory.com/vet.htm
Vet On-Line is the first and leading independent, peer-reviewed, free veterinary journal available on the Internet. Vet On-Line was launched in 1995 and is part of Priory Medical Journals, which offers many paper-based journals and their electronic counterparts. The aim is to build a lively, informative, and interactive journal for professionals and lay readers alike and to foster an interest in comparative medicine, through the association with Priory Medical Journals.

Journal of Applied Poultry Research (JAPR)
Coeditors in Chief
Don Bell
University of California (ret.)
don.bell@ucr.edu
William Weaver, Jr.
Pennsylvania State University (ret.)
bweaver@visi.net
www.psa.uiuc.edu/japr/japr.html
The purpose of this journal is to provide practical, reliable, and timely information to those whose livelihoods are derived from the commercial production of poultry and those whose research benefits from this sector; to address topics of near-term application based on appropriately designed studies and critical observations; to encourage scientific approaches to practical problem solving; and to present information comprehensible to a broad readership. Subscription and author guidelines are available online.

Journal of Dinosaur Paleontology
www.dinosauria.com/jdp/jdp.htm
The journal and Web site includes Dinosauria On-Line on behalf of the Dinosaur Mailing List and maintains the Dinosauria On-Line Dinosaur Omnipedia by the list and the public at large. The Dinosaur Mailing List is an e-mail based newsgroup, owned by Mickey Rowe, for scientific discussions about

dinosaurs. Dino-Dispatches are short articles that attempt to bring together information from technical publications, popular media, Internet resources, and personal communications to provide an up-to-date review of new discoveries and new thinking on topics related to dinosaurs and other Mesozoic animals. Dinosaur Picture Gallery may be available to those with any high-quality dinosaur drawings, or photographs of skeletons or digs; feel free to contact me about displaying them here. Both professional and amateur artists and photographers are welcome, as are nondinosaurian images (as long as they are animals, ancient and dead).

Journal of General Virology
Professor G. L. Smith, Reviews Editor
Wright-Fleming Institute
Imperial College School of Medicine
St Mary's Campus
Norfolk Place
London, W2 1PG, UK
Fax: +44-20-7594-3973
E-mail: glsmith@ic.ac.uk
vir.sgmjournals.org; www.sgm.ac.uk/JGVDirect
Presents research on aspects of animal, plant, insect, bacterial, and fungal viruses along with the transmissible spongiform encephalopathies. *Journal of General Virology* aims to publish papers that describe original research in virology and contribute significantly to their field. It is concerned particularly with fundamental studies. Papers must be in English. Standard papers, short communications, and review articles are published. Subscription and author guidelines are available online.

Journal of Indian Bird Records and Conservation
www.angelfire.com/fl/indianbirds
Provides periodic information on Indian birds, conservation, habitat status, and records of bird lists from the Indian subcontinent. Maintains a list for discussion at indianbirds.listbot.com and gratis Internet-based *Journal for Indian Birds.* Includes papers and notes about conservation issues in the subcontinent. A group of nearly twenty editors and reviewers from around the world participate. The *Journal of Indian Bird Records and Conservation* is a gratis and pioneering Internet-based ornithological publication of the Harini Nature Conservation Foundation. The journal welcomes original articles, scientific papers, field checklists, sighting records, habitat notes, and conservation recommendations about bird species known from the Indian subcontinent (India, Pakistan, Nepal, Bhutan, Bangla Desh, Sri Lanka, and Maldives). The journal would also function as a net-based archive of information about the birds of the Indian subcontinent.

Journal of Insect Systematics and Evolution
Dr. Verner Michelsen, Managing Editor
Zoological Museum
Universitetsparken 15, DK2100
Copenhagen Ø, Denmark
Ph.: +45 35352531
Fax: +45 35321010
E-mail: vmichelsen@zmuc.ku.dk
www.zmuc.dk/EntoWeb/InSysEvol/startcont.htm
An international journal of systematic entomology.

Journal of Molluscan Studies
Journals Marketing
Oxford University Press
2001 Evans Road
Cary, NC 27513
Fax: 919-677-1714
mollus.oupjournals.org/misc/ifora.shtml
Covers the biology of mollusks, molecular genetics, cladistic phylogenetics, ecophysiology, and ecological, behavioral, and systematic malacology

Journal of New Zealand Birds
nzbirds.com/NZBirdsJournal_2html.html
Publishes articles on their conservation and environment.

Journal of Oregon Ornithology (JOO)
Range Bayer, Editor
P.O. Box 1467
Newport, OR 97365
E-mail: rbayer@orednet.org
www.oregonvos.net/~rbayer/j/joomenu.htm
Documents the biology of birds in Oregon that would probably not otherwise be published.

The Journal of Research on Lepidoptera
Rudi Mattoni, Editor
The Lepidoptera Research Foundation, Inc.
9620 Heather Road
Beverly Hills, California 90210-1757
Ph.: 310-274-1052
Fax: 310-275-3290
E-mail: mattoni@ucla.edu
www.geog.ucla.edu/~longcore/jrl.html
An international peer-reviewed scientific journal featuring research on the biology, ecology, distribution, and systematics of the Lepidoptera. Subscription and author guidelines are available online.

Journal of the American Animal Hospital Association
Douglas Novick, D.V.M., President
10940 Lucky Oak Court
Cupertino, CA 95014
E-mail: dnovick@iknowledgenow.com
www.iknowledgenow.com/info.cfm?msg=aboutus
Offers full-text articles and conference proceedings.

Journal of the American Biological Safety Association
Editor
1202 Allanson Road
Mundelein, IL 60060
Ph.: 847-949-1517
Fax: 847-566-4580
www.absa.org/images/ABSA-logo3.gif
The official publication of the American Biological Safety Association (ABSA), which distributes a quarterly newsletter and conducts an annual Biological Safety Conference to inform members of regulatory initiatives, hazard recognition and management issues, risk communications, current biosafety publications, upcoming meetings, training opportunities, and employment opportunities. In addition, ABSA produces an annual membership directory to stimulate networking and provides a technical Review Committee to members preparing materials for publication.

Journal of Veterinary Medical Education
Dr. Richard B. Talbot, Editor
Journal of Veterinary Medical Education
VA-MD College of Veterinary Medicine
Virginia Polytechnic Institute and State University
Blacksburg, VA 24061
scholar.lib.vt.edu/ejournals/JVME/V21-1/tofc.html
Official publication of the Association of American Veterinary Medical Colleges.

Lab Animal
labanimal@natureny.com
www.labanimal.com
A peer-reviewed journal for professionals in animal research, emphasizing proper management and care. Editorial features include: new animal models of disease; breeds and breeding practices; lab animal care and nutrition; new research techniques; personnel and facility management; facility design; new lab equipment; education and training; diagnostic activities; clinical chemistry; toxicology; genetics; and embryology, as they relate to laboratory animal science. *Lab Animal* publishes timely and informative editorial material, reaching both the academic research world and

applied research industries, including: genetic engineering, human therapeutics, and pharmaceutical companies.

National Reference Center for Bioethics Literature (NRCBL) www.georgetown.edu/research/nrcbl

Web site describes the specialized list of books, journals, newspaper articles, legal materials, regulations, codes, government publications, and other relevant documents concerned with issues in biomedical and professional ethics. The library holdings represent the world's largest collection related to ethical issues in medicine and biomedical research. This collection functions both as a reference library for the public and as an in-depth research resource for scholars from the United States and abroad.

The New Anatomist

Mark H. Paalman, Ph.D., Managing Editor
John Wiley & Sons, Inc.
605 Third Avenue
New York, NY 10158-0012
Ph.: 410-990-9020
Fax: 410-990-9004
E-mail: mpaalman@wiley.com

This journal is a section of *The Anatomical Record* and one of the official publications of the American Association of Anatomists. *The New Anatomist* is a bimonthly, magazine-style section publication that uses the disciplines of anatomy to connect the biological fields of cell biology, development, physical anthropology, and neuroscience. Its articles, reviews, and tutorials focus on topics of interest to anatomists and bioscientists alike, from cutting-edge research to general science and technology breakthroughs, all highlighted by lively and accessible illustrations. This unique journal also provides readers with news and views of the field from scientific, social, and political arenas and a forum for debate on controversial issues.

Online Journal of Veterinary Research (OJVR)

E-mail: guerrin@usq.edu.au
www.uq.edu.au/~csvguerr/legal.htm

Full text peer-reviewed electronic journal; veterinary pathobiology, toxicology, and pharmacokinetics.

Painted Meadows Horse Journal

ursu@crosswinds.net
www.crosswinds.net/~ursu/pmhj.html

Journal devoted to the lifestyle of equestrians and their families. Formerly known as *Painted Meadows Horse Journal*, it is now known as the *Ursu Horse Journal*.

Paper Dinosaurs, 1824-1969

Linda Hall Library
5109 Cherry Street
Kansas City, MO 64110
E-mail: ashwortb@lhl.lib.mo.us
www.lhl.lib.mo.us/pubserv/hos/dino/welcome.htm

Virtual catalog of an exhibition of rare books and journals illustrating the early history of dinosaur discovery and restoration.

Perception Online

l.sackett@bristol.ac.uk (Editor)
www.perceptionweb.com

A scholarly journal reporting experimental results and theoretical ideas ranging over the fields of human, animal, and machine perception. Topics covered include physiological mechanisms and clinical neurological disturbances; psychological data on pattern and object perception in animals and man; the role of experience in developing perception; skills, such as driving and flying; effects of culture on perception and aesthetics; errors, illusions, and perceptual phenomena occurring in controlled conditions, with emphasis on their theoretical significance; cognitive experiments and theories relating knowledge to perception; development of categories and generalizations; strategies for interpreting sensory patterns in terms of objects by organisms and machines; special problems associated with perception of pictures and symbols; verbal and nonverbal skills; reading; philosophical implications of experiments and theories of perception for epistemology, aesthetics, and art. Papers may be full experimental reports or preliminary results, accounts of new phenomena or effects, or theoretical discussions or comments. Descriptions of novel apparatus and techniques are also acceptable.

Physiological and Biochemical Zoology
Editorial Office
Department of Environmental, Population, and Organismic Biology
Campus Box 334
University of Colorado
Boulder, CO 80309-0334
Ph.: 303-735-0297
Fax: 303-735-1811
physzoo@spot.colorado.edu
Presents current research in environmental, adaptational, and comparative physiology and biochemistry. Subscription and author guidelines are available online.

Placenta
www.harcourt-international.com/journals/plac/default.cfm?
Publishes full-length papers and short communications of high scientific quality on all aspects of human and animal placenta.

Poultry Science Association
1111 N. Dunlap Avenue
Savoy, IL 61874
Ph.: 217-356-3182
Fax: 217-398-4119
www.psa.uiuc.edu/japr/japr.html
Official publication of the Poultry Science Association.

Recent Ornithological Literature Online (ROL)
www.nmnh.si.edu/BIRDNET/ROL
A serial compilation of citations and abstracts from the worldwide scientific literature that pertains to birds and the science of ornithology. The ROL deals chiefly with periodicals, but also announces new and renamed journals and provides abstracts of conference proceedings, reports, doctoral dissertations, and other serial publications. Scientists, who voluntarily scan journals for ornithological articles, generally according to their geographic region and special scientific interests, prepare the entries. Papers dealing exclusively with domestic birds and their husbandry are excluded, unless applicable to nondomestic species.

Small Farmer's Journal
P.O. Box 1627
Sisters, OR 97759-1627
Ph.: 800-876-2893 or 541-549-2064
Fax: 541-549-4403
E-mail: comments@smallfarmersjournal.com
www.smallfarmersjournal.com
International quarterly, strongly supporting independent family farms. Also offers information on the use of animal-power for farming.

Wildlife Journal
40 Monument Circle
Indianapolis, IN 46204
Ph.: 800-733-8273
www.wildlifejournal.com/Subscribe.htm
For those who appreciate and enjoy wildlife.

The Wilson Bulletin
www.ummz.lsa.umich.edu/birds/wos.html
Official publication of the Wilson Ornithological Society, a worldwide organization of professional ornithologists and serious amateurs interested in birds. The Wilson Society, founded in 1888, is a worldwide organization of nearly 2,500 people who share a curiosity about birds. Named in honor of Alexander Wilson, the Father of American Ornithology, the Society publishes

this quarterly journal of ornithology, and holds annual meetings.

Yellowstone Journal
Ph.: 800-656-8762 in the United States or (307) 332-3111
www.yellowstonepark.com
The online edition of the journal, published several times a year, provides current, in-depth coverage about wolves, grizzlies, other wild animals, recreation, geysers, fires, the environment, and more.

Zootecnica International
E-mail: zootecnica@zootecnica.it
www.zootecnica.it
Journal specializing in poultry science and breeding technology.

—Mary E. Carey

ORGANIZATIONS

AgBiotechNet
www.agbiotechnet.com
Web site offers an online service for agricultural biotechnology, providing information for plant and animal biotechnology including cloning, genomics, genetic engineering, in vitro culture, biosafety, intellectual property rights and all key issues in agricultural biotechnology through news, reviews, abstracts, reports, links, book chapters. Also includes news, book chapters, reports, abstracts, and conference materials on agricultural biotechnology, including genetic engineering of plants and animals.

All for Animals
www.allforanimals.com
Web site offers directory of animal rescue groups, animal rights groups, companies against animal testing, pet goods and services, nature and wildlife. This organization offers a free newsletter to assist readers in learning more about animal testing, cruelty-free companies, vegetarianism, cruelty-free investments, coexisting with wildlife, the links between animal abuse and domestic violence, and much more. Each issue of the newsletter features three sections: Cruelty-free Tip of the Month, Did You Know? and the Web site of the Month.

Alliance for the Wild Rockies (AWR)
P.O. Box 8737
Missoula, MT
Ph.: 406-721-5420
Fax: 405-721-9917
www.wildrockiesalliance.org/index.html
Web site describes works to protect rivers, wilderness ecosystems, biological corridors, and native wildlife in the northern Rockies. AWR formed to meet the challenge of saving the Northern Rockies Bioregion from habitat destruction. Multiple individuals, business owners, and organizations take a bioregional approach to protect and restore this region. A membership-based nonprofit organization, the board and advisors include some of the nation's top scientists and conservationists. The term "bioregion" refers to a physiographic region of wildlands that are biologically connected. The Wild Rockies Bioregion includes wildlands in parts of Idaho, Montana, Wyoming, Oregon, Washington, Alberta, and British Columbia.

American Association for Laboratory Animal Science (AALAS)
National Office:
AALAS
9190 Crestwyn Hills Drive
Memphis, TN 38125
www.aalas.org/opening.htm
Web site describes this organization as dedicated to the humane care and treatment of laboratory animals, and to quality research that leads to scientific gains benefiting both humankind and animals.

American Association of Anatomists (AAA)
9650 Rockville Pike
Bethesda, MD 20814-3998
Ph.: 301-571-8314
Fax: 301-571-0619
E-mail: exec@anatomy.org
www.anatomy.org/anatomy
Web site describes this scientific society representing biomedical researchers who are interested in structural biology.

American Association of Clinical Anatomists (AACA)
Lawrence M. Ross, M.D., Ph.D., Secretary AACA
Ph.: 713-500-6169

Fax: 713-500-0621
www.clinicalanatomy.org
Web site describes the AACA, an organization that advances the science and art of clinical anatomy. It encourages research and publication in the field and maintaining high standards in the teaching of anatomy. Clinical anatomy is defined as anatomy in all its aspects—gross, histologic, developmental, and neurologic, as applied to clinical practice, the application of anatomic principles to the solution of clinical problems and/or the application of clinical observations to expand anatomic knowledge.

American Biological Safety Association (ABSA)
1202 Allanson Road
Mundelein, IL 60060
Ph.: 847-949-1517
Fax: 847-566-4580
E-mail: absa@absa.org
www.absa.org
Web site offers information on all aspects of biosafety and biological safety, including documents, regulations, and guidelines for members and non-members of the ABSA.

American Society for the Prevention of Cruelty to Animals (ASPCA)
424 East 92d Street
New York, NY 10128
Ph.: 1-800-426-4435 (Animal Poison Control Center)
www.aspca.org/body_index.asp
Web site describes this organization dedicated to alleviating pain, fear, and suffering in animals. A free newsletter is available through this site. The ASPCA Animal Poison Control Center is the first and only nonprofit animal-dedicated poison control center in North America. The Center consults with animal owners, veterinarians, and others about animal poisonings and other toxicology-related issues. Licensed veterinarians and board-certified veterinary toxicologists answer center phones twenty-four hours a day.

American Veterinary Medical Association (AVMA)
Headquarters:
1931 North Meacham Road, Suite 100
Schaumburg, IL 60173
Ph.: 847-925-8070
Fax: 847-925-1329
E-mail: avmainfo@avma.org
Governmental Relations Division:
1101 Vermont Avenue NW
Washington, DC 20005
Ph.: 202-789-0007
Fax: 202-842-4360
E-mail: avmagrd@avma.org
www.avma.org/resources/default.asp
Web site includes information on publications, client service handbooks, disasters and emergency preparedness, pet ownership sourcebook, public health, and veterinary schools. The site also includes a Care for Pets Index featuring information about veterinarians and how to select them, managing the loss of a pet, buying a pet, animal safety, and general animal health.

American Zoo and Aquarium Association (AZA)
8403 Colesville Road
Suite 710
Silver Spring, MD 20910-3314
Ph.: 301-562-0777
Fax: 301-562-0888
www.aza.org
Web site describes this organization with a mission to promote the welfare of zoological parks and aquariums and their advancement as public educational institutions, as scientific centers, as natural science and wildlife exhibition and conservation agencies, and as cultural recreational establishments dedicated to the enrichment of human and natural resources.

Conservation International
1919 M Street NW, Suite 600
Washington, DC 20036
Fax: 202-912-1000
www.conservation.org
Web site provides information about biodiversity conservation in the world's endangered ecosystems, including a map of global biodiversity hotspots, profiles of hotspots, and many other resources. The site also provides access to video news releases. Fact sheets designed to make information and resources available to journalists in television, radio, and print media are also available.

4-H Clubs
FourHWeb Project:
www.4-h.org/fourhweb/index.htm
Site brings together all 4-H clubs and resources on the web in one location.
National 4-H Council:
www.fourhcouncil.edu
Site describes this nonprofit organization dedicated to building partnerships for community youth development that value and involve youth in solving issues critical to their lives, their families, and society.
National 4-H Headquarters:
www.4h-usa.org
Families, 4-H, and Nutrition:
CSREES/USDA
Stop 2225
1400 Independence Avenue SW
Washington, DC 20250-2225
Ph.: 202-720-2908
National 4-H Web:
www.4-h.org
Web site created and maintained by the National 4-H Youth Technology Corps. Offers both IRC and Java Chat.

Humane America Animal Foundation
P.O. Box 7
Redondo Beach, CA 90277
Ph.: 310-263-2930
Fax: 310-263-2937
E-mail: info@humaneamerica.org
www.humaneamerica.org/images/f_ha.gif
Web site includes information on Humane America's 1-800-Save-a-Pet.com program, designed to assist communities across the country in reducing their overpopulation of healthy companion animals to the point that euthanization is no longer necessary as a means of population control.

International Commission on Zoological Nomenclature (ICZN)
c/o The Natural History Museum
Cromwell Road
London, SW7 5BD, UK
Ph.: +44 (0)20 7942 5653
E-mail: iczn@nhm.ac.uk
www.iczn.org
Web site describes the official body responsible for providing and regulating the system for ensuring that every animal has a unique and universally accepted scientific name.

International Federation of Associations of Anatomists (IFAA)
E-mail: iwhitmore@argonet.co.uk
www.ifaa.lsumc.edu
Web site describes this international body for anatomical associations. The IFAA is composed of the following organizations: the Anatomical Society of Great Britain and Ireland, the American Association of Anatomists, the American Association of Clinical Anatomists, the Anatomische Gesellschaft, and the Spanish Society of Anatomy.

International Species Information System (ISIS)
E-mail: ivan@isis.org
www.worldzoo.org
Web site offers information about this organization that helps zoological institutes manage their living collections by providing software for records keeping and collection

management, and then pools this information. ISIS is an international nonprofit membership organization that serves nearly 550 zoological institutional members, from fifty-four countries, worldwide. There is a high level of global cooperation by mostly city-based facilities, presently including about half of the world's recognized zoos and aquariums. ISIS supports conservation and preservation of species by helping member facilities manage their living collections. Information is available on 286,000 living specimens of 7,500 species, along with an additional 1,413,000 of their ancestors. Most of these specimens were bred in member facilities. ISIS cooperates closely with many national and regional associations of zoos and aquaria, and hosts the Secretariat of the World Zoo Organization.

International Wildlife Rehabilitation Council
4437 Central Place, Suite B-4
Suisun, CA 94585
www.iwrc-online.org
Web site describes the work of this organization in protecting animals and their habitat through wildlife rehabilitation.

Internet Zoological Society
E-mail: e-mail@izoo.org
www.izoo.org
Web site describes this organization established to help support those organizations and individuals that deal directly with education about, and conservation and rehabilitation of, wild animals and ecosystems.

Jane Goodall Center for Excellence in Environmental Studies
WCSU Westside
Room 134A
181 White Street
Danbury, CT 06810
Ph.: 203-837-8726
E-mail: jgiinformation@janegoodall.org
www.janegoodall.org
Web site describes organization that studies issues relevant to the environment, both natural and human-centered. A collaboration between the Jane Goodall Institute and Western Connecticut State University.

Kids Go Wild
2300 Southern Boulevard
Bronx, NY 10460
Ph.: 718-220-5100
wcs.org/9822
Web site where children can learn about saving wild animals and the environment, conservation education, and events sponsored by the Wildlife Conservation Society.

Laboratory Animal Management Association (LAMA)
P.O. Box 877
Killingworth, CT 06419
E-mail: doc@animalvillage.com
www.lama-online.org
Web site describes LAMA, dedicated to enhancing the quality of management and care of laboratory animals throughout the world.

National Snaffle Bit Association Office (NSBA)
4815 S. Sheridan, Suite 109
Tulsa, OK 74145
Ph.: 918-270-1469
Fax: 918-270-1471
E-mail: nsbaoffice@aol.com
www.premierpub.com/futurity/nsba.htm
Web site describes the purpose of the NSBA, which is to define, promote, and improve the quality of the pleasure horse; to promote exhibits, events, and contests in expositions and shows; to promote the training of pleasure horses; to promote interest in pleasure horses among younger horsemen; and to use and encourage the use of the standard rules for holding and judging contests of the pleasure horse.

National Wildlife Federation
Ph.: 1-800-822-9919
www.nwf.org/includes/2ndlevel
Web site describes this large, member-supported conservation group. Includes educational resources such as Animal Tracks, Campus Ecology, National Wildlife Week, and ordering information for the following magazines: *National Wildlife, International Wildlife, Ranger Rick, Your Back Yard,* and *Wild Animal Baby.*

Species Survival Commission (SSC)
www.iucn.org/themes/ssc/siteindx.htm
Web site describes this volunteer commission of the IUCN, whose mission is to conserve biological diversity through studying and managing species and their habitats.

U.S. Fish and Wildlife Service (FWS)
www.fws.gov
Web site is the official site of the FWS, whose mission is to conserve, protect, and enhance the nation's fish and wildlife and their habitats for the continuing benefit of people. Information can be found on conserving wildlife and habitats including fish, birds, endangered species, and refuges. Information on how to work with others around issues of sports and recreation, landowners, partnerships, and grants is provided. An office directory by state and region is available through this site, as well as information about fishing and other wildlife permits.

Wildlife Conservation Society (WCS)
2300 Southern Boulevard
Bronx, NY 10460
Ph.: 718-220-5100
www.wcs.org
Web site offers information about the organization's conservation activities. Also described are other entities run by the WCS including the Bronx Zoo, the New York Aquarium, and works to save wildlife and wild lands throughout the world. The Bronx Zoo web site features pictures, maps, and details about the largest metropolitan wildlife conservation park in the United States. The New York's Aquarium for Wildlife Conservation's web site features information on fish and marine creatures, including beluga whales, sharks, walruses, and dolphins. Also included under the umbrella of the WCS is the Hornocker Wildlife Institute, which conducts long-term research on threatened species and sensitive ecological systems. The institute site also offers access to specialists in carnivore research.

The Wildlife Society
5410 Grosvenor Lane
Bethesda, MD 20814
Ph.: 301-897-9770
Fax: 301-530-2471
E-mail: tws@wildlife.org
www.wildlife.org/index.html
Web site provides information on research scientists, educators, communications specialists, conservation law enforcement officers, resource managers, administrators, and students from more than sixty countries.

Wildlife Web: Conservation
www.selu.com/bio/wildlife/text/conservation.html
Web site is a directory to conservation sites.

World Wide Fund for Nature (WWF)
www.panda.org/resources/publications
Web site includes access to publications and other resources, including the Living Planet series, featuring information on climate change, endangered seas, forests, species, toxins, and water. The multimedia series includes a photography collection, video library, and a "Just for Kids" section.

—*Mary E. Carey*

MAGILL'S ENCYCLOPEDIA OF SCIENCE

ANIMAL LIFE

ALPHABETICAL LIST

Volume 1

Volume 2

Volume 3

Volume 4

CATEGORY LIST

Amphibians
Amphibians
Cold-blooded animals
Frogs and toads
Metamorphosis
Salamanders and newts
Vertebrates

Anatomy
Anatomy
Antennae
Beaks and bills
Bone and cartilage
Brain
Cell types
Claws, nails, and hooves
Digestive tract
Ears
Endoskeletons
Exoskeletons
Eyes
Feathers
Fins and flippers
Fur and hair
Heart
Horns and antlers
Hydrostatic skeletons
Kidneys and other excretory structures
Lungs, gills, and tracheas
Muscles in invertebrates
Muscles in vertebrates
Noses
Scales
Sense organs
Shells
Skin
Tails
Teeth, fangs, and tusks
Tentacles
Wings

Arthropods
Arachnids
Arthropods
Centipedes and millipedes
Cold-blooded animals
Crabs and lobsters
Crustaceans
Exoskeletons
Horseshoe crabs
Invertebrates
Scorpions
Spiders
Vertebrates

Behavior
Adaptations and their mechanisms
Altruism
Camouflage
Cannibalism
Carnivores
Communication
Communities
Competition
Copulation
Courtship
Death and dying
Defense mechanisms
Displays
Domestication

Birds

Carnivores

Cell Biology

Neutral mutations and evolutionary clocks
Nonrandom mating, genetic drift, and mutation
Nutrient requirements
Osmoregulation
pH maintenance
Regeneration
Reproduction
Sense organs
Water balance in vertebrates

Classification

Aardvarks
Allosaurus
American pronghorns
Amphibians
Animal kingdom
Antelope
Ants
Apatosaurus
Arachnids
Archaeopteryx
Armadillos, anteaters, and sloths
Arthropods
Baboons
Bats
Bears
Beavers
Bees
Beetles
Birds
Brachiosaurus
Butterflies and moths
Camels
Cats
Cattle, buffalo, and bison
Centipedes and millipedes
Chameleons
Cheetahs
Chickens, turkeys, pheasant, and quail
Chimpanzees
Chordates, lower
Clams and oysters
Cockroaches
Coral
Crabs and lobsters
Cranes
Crocodiles
Crustaceans
Deer
Dinosaurs
Dogs, wolves, and coyotes
Dolphins, porpoises, and toothed whales
Donkeys and mules
Ducks
Eagles
Echinoderms
Eels
Elephant seals
Elephants
Elk
Fish
Flamingos
Flatworms
Flies
Foxes
Frogs and toads
Geese
Giraffes
Goats
Gophers
Gorillas
Grasshoppers
Grizzly bears
Hadrosaurs
Hawks
Hippopotamuses
Hominids
Horses and zebras
Horseshoe crabs
Hummingbirds
Hyenas
Hyraxes
Ichthyosaurs
Insects
Invertebrates
Jaguars
Jellyfish
Kangaroos
Koalas
Lampreys and hagfish
Lemurs
Leopards
Lions
Lizards
Lungfish
Mammals
Mammoths
Manatees
Meerkats
Mice and rats
Moles
Mollusks
Monkeys
Moose
Mosquitoes
Mountain lions
Neanderthals
Octopuses and squid
Opossums
Orangutans
Ostriches and related birds
Otters
Owls
Pandas
Parrots
Pelicans
Penguins
Pigs and hogs
Platypuses
Polar bears
Porcupines
Praying mantis
Primates
Protozoa
Pterosaurs
Rabbits, hares, and pikas
Raccoons and related mammals
Reindeer
Reptiles
Rhinoceroses
Rodents

Embryology
Ethology
Human evolution analysis
Marine biology
Paleoecology
Paleontology
Phylogeny
Physiology
Population analysis
Population genetics
Systematics
Veterinary medicine
Zoology

Fish
Cold-blooded animals
Deep-sea animals
Eels
Fins and flippers
Fish
Lakes and rivers
Lampreys and hagfish
Lungfish
Marine animals
Marine biology
Reefs
Salmon and trout
Scales
Seahorses
Sharks and rays
Tidepools and beaches
Vertebrates
Whale sharks
White sharks

Genetics
Gene flow
Genetics
Hardy-Weinberg law of genetic equilibrium
Homo sapiens and human diversification
Human evolution analysis
Morphogenesis
Multicellularity
Mutations
Natural selection
Neutral mutations and evolutionary clocks
Nonrandom mating, genetic drift, and mutation

Geography
Biogeography
Chaparral
Deep-sea animals
Deserts
Fauna: Africa
Fauna: Antarctica
Fauna: Arctic
Fauna: Asia
Fauna: Australia
Fauna: Caribbean
Fauna: Central America
Fauna: Europe
Fauna: Galápagos Islands
Fauna: Madagascar
Fauna: North America
Fauna: Pacific Islands
Fauna: South America
Forests, coniferous
Forests, deciduous
Grasslands and prairies
Habitats and biomes
Lakes and rivers
Marine animals
Mountains
Rain forests
Reefs
Savannas
Swamps and marshes
Tidepools and beaches
Tundra
Urban and suburban wildlife

Habitats
Chaparral
Deep-sea animals
Deserts
Forests, coniferous
Forests, deciduous
Grasslands and prairies
Habitats and biomes
Lakes and rivers
Marine animals
Mountains
Rain forests
Reefs
Savannas
Swamps and marshes
Tidepools and beaches
Tundra
Urban and suburban wildlife

Herbivores
American pronghorns
Antelope
Apatosaurus
Beavers
Brachiosaurus
Camels
Cattle, buffalo, and bison
Deer
Donkeys and mules
Elephants
Elk
Food chains and food webs
Giraffes
Goats
Gophers
Hadrosaurs
Herbivores
Herds
Hippopotamuses
Horns and antlers
Horses and zebras
Hyraxes
Kangaroos
Koalas
Mammoths
Manatees
Mice and rats
Moles
Moose
Pandas
Pigs and hogs
Platypuses
Porcupines

Human Origins

Insects

Invertebrates

Mammals

Monkeys
Monotremes
Moose
Mountain lions
Neanderthals
Orangutans
Otters
Pandas
Pigs and hogs
Placental mammals
Platypuses
Polar bears
Porcupines
Primates
Rabbits, hares, and pikas
Raccoons and related mammals
Reindeer
Reproductive system of female mammals
Reproductive system of male mammals
Rhinoceroses
Rodents
Ruminants
Seals and walruses
Sheep
Shrews
Skunks
Squirrels
Tasmanian devils
Tigers
Ungulates
Vertebrates
Warm-blooded animals
Weasels and related mammals
Whales, baleen

Marine Biology
Clams and oysters
Coral
Crabs and lobsters
Crustaceans
Deep-sea animals
Dolphins, porpoises, and toothed whales
Echinoderms
Eels
Elephant seals
Fins and flippers
Fish
Horseshoe crabs
Ichthyosaurs
Jellyfish
Lakes and rivers
Lampreys and hagfish
Lungfish
Manatees
Marine animals
Marine biology
Mollusks
Octopuses and squid
Otters
Penguins
Reefs
Salmon and trout
Scales
Seahorses
Seals and walruses
Sharks and rays
Sponges
Starfish
Tentacles
Tidepools and beaches
Turtles and tortoises
Whale sharks
Whales, baleen
White sharks
Zooplankton

Marsupials
Fauna: Australia
Kangaroos
Koalas
Marsupials
Opossums
Reproduction
Reproductive strategies
Reproductive system of female mammals
Tasmanian devils

Omnivores
Baboons
Bears
Chimpanzees
Food chains and food webs
Gorillas
Hominids
Lemurs
Monkeys
Omnivores
Orangutans
Raccoons and related mammals
Shrews
Skunks
Weasels and related mammals

Physiology
Adaptations and their mechanisms
Aging
Asexual reproduction
Bioluminescence
Birth
Camouflage
Circulatory systems of invertebrates
Circulatory systems of vertebrates
Cleavage, gastrulation, and neurulation
Cold-blooded animals
Determination and differentiation
Digestion
Digestive tract
Diseases
Endocrine systems of invertebrates
Endocrine systems of vertebrates
Endoskeletons
Estrus
Exoskeletons
Fertilization
Gametogenesis
Gas exchange

Population Biology

Prehistoric Animals

Primates

Reproduction and Development

Hormones and behavior
Hormones in mammals
Imprinting
Infanticide
Lactation
Life spans
Marsupials
Mating
Metamorphosis
Molting and shedding
Monotremes
Morphogenesis
Nesting
Offspring care
Pair-bonding
Parthenogenesis
Pheromones
Placental mammals
Pregnancy and prenatal development
Regeneration
Reproduction
Reproductive strategies
Reproductive system of female mammals
Reproductive system of male mammals
Sex differences: Evolutionary origins
Sexual development
Vocalizations

Reptiles
Allosaurus
Apatosaurus
Archaeopteryx
Brachiosaurus
Chameleons
Cold-blooded animals
Crocodiles
Dinosaurs
Hadrosaurs
Ichthyosaurs
Lizards
Pterosaurs
Reptiles
Sauropods
Scales
Snakes
Stegosaurs
Triceratops
Turtles and tortoises
Tyrannosaurus
Velociraptors
Vertebrates

Scientific Methods
Breeding programs
Cloning of extinct or endangered species
Demographics
Fossils
Hardy-Weinberg law of genetic equilibrium
Human evolution analysis
Mark, release, and recapture methods
Paleoecology
Paleontology
Population analysis
Population genetics
Systematics
Veterinary medicine

INDEX